Introduction to X-ray Powder Diffractometry

CHEMICAL ANALYSIS

A SERIES OF MONOGRAPHS ON
ANALYTICAL CHEMISTRY AND ITS APPLICATIONS

Editor
J. D. WINEFORDNER

VOLUME 138

A WILEY-INTERSCIENCE PUBLICATION

JOHN WILEY & SONS, INC.

New York / Chichester / Brisbane / Toronto / Singapore

Introduction to X-ray Powder Diffractometry

RON JENKINS

International Centre for Diffraction Data
Newtown Square, Pennsylvania

ROBERT L. SNYDER

New York State College of Ceramics
Alfred University
Alfred, New York

A WILEY-INTERSCIENCE PUBLICATION

JOHN WILEY & SONS, INC.

New York / Chichester / Brisbane / Toronto / Singapore

This text is printed on acid-free paper.

Copyright © 1996 by John Wiley & Sons, Inc.

All rights reserved. Published simultaneously in Canada.

Reproduction or translation of any part of this work beyond that permitted by Section 107 or 108 of the 1976 United States Copyright Act without the permission of the copyright owner is unlawful. Requests for permission or further information should be addressed to the Permissions Department, John Wiley & Sons, Inc., 605 Third Avenue, New York, NY 10158-0012.

Library of Congress Cataloging in Publication Data
Jenkins, Ron, 1932–
  Introduction to X-ray powder diffractometry / Ron Jenkins and Robert L. Snyder.
   p. cm.—(Chemical analysis ; v. 138)
  "A Wiley-Interscience publication."
  Includes index.
  ISBN 0-471-51339-3 (cloth : alk. paper)
  1. X-rays—Diffraction—Technique. 2. X-ray diffractometer.
3. Powders—Optical properties—Measurement. I. Snyder, R. L. (Robert L.), 1941– II. Title. III. Series.
QC482.D5J46 1996
548'.83—dc20   96-12039
                         CIP

Printed in the United States of America

10 9 8 7 6 5 4 3 2 1

Dedicated to

J. L. de Vries
and
W. Parrish

CONTENTS

PREFACE	xvii
CUMULATIVE LISTING OF VOLUMES IN SERIES	xix
CHAPTER 1. CHARACTERISTICS OF X-RADIATION	**1**
1.1. Early Development of X-ray Diffraction	1
1.2. Origin of X-radiation	2
1.3. Continuous Radiation	3
1.4. Characteristic Radiation	5
1.4.1. The Photoelectric Effect	5
1.4.2. The Auger Effect	5
1.4.3. Fluorescent Yield	7
1.4.4. Selection Rules	7
1.4.5. Nondiagram Lines	11
1.4.6. Practical Form of the Copper K Spectrum	12
1.5. Scattering of X-rays	14
1.5.1. Coherent Scatter	15
1.5.2. Compton Scatter	15
1.6. Absorption of X-rays	16
1.7. Safety Considerations	19
References	21
CHAPTER 2. THE CRYSTALLINE STATE	**23**
2.1. Introduction to the Crystalline State	23
2.2. Crystallographic Symmetry	26
2.2.1. Point Groups and Crystal Systems	28
2.2.2. The Unit Cell and Bravais Lattices	30

2.2.3. Reduced Cells		31
2.2.4. Space Groups		34
2.3. Space Group Notation		35
2.3.1. The Triclinic or Anorthic Crystal System		35
2.3.2. The Monoclinic Crystal System		35
2.3.3. The Orthorhombic Crystal System		37
2.3.4. The Tetragonal Crystal System		37
2.3.5. The Hexagonal and Trigonal Crystal Systems		38
2.3.6. The Cubic Crystal System		38
2.3.7. Equivalent Positions		39
2.3.8. Special Positions and Site Multiplicity		40
2.4. Space Group Theory		41
2.5. Crystallographic Planes and Miller Indices		43
References		44
CHAPTER 3. DIFFRACTION THEORY		**47**
3.1. Diffraction of X-rays		47
3.2. The Reciprocal Lattice		49
3.3. The Ewald Sphere of Reflection		54
3.4. Origin of the Diffraction Pattern		57
3.4.1. Single Crystal Diffraction		57
3.4.2. The Powder Diffraction Pattern		58
3.5. The Location of Diffraction Peaks		60
3.6. Intensity of Diffraction Peaks		64
3.6.1. Electron Scattering		64
3.6.2. The Atomic Scattering Factor		65
3.6.3. Anomalous Scattering		67
3.6.4. Thermal Motion		68
3.6.5. Scattering of X-rays by a Crystal: The Structure Factor		70
3.7. The Calculated Diffraction Pattern		75
3.7.1. Factors Affecting the Relative Intensity of Bragg Reflections		76
3.7.2. The Intensity Equation		80

	3.8. Calculation of the Powder Diffraction Pattern of KCl	82
	3.9. Anisotropic Distortions of the Diffraction Pattern	85
	3.9.1. Preferred Orientation	85
	3.9.2. Crystallite Size	89
	3.9.3. Residual Stress and Strain	91
	References	94
CHAPTER 4.	SOURCES FOR THE GENERATION OF X-RADIATION	97
	4.1. Components of the X-ray Source	97
	4.2. The Line-Voltage Supply	98
	4.3. The High-Voltage Generator	99
	4.3.1. Selection of Operating Conditions	102
	4.3.2. Source Stability	104
	4.4. The Sealed X-ray Tube	105
	4.4.1. Typical X-ray Tube Configuration	106
	4.4.2. Specific Loading	109
	4.4.3. Care of the X-ray Tube	113
	4.5. Effective Line Width	114
	4.6. Spectral Contamination	116
	4.6.1. X-ray Tube Life	117
	4.7. The Rotating Anode X-ray Tube	118
	References	120
CHAPTER 5.	DETECTORS AND DETECTION ELECTRONICS	121
	5.1. X-ray Detectors	121
	5.2. Desired Properties of an X-ray Detector	122
	5.2.1. Quantum-Counting Efficiency	122
	5.2.2. Linearity	123
	5.2.3. Energy Proportionality	125
	5.2.4. Resolution	126
	5.3. Types of Detector	127
	5.3.1. The Gas Proportional Counter	128

	5.3.2. Position-Sensitive Detectors	130
	5.3.3. The Scintillation Detector	131
	5.3.4. The Si(Li) Detector	132
	5.3.5. Other X-ray Detectors	135
5.4.	Pulse Height Selection	136
5.5.	Counting Circuits	138
	5.5.1. The Ratemeter	139
5.6.	Counting Statistics	140
5.7.	Two-Dimensional Detectors	142
	References	148

CHAPTER 6. PRODUCTION OF MONOCHROMATIC RADIATION — 151

- 6.1. Introduction — 151
- 6.2. Angular Dispersion — 153
- 6.3. Makeup of a Diffractogram — 154
  - 6.3.1. Additional Lines in the Diffractogram — 155
  - 6.3.2. Reduction of Background — 157
- 6.4. The β-Filter — 158
  - 6.4.1. Thickness of the β-Filter — 159
  - 6.4.2. Use of Pulse Height Selection to Supplement the β-Filter — 160
  - 6.4.3. Placement of the β-Filter — 162
- 6.5. The Proportional Detector and Pulse Height Selection — 162
- 6.6. Use of Solid State Detectors — 163
- 6.7. Use of Monochromators — 164
  - 6.7.1. The Diffracted-Beam Monochromator — 167
  - 6.7.2. The Primary-Beam Monochromator — 170
- 6.8. Comparison of Monochromatization Methods — 170
- References — 172

CHAPTER 7. INSTRUMENTS FOR THE MEASUREMENT OF POWDER PATTERNS — 173

- 7.1. Camera Methods — 173
  - 7.1.1. The Debye–Scherrer/Hull Method — 173

	7.1.2. The Gandolfi Camera	174
	7.1.3. The Guinier Camera	177
7.2.	The Powder Diffractometer	178
7.3.	The Seemann–Bohlin Diffractometer	180
7.4.	The Bragg–Brentano Diffractometer	180
7.5.	Systematic Aberrations	187
	7.5.1. The Axial-Divergence Error	187
	7.5.2. The Flat-Specimen Error	191
	7.5.3. Error Due to Specimen Transparency	193
	7.5.4. Error Due to Specimen Displacement	194
7.6.	Selection of Goniometer Slits	195
	7.6.1 Effect of Receiving Slit Width	195
	7.6.2. Effect of the Divergence Slit	197
	References	202

CHAPTER 8. ALIGNMENT AND MAINTENANCE OF POWDER DIFFRACTOMETERS — 205

8.1.	Principles of Alignment	205
	8.1.1. The Rough xyz Alignment	206
	8.1.2. Setting the Takeoff Angle	208
	8.1.3. Setting the Mechanical Zero	210
	8.1.4. Setting the 2:1	212
	8.1.5. Aligning of the Divergence Slit	213
	8.1.6. Tuning of the Monochromator	214
8.2.	Routine Alignment Checks	216
8.3.	Evaluation of the Quality of Alignment	222
8.4.	Troubleshooting	226
	References	229

CHAPTER 9. SPECIMEN PREPARATION — 231

9.1.	General Considerations	231
9.2.	Compositional Variations Between Sample and Specimen	233
9.3.	Absorption Problems	234
9.4.	Problems in Obtaining a Random Specimen	235
	9.4.1. Particle Inhomogeneity	235

9.4.2. Crystal Habit and Preferred Orientation   236
9.4.3. Particle Statistics   240
9.5. Particle Separation and Size Reduction Methods   244
9.6. Specimen Preparation Procedures   244
    9.6.1. Use of Standard Mounts   246
    9.6.2. Back and Side Loading   247
    9.6.3. Top Loading   249
    9.6.4. The Zero Background Holder Method   249
    9.6.5. Spray-Drying   251
    9.6.6. Use of Aerosols   253
9.7. Measurement of the Prepared Specimen   254
    9.7.1. Specimen Displacement   254
    9.7.2. Mechanical Methods for Randomizing   255
    9.7.3. Handling of Small Samples   257
    9.7.4. Special Samples   257
References   258

CHAPTER 10. ACQUISITION OF DIFFRACTION DATA   261

10.1. Introduction   261
10.2. Steps in Data Acquisition   261
10.3. Typical Data Quality   264
10.4. Selection of the d-Spacing Range of the Pattern   265
    10.4.1. Choice of the 2θ Range   266
    10.4.2. Choice of Wavelength   266
10.5. Manual Powder Diffractometers   270
    10.5.1. Synchronous Scanning   270
    10.5.2. Use of Ratemeters   270
    10.5.3. Step Scanning   272
10.6. Automated Powder Diffractometers   274
    10.6.1. Step Scanning with the Computer   277
    10.6.2. Choice of Step Width   279
    10.6.3. Open-Loop and Absolute Encoders   280
10.7. Use of Calibration Standards   281
    10.7.1. External 2θ Standards   282

10.7.2. Internal 2θ and d-Spacing Standards	283
10.7.3. Quantitative Analysis Standards	283
10.7.4. Sensitivity Standards	284
10.7.5. Line Profile Standards	285
References	285

CHAPTER 11. REDUCTION OF DATA FROM AUTOMATED POWDER DIFFRACTOMETERS — 287

11.1. Data Reduction Procedures	287
11.2. Range of Experimental Data to Be Treated	287
11.2.1. Computer Reduction of Data	288
11.3. Steps in Data Treatment	291
11.3.1. Use of Data Smoothing	292
11.3.2. Background Subtraction	297
11.3.3. Treatment of the α_2	299
11.3.4. Peak Location Methods	300
11.4. Conversion Errors	305
11.5. Calibration Methods	308
11.5.1. 2θ Correction Using an External Standard	308
11.5.2. 2θ and d-Spacing Correction Using an Internal Standard	309
11.5.3. Sensitivity Correction Using an External Intensity Standard	309
11.6. Evaluation of Data Quality	310
11.6.1. Use of Figures of Merit	310
11.6.2. Use of Figures of Merit for Instrument Performance Evaluation	312
11.6.3. Use of Figures of Merit for Data Quality Evaluation	313
11.6.4. Use of Figures of Merit in Indexing of Powder Patterns	316
References	317

CHAPTER 12. QUALITATIVE ANALYSIS — 319

12.1. Phase Identification by X-ray Diffraction	319
12.1.1. Quality of Experiment Data	322

	12.2. Databases	323
	12.2.1. The Powder Diffraction File	324
	12.2.2. The Crystal Data File	326
	12.2.3. The Elemental and Interplanar Spacing Index (EISI)	327
	12.2.4. The Metals and Alloys Index	328
	12.3. Media on Which ICDD Databases Are Supplied	329
	12.3.1. Historical Evolution of Database Media	329
	12.3.2. Computer-Readable Products	330
	12.3.3. The CD–ROM System	331
	12.4. Manual Search/Matching Methods	332
	12.4.1. The Alphabetic Method	333
	12.4.2. The Hanawalt Search Method	335
	12.4.3. The Fink Search Method	339
	12.5. Limitations with the Use of Paper Search Manuals	344
	12.6. Boolean Search Methods	345
	12.7. Fully Automated Search Methods	347
	12.7.1. First-Generation Programs	347
	12.7.2. Second-Generation Search/Match Algorithms	348
	12.7.3. Commercial Search/Match Programs	348
	12.7.4. Third-Generation Search/Match Algorithms	349
	12.8. Effectiveness of Search/Matching Using the Computer	350
	References	351
CHAPTER 13.	**QUANTITATIVE ANALYSIS**	**355**
	13.1. Historical Development of Quantitative Phase Analysis	355
	13.2. Measurement of Line Intensities	356
	13.3. Foundation of Quantitative Phase Analysis	361
	13.4. The Absorption–Diffraction Method	362
	13.4.1. Use of Klug's Equation	365

	13.4.2.	Use of Measured Mass Attenuation Coefficients	367
	13.4.3.	Use of Mass Attenuation Coefficients Derived from Elemental Chemistry	368
13.5.	Method of Standard Additions		369
13.6.	The Internal Standard Method of Quantitative Analysis		370
	13.6.1.	$I/I_{corundum}$ and the Reference Intensity Ratio Method	372
	13.6.2.	The Generalized Reference Intensity Ratio	372
	13.6.3.	Quantitative Analysis with RIRs	373
	13.6.4.	The Normalized RIR Method	373
	13.6.5.	Constrained XRD Phase Analysis: Generalized Internal Standard Method	374
13.7.	Quantitative Phase Analysis Using Crystal Structure Constraints		376
13.8.	Quantitative Methods Based on Use of the Total Pattern		378
	13.8.1.	The Rietveld Method	378
	13.8.2.	Full-Pattern Fitting with Experimental Patterns	383
13.9.	Detection of Low Concentrations		384
	References		386

APPENDIX A:	**COMMON X-RAY WAVELENGTHS**	**389**
APPENDIX B:	**MASS ATTENUATION COEFFICIENTS**	**390**
APPENDIX C:	**ATOMIC WEIGHTS AND DENSITIES**	**391**
APPENDIX D:	**CRYSTALLOGRAPHIC CLASSIFICATION OF THE 230 SPACE GROUPS**	**392**
INDEX		**397**

PREFACE

The purpose of this book is to act as an introductory text for users of X-ray powder diffractometers and diffractometry. We have worked to supply the fundamental information on the diffractometer, including consideration of its components, alignment, calibration, and automation. Much of this information is being presented in textbook form for the first time here. The goal of this book is to act as an introduction to students of materials science, mineralogy, chemistry, and physics. This volume contains all of the fundamentals required to appreciate the theory and practice of powder diffraction, with a strong emphasis on the two most important applications: qualitative and quantitative analysis. The treatment of advanced applications of powder diffraction would have both distracted the introductory user and more than doubled the size of this volume. Therefore we decided to develop *Applications of Powder Diffraction* as a companion volume to this one. The pair of books are designed to bring an introductory user to appreciate all of the applications at the current state of the art. A reader content to understand and use the most common applications should find everything required in the present volume.

A book such as this can never be considered simply the work of two people. Science is a discipline that is built on the inspiration of a few and the mistakes of many. The X-ray powder diffraction field is certainly no exception to this general rule. To this extent we would both like to thank not just the few that have inspired us but also the many who have accepted our mistakes and shortcomings. We are especially grateful to those who have taken time from their busy schedules to review the manuscript at the various stages of its preparation. In this context we are especially indebted to Tom Blanton, Greg Hamill, Jim Kaduk, Greg McCarthy, and Paul Predecki. Special thanks go to Chan Park for his painstaking help in preparing some of the difficult figures. We are pleased to acknowledge the help of those who patiently read the final manuscript: Mario Fornoff, Paden Dismore, Chan Park, and Mike Haluska. Our special thanks also go to Leo Zwell and Zhouhui Yang, who provided invaluable help in locating some of the more obscure original references. We would also like to thank a full generation of graduate students of the New York State College of Ceramics at Alfred University, who have contributed ideas that have helped simplify the explanations of various sections of this book. Anything that does not strike the reader as brilliantly clear, however, may safely be blamed on the authors. The two women whose judgment of human nature was so poor as to marry each of us also need to be acknowledged. Our marriages have survived the first volume of this endeavor; perhaps a bit of time in Cancún will get us through the companion volume!

We have chosen to dedicate this book to two people who have made a dramatic impact on our professional lives, Dr. J. L. (Hans) de Vries and Dr. W. (Bill) Parrish (now deceased). These two men did more than any others to promote the budding field of X-ray powder diffractometry in the early 1950s when the two of us were still in the salad days of our youth. We have learned much at the feet of our masters and gratefully acknowledge their patience and understanding.

RON JENKINS
ROBERT L. SNYDER

Newtown Square, Pennsylvania
Alfred, New York
May 1996

CHEMICAL ANALYSIS

A SERIES OF MONOGRAPHS ON ANALYTICAL CHEMISTRY AND ITS APPLICATIONS

J. D. Winefordner, *Series Editor*

Vol. 1. **The Analytical Chemistry of Industrial Poisons, Hazards, and Solvents.** *Second Edition.* By the late Morris B. Jacobs
Vol. 2. **Chromatographic Adsorption Analysis.** By Harold H. Strain (*out of print*)
Vol. 3. **Photometric Determination of Traces of Metals.** *Fourth Edition*
Part 1: General Aspects. By E. B. Sandell and Hiroshi Onishi
Part IIA: Individual Metals, Aluminum to Lithium. By Hiroshi Onishi
Part IIB: Individual Metals, Magnesium to Zirconium. By Hiroshi Onishi
Vol. 4. **Organic Reagents Used in Gravimetric and Volumetric Analysis.** By John F. Flagg (*out of print*)
Vol. 5. **Aquametry: A Treatise on Methods for the Determination of Water.** *Second Edition* (*in three parts*). By John Mitchell, Jr. and Donald Milton Smith
Vol. 6. **Analysis of Insecticides and Acaricides.** By Francis A. Gunther and Roger C. Blinn (*out of print*)
Vol. 7. **Chemical Analysis of Industrial Solvents.** By the late Morris B. Jacobs and Leopold Schetlan
Vol. 8. **Colorimetric Determination of Nonmetals.** *Second Edition.* By the late David F. Boltz and James A. Howell
Vol. 9. **Analytical Chemistry of Titanium Metals and Compounds.** By Maurice Codell
Vol. 10. **The Chemical Analysis of Air Pollutants.** By the late Morris B. Jacobs
Vol. 11. **X-Ray Spectrochemical Analysis.** *Second Edition.* By L. S. Birks
Vol. 12. **Systematic Analysis of Surface-Active Agents.** *Second Edition.* By Milton J. Rosen and Henry A. Goldsmith
Vol. 13. **Alternating Current Polarography and Tensammetry.** By B. Breyer and H. H. Bauer
Vol. 14. **Flame Photometry.** By R. Herrmann and J. Alkemade
Vol. 15. **The Titration of Organic Compounds** (*in two parts*). By M. R. F. Ashworth
Vol. 16. **Complexation in Analytical Chemistry: A Guide for the Critical Selection of Analytical Methods Based on Complexation Reactions.** By the late Anders Ringbom
Vol. 17. **Electron Probe Microanalysis.** *Second Edition.* By L. S. Birks
Vol. 18. **Organic Complexing Reagents: Structure, Behavior, and Application to Inorganic Analysis.** By D. D. Perrin

Vol.	19.	**Thermal Analysis.** *Third Edition.* By Wesley Wm. Wendlandt
Vol.	20.	**Amperometric Titrations.** By John T. Stock
Vol.	21.	**Reflectance Spectroscopy.** By Wesley Wm. Wendlandt and Harry G. Hecht
Vol.	22.	**The Analytical Toxicology of Industrial Inorganic Poisons.** By the late Morris B. Jacobs
Vol.	23.	**The Formation and Properties of Precipitates.** By Alan G. Walton
Vol.	24.	**Kinetics in Analytical Chemistry.** By Harry B. Mark, Jr. and Garry A. Rechnitz
Vol.	25.	**Atomic Absorption Spectroscopy.** *Second Edition.* By Morris Slavin
Vol.	26.	**Characterization of Organometallic Compounds** (*in two parts*). Edited by Minoru Tsutsui
Vol.	27.	**Rock and Mineral Analysis.** *Second Edition.* By Wesley M. Johnson and John A. Maxwell
Vol.	28.	**The Analytical Chemistry of Nitrogen and Its Compounds** (*in two parts*). Edited by C. A. Streuli and Philip R. Averell
Vol.	29.	**The Analytical Chemistry of Sulfur and Its Compounds** (*in three parts*). By J. H. Karchmer
Vol.	30.	**Ultramicro Elemental Analysis.** By Günther Tölg
Vol.	31.	**Photometric Organic Analysis** (*in two parts*). By Eugene Sawicki
Vol.	32.	**Determination of Organic Compounds: Methods and Procedures.** By Frederick T. Weiss
Vol.	33.	**Masking and Demasking of Chemical Reactions.** By D. D. Perrin
Vol.	34.	**Neutron Activation Analysis.** By D. De Soete, R. Gijbels, and J. Hoste
Vol.	35.	**Laser Raman Spectroscopy.** By Marvin C. Tobin
Vol.	36.	**Emission Spectrochemical Analysis.** By Morris Slavin
Vol.	37.	**Analytical Chemistry of Phosphorus Compounds.** Edited by M. Halmann
Vol.	38.	**Luminescence Spectrometry in Analytical Chemistry.** By J. D. Winefordner, S. G. Schulman and T. C. O'Haver
Vol.	39.	**Activation Analysis with Neutron Generators.** By Sam S. Nargolwalla and Edwin P. Przybylowicz
Vol.	40.	**Determination of Gaseous Elements in Metals.** Edited by Lynn L. Lewis, Laben M. Melnick, and Ben D. Holt
Vol.	41.	**Analysis of Silicones.** Edited by A. Lee Smith
Vol.	42.	**Foundations of Ultracentrifugal Analysis.** By H. Fujita
Vol.	43.	**Chemical Infrared Fourier Transform Spectroscopy.** By Peter R. Griffiths
Vol.	44.	**Microscale Manipulations in Chemistry.** By T. S. Ma and V. Horak
Vol.	45.	**Thermometric Titrations.** By J. Barthel
Vol.	46.	**Trace Analysis: Spectroscopic Methods for Elements.** Edited by J. D. Winefordner
Vol.	47.	**Contamination Control in Trace Element Analysis.** By Morris Zief and James W. Mitchell
Vol.	48.	**Analytical Applications of NMR.** By D. E. Leyden and R. H. Cox
Vol.	49.	**Measurement of Dissolved Oxygen.** By Michael L. Hitchman
Vol.	50.	**Analytical Laser Spectroscopy.** Edited by Nicolò Omenetto
Vol.	51.	**Trace Element Analysis of Geological Materials.** By Roger D. Reeves and Robert R. Brooks

CHEMICAL ANALYSIS: A SERIES OF MONOGRAPHS

Vol. 52. **Chemical Analysis by Microwave Rotational Spectroscopy.** By Ravi Varma and Lawrence W. Hrubesh
Vol. 53. **Information Theory As Applied to Chemical Analysis.** By Karel Eckschlager and Vladimir Štěpánek
Vol. 54. **Applied Infrared Spectroscopy: Fundamentals, Techniques, and Analytical Problem-solving.** By A. Lee Smith
Vol. 55. **Archaeological Chemistry.** By Zvi Goffer
Vol. 56. **Immobilized Enzymes in Analytical and Clinical Chemistry.** By P. W. Carr and L. D. Bowers
Vol. 57. **Photoacoustics and Photoacoustic Spectroscopy.** By Allan Rosencwaig
Vol. 58. **Analysis of Pesticide Residues.** Edited by H. Anson Moye
Vol. 59. **Affinity Chromatography.** By William H. Scouten
Vol. 60. **Quality Control in Analytical Chemistry.** *Second Edition.* By G. Kateman and L. Buydens
Vol. 61. **Direct Characterization of Fineparticles.** By Brian H. Kaye
Vol. 62. **Flow Injection Analysis.** By J. Ruzicka and E. H. Hansen
Vol. 63. **Applied Electron Spectroscopy for Chemical Analysis.** Edited by Hassan Windawi and Floyd Ho
Vol. 64. **Analytical Aspects of Environmental Chemistry.** Edited by David F. S. Natusch and Philip K. Hopke
Vol. 65. **The Interpretation of Analytical Chemical Data by the Use of Cluster Analysis.** By D. Luc Massart and Leonard Kaufman
Vol. 66. **Solid Phase Biochemistry: Analytical and Synthetic Aspects.** Edited by William H. Scouten
Vol. 67. **An Introduction to Photoelectron Spectroscopy.** By Pradip K. Ghosh
Vol. 68. **Room Temperature Phosphorimetry for Chemical Analysis.** By Tuan Vo-Dinh
Vol. 69. **Potentiometry and Potentiometric Titrations.** By E. P. Serjeant
Vol. 70. **Design and Application of Process Analyzer Systems.** By Paul E. Mix
Vol. 71. **Analysis of Organic and Biological Surfaces.** Edited by Patrick Echlin
Vol. 72. **Small Bore Liquid Chromatography Columns: Their Properties and Uses.** Edited by Raymond P. W. Scott
Vol. 73. **Modern Methods of Particle Size Analysis.** Edited by Howard G. Barth
Vol. 74. **Auger Electron Spectroscopy.** By Michael Thompson, M. D. Baker, Alec Christie, and J. F. Tyson
Vol. 75. **Spot Test Analysis: Clinical, Environmental, Forensic and Geochemical Applications.** By Ervin Jungreis
Vol. 76. **Receptor Modeling in Environmental Chemistry.** By Philip K. Hopke
Vol. 77. **Molecular Luminescence Spectroscopy: Methods and Applications** (*in three parts*). Edited by Stephen G. Schulman
Vol. 78. **Inorganic Chromatographic Analysis.** By John C. MacDonald
Vol. 79. **Analytical Solution Calorimetry.** Edited by J. K. Grime
Vol. 80. **Selected Methods of Trace Metal Analysis: Biological and Environmental Samples.** By Jon C. Van Loon

Vol. 81. **The Analysis of Extraterrestrial Materials.** By Isidore Adler
Vol. 82. **Chemometrics.** By Muhammad A. Sharaf, Deborah L. Illman, and Bruce R. Kowalski
Vol. 83. **Fourier Transform Infrared Spectrometry.** By Peter R. Griffiths and James A. de Haseth
Vol. 84. **Trace Analysis: Spectroscopic Methods for Molecules.** Edited by Gary Christian and James B. Callis
Vol. 85. **Ultratrace Analysis of Pharmaceuticals and Other Compounds of Interest.** By S. Ahuja
Vol. 86. **Secondary Ion Mass Spectrometry: Basic Concepts, Instrumental Aspects, Applications and Trends.** By A. Benninghoven, F. G. Rüdenauer, and H. W. Werner
Vol. 87. **Analytical Applications of Lasers.** Edited by Edward H. Piepmeier
Vol. 88. **Applied Geochemical Analysis.** By C. O. Ingamells and F. F. Pitard
Vol. 89. **Detectors for Liquid Chromatography.** Edited by Edward S. Yeung
Vol. 90. **Inductively Coupled Plasma Emission Spectroscopy: Part I: Methodology, Instrumentation, and Performance; Part II: Applications and Fundamentals.** Edited by J. M. Boumans
Vol. 91. **Applications of New Mass Spectrometry Techniques in Pesticide Chemistry.** Edited by Joseph Rosen
Vol. 92. **X-Ray Absorption: Principles, Applications, Techniques of EXAFS, SEXAFS, and XANES.** Edited by D. C. Konnigsberger
Vol. 93. **Quantitative Structure–Chromatographic Retention Relationships.** Edited by Roman Kaliszan
Vol. 94. **Laser Remote Chemical Analysis.** Edited by Raymond M. Measures
Vol. 95. **Inorganic Mass Spectrometry.** Edited by F. Adams, R. Gijbels, and R. Van Grieken
Vol. 96. **Kinetic Aspects of Analytical Chemistry.** By Horacio A. Mottola
Vol. 97. **Two-Dimensional NMR Spectroscopy.** By Jan Schraml and Jon M. Bellama
Vol. 98. **High Performance Liquid Chromatography.** Edited by Phyllis R. Brown and Richard A. Hartwick
Vol. 99. **X-Ray Fluorescence Spectrometry.** By Ron Jenkins
Vol. 100. **Analytical Aspects of Drug Testing.** Edited by Dale G. Deutsch
Vol. 101. **Chemical Analysis of Polycyclic Aromatic Compounds.** Edited by Tuan Vo-Dinh
Vol. 102. **Quadrupole Storage Mass Spectrometry.** By Raymond E. March and Richard J. Hughes
Vol. 103. **Determination of Molecular Weight.** Edited by Anthony R. Cooper
Vol. 104. **Selectivity and Detectability Optimizations in HPLC.** By Satinder Ahuja
Vol. 105. **Laser Microanalysis.** By Lieselotte Moenke-Blankenburg
Vol. 106. **Clinical Chemistry.** Edited by E. Howard Taylor
Vol. 107. **Multielement Detection Systems for Spectrochemical Analysis.** By Kenneth W. Busch and Marianna A. Busch
Vol. 108. **Planar Chromatography in the Life Sciences.** Edited by Joseph C. Touchstone
Vol. 109. **Fluorometric Analysis in Biomedical Chemistry: Trends and Techniques Including HPLC Applications.** By Norio Ichinose, George Schwedt, Frank Michael Schnepel, and Kyoko Adochi
Vol. 110. **An Introduction to Laboratory Automation.** By Victor Cerdá and Guillermo Ramis

CHEMICAL ANALYSIS: A SERIES OF MONOGRAPHS

Vol. 111. **Gas Chromatography: Biochemical, Biomedical, and Clinical Applications.** Edited by Ray E. Clement
Vol. 112. **The Analytical Chemistry of Silicones.** Edited by A. Lee Smith
Vol. 113. **Modern Methods of Polymer Characterization.** Edited by Howard G. Barth and Jimmy W. Mays
Vol. 114. **Analytical Raman Spectroscopy.** Edited by Jeannette Graselli and Bernard J. Bulkin
Vol. 115. **Trace and Ultratrace Analysis by HPLC.** By Satinder Ahuja
Vol. 116. **Radiochemistry and Nuclear Methods of Analysis.** By William D. Ehmann and Diane E. Vance
Vol. 117. **Applications of Fluorescence in Immunoassays.** By Ilkka Hemmila
Vol. 118. **Principles and Practice of Spectroscopic Calibration.** By Howard Mark
Vol. 119. **Activation Spectrometry in Chemical Analysis.** By S. J. Parry
Vol. 120. **Remote Sensing by Fourier Transform Spectrometry.** By Reinhard Beer
Vol. 121. **Detectors for Capillary Chromatography.** Edited by Herbert H. Hill and Dennis McMinn
Vol. 122. **Photochemical Vapor Deposition.** By J. G. Eden
Vol. 123. **Statistical Methods in Analytical Chemistry.** By Peter C. Meier and Richard Zund
Vol. 124. **Laser Ionization Mass Analysis.** Edited by Akos Vertes, Renaat Gijbels, and Fred Adams
Vol. 125. **Physics and Chemistry of Solid State Sensor Devices.** By Andreas Mandelis and Constantinos Christofides
Vol. 126. **Electroanalytical Stripping Methods.** By Khjena Z. Brainina and E. Neyman
Vol. 127. **Air Monitoring by Spectroscopic Techniques.** Edited by Markus W. Sigrist
Vol. 128. **Information Theory in Analytical Chemistry.** By Karel Eckschlager and Klaus Danzer
Vol. 129. **Flame Chemiluminescence Analysis by Molecular Emission Cavity Detection.** Edited by David Stiles, Anthony Calokerinos, and Alan Townshend
Vol. 130. **Hydride Generation Atomic Absorption Spectrometry.** Edited by Jiri Dedina and Dimiter L. Tsalev
Vol. 131. **Selective Detectors: Environmental, Industrial, and Biomedical Applications.** Edited by Robert E. Sievers
Vol. 132. **High-Speed Countercurrent Chromatography.** Edited by Yoichiro Ito and Walter D. Conway
Vol. 133. **Particle-Induced X-Ray Emission Spectrometry.** By Sven A. E. Johansson, John L. Campbell, and Klass G. Malmqvist
Vol. 134. **Photothermal Spectroscopy Methods for Chemical Analysis.** By Stephen Bialkowski
Vol. 135. **Element Speciation in Bioinorganic Chemistry.** Edited by Sergio Caroli
Vol. 136. **Laser-Enhanced Ionization and Spectrometry.** Edited by John C. Travis and Gregory C. Turk
Vol. 137. **Fluorescence Imaging Spectroscopy and Microscopy.** Edited by Xue Feng Wang and Brian Herman
Vol. 138. **Introduction to X-ray Powder Diffractometry.** By Ron Jenkins and Robert L. Snyder

CHAPTER

1

CHARACTERISTICS OF X-RADIATION

1.1. EARLY DEVELOPMENT OF X-RAY DIFFRACTION

Following the discovery of X-rays by W. C. Röntgen in 1895, three major branches of science have developed from the use of this radiation. The first and oldest of these is X-ray radiography, which makes use of the fact that the relative absorption of X-rays by matter is a function of the average atomic number and density of the matter concerned. From this has developed the whole range of diagnostic methods for medical and industrial use. Early attempts to confirm the dual nature of X-rays, i.e., their particle and wave character, were frustrated by experimental difficulties involved with the handling of the very short wavelengths in question. Not until the classic work of Max von Laue in 1912 was the wave character confirmed by diffraction experiments from a single crystal. From this single experiment has developed the field of *X-ray crystallography*, of which X-ray powder diffractometry is one important member. X-ray crystallography, using single crystals or powder, is mainly concerned with structure analysis. The third technique, *X-ray spectrometry*, also has its fundamental roots in the early part of this century, but routine application of X-ray fluorescence spectrometry has only developed over the last 20 to 30 years.

The purpose of this work is to discuss *X-ray powder diffractometry*. Powder diffractometry is mainly used for the identification of compounds by their diffraction patterns. The first X-ray powder diffractometer was developed in 1935 by Le Galley [1], but, due mainly to the lack of parafocusing conditions, the instrument gave relatively poor intensities. In 1945 Parrish and Gordon [2] developed a Geiger-counter *spectrometer*[1] for the precision cutting of quartz oscillator plates used in frequency control for military communication equipment. At the same time, Friedman [3] was working on X-ray spectrometer techniques at the U.S. Naval Research Laboratory in Washington,

---

[1]Confusion may occur in the study of early papers in this field since the term *X-ray spectrometer* referred to a system incorporating a crystal diffracting medium. The modern term *spectrometer* refers to an instrument employing a crystal or grating to separate a polychromatic beam of radiation into its constituent wavelengths. A *diffractometer* utilizes a monochromatic beam of radiation to yield information about d-spacings and intensities from a single crystal or crystalline powder.

DC. The modern parafocusing X-ray powder diffractometer was based on these ideas, and the first commercial equipment was introduced by North American Philips in 1947. The latest versions of the powder diffractometer differ little in their construction and geometry, but considerable advances have been made in detection and counting systems, automation, and in the X-ray tubes themselves.

1.2. ORIGIN OF X-RADIATION

X-rays are relatively short-wavelength, high-energy beams of electromagnetic radiation. When an X-ray beam is viewed as a wave, one can think of it as a sinusoidal oscillating electric field with, at right angles to it, a similarly varying magnetic field changing with time. Another description of X-rays is as particles of energy called *photons*. All electromagnetic radiation is characterized either by its wave character using its *wavelength* λ (i.e., the distance between peaks) or its *frequency* ν (the number of peaks that pass a point in unit time) or by means of its *photon energy E*. The following equations represent the relationships between these quantities:

$$\nu = \frac{c}{\lambda}, \tag{1.1}$$

$$E = h\nu, \tag{1.2}$$

where c is the speed of light and h is Planck's constant. The X-ray region is normally considered to be that part of the electromagnetic spectrum lying between 0.1 and 100 Å(1 Å = 10^{-10} m), being bounded by the γ-ray region to the short-wavelength side and the vacuum ultraviolet region to the long-wavelength side. In terms of energy, the X-ray region covers the range from about 0.1 to 100 keV. From a combination of Equations 1.1 and 1.2, it follows that the energy equivalent of an X-ray photon is

$$E = \frac{hc}{\lambda}. \tag{1.3}$$

Insertion of the appropriate values for the fundamental constants gives

$$E = \frac{4.135 \times 10^{-15} \text{eV s} \times 3 \times 10^{18} \text{Å/s}}{\lambda(\text{Å})} \tag{1.4}$$

or

$$E = \frac{12.398}{\lambda}, \tag{1.5}$$

where E is in keV and λ in angstroms. As an example the Cu $K\alpha_1$, $K\alpha_2$ doublet has an energy of about 8.05 keV, corresponding to a wavelength of $12.398/8.046 = 1.541$ Å.

In the early days of crystallography there was no standard value for—or way to determine—the wavelength of any particular X-ray photon. A practical definition was made defining wavelength in terms of the cubic lattice parameter of calcite. These units are referred to as *kX units* and were used in the literature into the 1950s. The *angstrom* (Å) *unit* has always been the preferred measure of wavelength and is related to kX (the crystallographic unit) by 1 Å = 1.00025 kX units. Even though the latest recommendation from the International Union of Pure and Applied Chemistry (IUPAC) discourages use of the angstrom and encourages use of the *nanometer* (nm; 1×10^{-9} m), the powder diffraction community has fought for retention of the angstrom and this remains the common unit in use in the field today. For this reason, in this book we will use the angstrom unit. The common electron-volt energy unit is also not IUPAC approved in that the standard energy unit is the joule (J), which may be converted by 1 eV = 1.602×10^{-19} J.

1.3. CONTINUOUS RADIATION

X-radiation arises when matter is irradiated with a beam of high-energy charged particles or photons. When an element is bombarded with electrons the spectrum obtained is similar to that shown in Figure 1.1. The figure illustrates the main features of the spectrum that would be obtained from a copper anode (target) X-ray tube, operated at 8.5, 25, and 50 kV, respectively. It will be seen that the spectrum consists of a broad band of continuous radiation (bremsstrahlung, or white radiation) superimposed on which are discrete wavelengths of varying intensity. The continuous radiation is produced as the impinging high-energy electrons are decelerated by the atomic electrons of the target element. The continuum is typified by a minimum wavelength, λ_{min}, which is related to the maximum accelerating potential V of the electrons. Thus, as follows from Equation 1.5,

$$\lambda_{min} = \frac{hc}{V} = \frac{12.398}{V}. \tag{1.6}$$

Note from Figure 1.1 that as the operating voltage is increased from 8.5 to 25 to 50 kV, the λ_{min} value shifts to shorter wavelengths and the intensity of the continuum increases. The intensity distribution of the continuum reaches a maximum intensity at a wavelength of about 1.5 to 2 times λ_{min}. The wavelength distribution of the continuum can be expressed quantitatively in

4 CHARACTERISTICS OF X-RADIATION

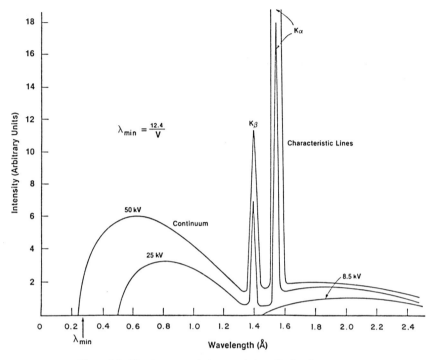

Figure 1.1. Continuous and characteristic radiation for copper.

terms of the excitation conditions by means of Kramers' formula [4]:

$$I_{(\lambda)}\,d\lambda = \frac{KiZ(\lambda_{\min}-1)}{\lambda^2}\,d\lambda. \tag{1.7}$$

Kramers' formula relates the intensity $I_{(\lambda)}$ from an infinitely thick target of atomic number Z with the applied current i where K is a constant. This expression does not correct for self-absorption by the target, which in practice leads to some modification of the intensity distribution.

It will also be seen from Figure 1.1 that somewhere between X-ray tube potentials of 8.5 and 25 kV sharp lines appear, superimposed on the continuum. These lines were shown by Moseley [5] to be *characteristic* wavelengths since their values differ for each unique target element. These characteristic lines will only appear when their equivalent excitation potential value V is exceeded. While the wavelengths of these characteristic lines are completely independent of the X-ray tube conditions, the intensities of the lines are very much dependent on the X-ray tube current i and voltage V; see Section 4.3.

1.4. CHARACTERISTIC RADIATION

1.4.1. The Photoelectric Effect

The processes whereby characteristic radiation is produced in an X-ray tube are based on interactions between the atomic electrons of the target and the incident particles. In the case described in Figure 1.2, the incident particles are high-voltage electrons. The incident particle can also be an X-ray photon, a γ-ray, or a proton. Each will produce similar effects if the energy of the particle is greater than the energy binding the electron to the nucleus. The atomic electron may be removed from its original atomic position leaving the atom in an *ionized* state. The free electron, called a *photoelectron*, will leave the atom with a kinetic energy $E - \phi_e$, i.e., equal to the difference between the energy E of the incident photon and the binding energy ϕ_e of the electron.

Figure 1.2 shows the basic processes involved in a photoelectric interaction. Figure 1.2a shows an atom with its various energy levels ϕ_K, ϕ_L, ϕ_M, etc., and incident upon it is a photon of energy E. Figure 1.2b shows the ejected photoelectron leaving the atom with an energy equal to $E - \phi_K$. Note that this process creates a vacancy in the atom, in this instance, with an equivalent energy of ϕ_K. One of the processes by which this vacancy can be filled is by transferring an outer orbital electron to fill its place. Such a transference is shown in Figure 1.2c, where an electron from the L level is transferred to the K vacancy. Associated with this electron transfer (and subsequent lowering of the ionized energy of the atom) will be the production of a fluorescent X-ray photon with an energy $E_{\text{X-ray}}$ equal to $\phi_K - \phi_L$. As will be shown later, this photon is called a $K\alpha$ photon.

1.4.2. The Auger Effect

An alternative deexcitation process, called the *Auger effect*, can also occur, and this effect is illustrated in Figure 1.2d. It may happen that the ionization of an inner shell electron produces a photon that in turn gets absorbed by an outer shell electron. Thus, the incident X-ray is absorbed by, for example, a K shell electron that leaves the atom. Next, an electron falls into the K shell, producing a $K\alpha$ photon. The $K\alpha$ photon, in turn, may be absorbed by an M electron, causing its ionization as an Auger electron. The kinetic energy of the emitted Auger electron is not just dependent on the energy of the initial X-ray photon (or particle) that ionized the K electron. Any incident particle with sufficient

6   CHARACTERISTICS OF X-RADIATION

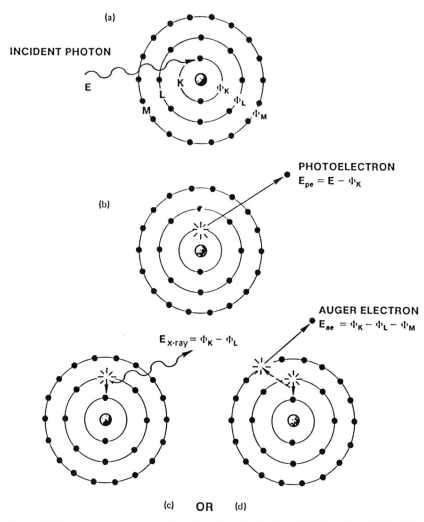

Figure 1.2. The Auger and photoelectric effects. From R. Jenkins, R. W. Gould, and D. Gedcke, *Quantitative X-Ray Spectrometry*, p. 16, Fig. 2–9. Dekker, New York, 1981. Reprinted by courtesy of Marcel Dekker Inc.

energy to create the initial vacancy can be responsible for the subsequent production of an Auger electron of unique energy. Study of the energy and intensities of Auger electrons, called *Auger spectroscopy*, allows measurement of the precise energy of the chemical bonds that involve the valence electrons.

CHARACTERISTIC RADIATION

1.4.3. Fluorescent Yield

It is apparent from the foregoing that there are two competing processes by which an ionized atom can regain its initial, or *ground*, state, these processes being the fluorescence of an X-ray photon and the Auger effect. The efficiency of the production of characteristic X-rays will be dependent upon the relative effectiveness of these two processes in a given atom. As the atomic number decreases, the production of Auger electrons becomes more probable and thus the production of K radiation falls off at low Z. The efficiency of a particular element producing fluorescent X-rays is quantified by the *fluorescent yield* ω. The fluorescent yield is the ratio of the number of photons produced from a given atomic shell to the number of equivalent shell vacancies created. For the production of K radiation from a specimen,

$$\omega_K = \frac{\text{number of } K \text{ photons produced}}{\text{number of } K \text{ shell vacancies created}}. \tag{1.8}$$

The probability of the production of an Auger electron is $1 - \omega$. Fluorescent yield values vary as the fourth power of atomic number and range from almost unity, for high atomic number elements, to 0.001, for low atomic numbers. For the wavelengths typically employed in powder diffraction, the K fluorescent yield values are about 0.5.

1.4.4. Selection Rules

Characteristic radiation arises from the rearrangement of the orbital electrons of the target element following the ejection of one or more electrons in the excitation process. The final resting place of the transferred electron determines the type of radiation, i.e., K, L, M, etc. Thus, ejection of a K electron leaves the atom in the highly energetic K^+ state. Transference of an electron from the L shell reduces the energy state from $K^+ \to L^+$ and the excess energy $(K^+ - L^+)$ is emitted as K X-radiation, in this instance $K\alpha$ radiation. Since every energy level has a unique binding energy, every element will have a unique set of binding energies and the energy state *differences* will also be unique (see, e.g., Table 1.1). There are obviously a great number of possibilities for electron transitions, particularly when one considers the various quantized states that each electron may have. However, in practice, X-ray spectra are far simpler than might appear at first sight, and just three selection rules cover the allowed transitions. These are $\Delta n \geq 1$, $\Delta l = 1$, and $\Delta J = 0$ *or* 1, where n is the *principal* or group quantum number, l the *angular* quantum number, and J the vector sum of the angular and *spin*(s) quantum numbers. It is common practice to refer to transition groups K, L_I, L_{II}, etc. (Table 1.2), which are simply built up

CHARACTERISTICS OF X-RADIATION

Table 1.1. Binding Energies for the K, L, and M Levels of Copper

Level	Binding Energy (keV)
K	8.978
L_{II}	0.953
L_{III}	0.933
M_{II}	0.078
M_{III}	0.075

Table 1.2. Siegbahn and IUPAC Notation for the K Series

Transition	Siegbahn	IUPAC	E (keV)	λ (Å)
KL_{III}	$K\alpha_1$	KL3	8.045	1.5406
KL_{II}	$K\alpha_2$	KL2	8.025	1.5444
KM_{III}	$K\beta_1$	KM3	8.903	1.3922
KM_{II}	$K\beta_3$	KM2	8.900	1.3922

Table 1.3. Construction of Transition Groups and Number of Electrons Allowed in Each State (Multiplicity)

Group	l	s	$J(=l+s)$	Multiplicity $(2J+1)$
K	0	+1/2	1/2	2
L_I	0	+1/2	1/2	2
L_{II}	1	−1/2	1/2	2
L_{III}	1	+1/2	3/2	4
M_I	0	+1/2	1/2	2
M_{II}	1	−1/2	1/2	2
M_{III}	1	+1/2	3/2	4
M_{IV}	2	−1/2	3/2	4
M_V	2	+1/2	5/2	6
N_I	0	+1/2	1/2	2
N_{II}	1	−1/2	1/2	2
N_{III}	1	+1/2	3/2	4
N_{IV}	2	−1/2	3/2	4
N_V	2	+1/2	5/2	6
O_I	3	−1/2	5/2	6
O_{II}	3	+1/2	7/2	8

by combining l and J quantum numbers. The construction of the transition groups is given in Table 1.3, which also shows the number of electrons allowed in each group. The familiar spectroscopic names of the various electron states derive from the values of l such that when $l = 0$, the state is called s; when $l = 1$, it is called p; when $l = 2$, it is called d; when $l = 3$, it is called f; and so on. The value of the J quantum number, which is the sum of l and s, determines the number of degenerate electron states for each energy level. For example, when l is 1 and $J = 1/2$, the two resulting energy states are called $p^{1/2}$; when l is 1 and $J = 3/2$, the four resulting states are referred to as $p^{3/2}$.

Figure 1.3 shows the usual transitions for the K spectrum of copper, giving both transition groups and quantum numbers. It will be seen that the copper K spectrum is quite simple, consisting of two α lines (called a doublet) from $2p^{1/2} \to 1s$ and $2p^{3/2} \to 1s$ transitions, and two β lines from $3p^{1/2} \to 1s$ and $3p^{3/2} \to 1s$ transitions. Since in the case of both doublets the energy difference between the lines in each pair is simply that due to the spin quantum number, the relative energy difference is very small. As will be seen in later sections, the energy difference between α_1 and α_2 being very small causes only partial separation of the two wavelengths to generally occur in diffraction. In practice the β_1, β_3 doublet is never resolved, but the α_1, α_2 is resolved at moderate-to-high diffraction angles. It should be noted that since copper does not have electrons in the $4p$ level the $4p \to 1s$ transition (the β_3 doublet) is absent. In the

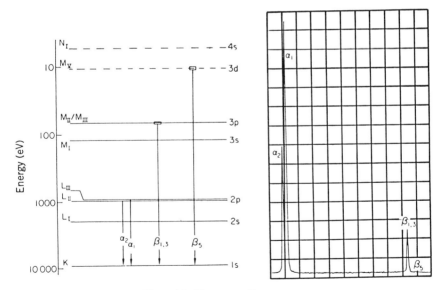

Figure 1.3. The copper $K\alpha$ spectrum.

case of higher atomic numbers such as silver (47) and molybdenum (42) the β_2 doublet is observed as an unresolved line. The relative intensities of the characteristic wavelengths are determined by the appropriate quantum mechanical transition probabilities. It is evident that a K shell with a missing electron represents a higher energy state than a similar hole in the corresponding L shell. The transition probability is a rather complex function of the difference in energy of the two levels concerned.

The relative intensity ratio of possible lines for an element is constant but may differ from one element to another. The greater the energy difference, the less probable the transition becomes and the less intense is the resulting line. Consequently the intensity of $K\alpha_1$, $K\alpha_2 > K\beta_1$, $\beta_3 > K\beta_2$. For a copper anode the ratio is about 5:1:0, and for molybdenum about 3:1:0.3. The relative intensity of α_1, α_2 (and also $K\beta_1$, β_3) is much simpler to predict since for these line pairs the Burger–Dorgelo rule [6] holds, stating that the intensity ratio is equal to the number of electrons that may make the transition. In the case of the $K\alpha_1:K\alpha_2$ ratio, there are four $p^{3/2}$ electrons (as shown in Table 1.3) giving rise to the $K\alpha_1$ line and two $p^{1/2}$ electrons giving rise to the $K\alpha_2$ line. Thus, the intensity ratio is 4:2 or 2:1.

The original nomenclature system for X-ray wavelengths was proposed by K. M. G. Siegbahn in the 1920s and is properly called the *Siegbahn notation*. Since the introduction of the Siegbahn notation a number of lines have been observed that have not been classified within the Siegbahn nomenclature, particularly for the M and N series. A further complication is that the Siegbahn notation is unsystematic and consequently rather difficult to learn. In recent years this problem has been addressed by the IUPAC, with the result that a new *IUPAC notation* has been recommended [7]. At the time of the publication of this book, the acceptance of the new IUPAC nomenclature among the X-ray community is still uncertain. However, for information, Tables 1.2 and 1.4 list typical Siegbahn and IUPAC equivalents.

Table 1.4. Nomenclature for the Copper K Series Wavelengths

Transition	Siegbahn	IUPAC
$2p^{3/2} \to 1s$	$K\alpha_1$	KL3
$2p^{1/2} \to 1s$	$K\alpha_2$	KL2
$3p^{3/2} \to 1s$	$K\beta_1$	KM3
$3p^{1/2} \to 1s$	$K\beta_3$	KM2
$2p^{3/2}(2p^{-1}) \to 1s$	$K\alpha_3$	KL3,3
$2p^{1/2}(2p^{-1}) \to 1s$	$K\alpha_4$	KL2,3
$2p^{3/2}(2s^{-1}) \to 1s$	$K\alpha_5$	KL3,1
$2p^{3/2}(2s^{-1}) \to 1s$	$K\alpha_6$	KL2,1

1.4.5. Nondiagram Lines

Not all observed characteristic X-ray lines can be satisfactorily described by the selection rules just outlined. Other lines occur following special conditions of ionization that generally fit into one of two categories—forbidden transitions and satellites. The origin of forbidden transitions is rather complex, but, in simple terms, forbidden transitions arise because outer orbital electrons are typically not distributed in absolutely unique and well-separated orbitals. For example, there is much hybridization of outer orbitals, meaning that an s electron may tend to show the character of a p electron and so on. Thus, transitions may occur that are close to obeying the selection rules, but in which the electron being transferred acts as if it had a different angular quantum number than expected. As an example, in the Cu K spectrum a weak $K\beta_5$ occurs from a $3d \rightarrow 1s$ transition. Since such a line corresponds to a Δl of 2 it is *forbidden* by the selection rules for normal lines.

Satellite lines occur from transitions involving removal of more than one electron from a target atom (dual ionization). Although the excitation/deexcitation process is fast ($\simeq 10^{-12}$ s), it is finite, and there is a probability that a second electron may be removed before the first vacancy is filled. Figure 1.4 shows the origin of the $K\alpha$ satellites in the Cu K spectrum. The

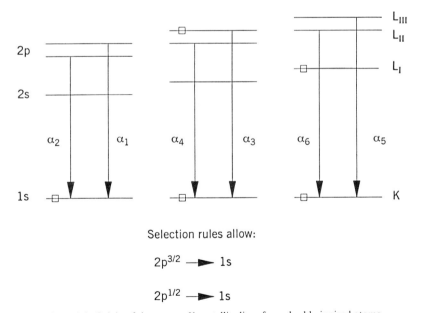

Selection rules allow:

$2p^{3/2} \longrightarrow 1s$

$2p^{1/2} \longrightarrow 1s$

Figure 1.4. Origin of the copper $K\alpha$ satellite lines from doubly ionized atoms.

left-hand diagram shows the usual situation with a K vacancy being filled by a $2p^{3/2} \to 1s$ and a $2p^{1/2} \to 1s$ transition, giving the $K\alpha_1$ and $K\alpha_2$ lines, respectively. The center diagram shows similar transitions, except that now there are two atomic vacancies, one in the K shell and one in the L_{III} level. Removal of the L_{III} electron decreases the total electron charge of the atom and the attraction of the charge by the nucleus of the atom. There is a consequent widening of the energy gap between the K and L levels. Hence, the two transitions give a pair of lines similar to the $K\alpha_1$ and $K\alpha_2$, but of shorter wavelength. These lines are called the α_3 and α_4. The right-hand diagram shows a similar circumstance in which the second vacancy is in the L_I level, which gives rise to the α_5 and α_6 lines. Thus, each of the $K\alpha_1$ and $K\alpha_2$ lines is actually a triplet, and there are actually six lines (i.e., two triplets) in what is usually called the $K\alpha_1$, $K\alpha_2$ doublet. A major difference between the satellites and the forbidden transitions is that the satellites occur close to the α_1, α_2 doublet, and even though they are not resolved by the normal monochromatization devices, they do play some part [8] in the profile-fitting process.

1.4.6. Practical Form of the Copper K Spectrum

From the foregoing discussion it will be clear that the characteristic α radiation emission from copper is much more complex than the simple α-doublet and β-doublet model generally employed in classical powder diffractometry. The relative intensities of the satellite lines in each line within each triplet differ somewhat in the α_1 and α_2 sets, which probably accounts for the higher degree of asymmetry of the α_2 relative to the α_1 typically observed. The largest energy gap within any of the triplets is only about 2.5 eV. Since the absolute energy resolution of the powder diffractometer using Cu $K\alpha$ radiation ranges from about 200 eV at $10° 2\theta$ to about 2.5 eV at $140° 2\theta$, the fine structure of the triplets is not resolved. However, as indicated in Figure 1.5, asymmetry is introduced in the α_2, which starts to become apparent at very high 2θ values. Even more important, where profile-fitting methods are employed, the effective "fitting resolution" is probably on the order of a few electronvolts, and here allowance must be made for wavelengths other than the α_1 and α_2 if accurate (<2% or so) fitting is required. For most practical purposes, however, in powder diffractometry, the copper K spectrum is considered to consist simply of two pairs of lines, the $K\alpha_1$, $K\alpha_2$ doublet occurring from $2p \to 1s$ transitions; and the $K\beta_1, \beta_3$ doublet from $3p \to 1s$ transitions. In most experimental work the β doublet intensity is typically reduced to less than a few percent of the α-doublet intensity by use of filtration or is removed by use of a crystal monochromator or an Si(Li) energy-resolving detector. In each case, what remains is essentially bichromatic radiation.

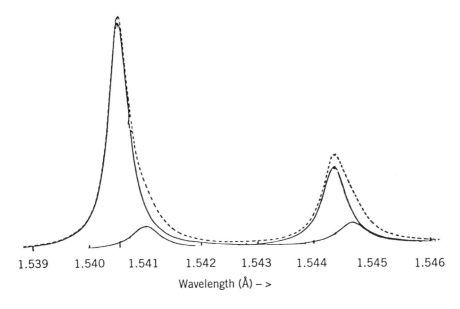

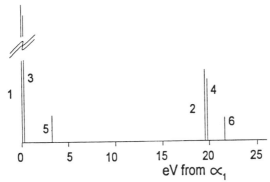

Figure 1.5. Full spectrum for copper K radiation.

The most commonly used values for the wavelengths of Cu $K\alpha_1$ and Cu $K\alpha_2$ are 1.54056 and 1.54439 Å, respectively. These values were reported by Bearden [9] and have been recommended in the International Union of Crystallography (IUCr) publication *International Critical Tables* [10]. The values are generally given in terms of a unit length in angstrom units, based on the energy of the W $K\alpha$ line of 59.31821 keV. There is a degree of uncertainty that arises because of the conversion of X units (Xu) to angstrom units. Bearden uses a value of 1.002056 for this conversion factor. Because of this uncertainty, minor differences will be found in other tables. As an example,

Cauchois and Senemaud [11] list a value of 1537.400 Xu with a conversion factor of 1.0020802 to give 1.540598 Å. This value has also been used by the National Bureau of Standards (NBS; now the National Institute of Standards and Technology, NIST) and has been widely used in the powder diffraction community for the last 15 years. The NBS value is based on the techniques used by Deslattes et al. [12]. There is clearly some inconsistency between these values, and Bearden et al. [13] have suggested a new value of 1537.370 Xu for Cu $K\alpha$. Most recently Härtwig et al. [14] have suggested 1.54059292 Å. These minor variations will not affect most X-ray powder diffraction measurements, and we recommend the use of 1.54060 Å for the Cu $K\alpha_1$ line.

1.5. SCATTERING OF X-RAYS

Electromagnetic radiation is a form of energy that can be described as an oscillating electric field **E** with an oscillating magnetic field **H** at right angles to it, as shown in Figure 1.6. The magnetic field will only interact with other magnetic fields and is therefore not generally important in considering the interactions of X-rays with matter. However, the oscillating electric field will couple to the charged electrons surrounding the atoms and cause them to accelerate and decelerate. Since an electron bound to an atom has an amount of energy fixed by the laws of quantum mechanics, the extra energy imparted to it from the acceleration must be reradiated or absorbed by perhaps stimulating a vibrational mode of the lattice. The phenomenon known as scattering occurs when any of the incident energy is reradiated.

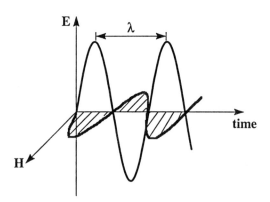

Figure 1.6. Electromagnetic radiation.

1.5.1. Coherent Scatter

Coherent scatter, or elastic scatter, can be thought of as a perfectly elastic collision between a photon and an electron. The photon changes direction after colliding with the electron but transfers none of its energy to the electron. The result is that the scattered photon leaves in a new direction but with the same phase and energy as that of the incident photon.

From the wave perspective one thinks of the incoming wave being instantaneously absorbed by an electron and reemitted (i.e., in a time interval short enough to fall within the uncertainty principle) as spherical waves. Thus, each electron on a scattering atom in a material will act as a Huygens scattering center. If the scattering atoms are arranged in an orderly manner, with a distance between each on the order of the wavelength of the radiation, then the phase relationships between scatterers will become periodic and interference diffraction effects will be observed at various angles of view (see Section 3.1).

1.5.2. Compton Scatter

It can also happen that the X-ray photon loses part of its energy in the collision process, especially where the electron is only loosely bound. In this case the scatter is said to be incoherent (*Compton scatter*) and the wavelength of the incoherently scattered photons will be longer than the coherently scattered wavelength. Compton scatter amounts to an inelastic collision between a photon and an electron. Part of the energy of the incident photon is absorbed by an electron, and the electron is ionized. However, instead of all of the remaining energy of the original photon converting to kinetic energy of the ionized photoelectron, some of it is reemitted as an X-ray photon of lower energy. Not only has the energy of this Compton photon been lowered, but it loses any phase relationship to the incident photon. For this reason the process is often called *incoherent scatter*. Since the Compton (phase) modified photons are emitted in arbitrary directions very few of them will reach the detector and, therefore, this is also a source of absorption. Compton scatter decreases in importance as the atomic number of the scatterer increases.

The total scatter σ is made up of both coherent and incoherent terms:

$$\sigma = Zf^2 + (1 - f^2) \qquad (1.9)$$

The first term is the coherent term, and the second is the incoherent term; f is called the atomic scattering factor and will be discussed in Section 3.6.2.

1.6. ABSORPTION OF X-RAYS

When a beam of X-radiation falls onto an absorber, a number of different processes may occur. The more important of these are illustrated in Figure 1.7. In this example, a monochromatic beam of radiation of wavelength λ and intensity I_0 is incident on an absorber of thickness t (with differential thickness dt) and density ρ. A certain portion, I, of the radiation may pass through the absorber. Where this happens the wavelength of the transmitted beam is unchanged and the intensity of this transmitted beam $I(\lambda)$ is given by

$$I(\lambda) = I_0(\lambda)\exp\left[-\left(\frac{\mu}{\rho}\right)\rho t\right], \quad (1.10)$$

where μ/ρ is the mass attenuation coefficient of the absorber for the wavelength λ and the density ρ. Equation 1.10 is very general and is called the mass-absorption law. In other parts of the spectrum the same equation will be called the Lambert–Beer law. The value of the X-ray mass attenuation coefficient μ/ρ in Equation 1.10 is a function both of the photoelectric absorption τ and the scatter σ:

$$\frac{\mu}{\rho} = f(\tau) + f(\sigma). \quad (1.11)$$

The scatter term contains contributions from coherent and incoherent scatter.

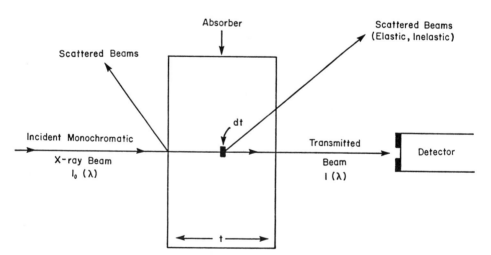

Figure 1.7. Absorption processes for X-rays.

However, τ is generally large in comparison with σ and generally $\mu/\rho \simeq f(\tau)$. For this reason, the *mass attenuation coefficient* is often referred to as the *mass absorption coefficient*. The mass attenuation coefficient is independent of the physical state of a material (i.e., solid, liquid or gas) and depends only on the wavelength of the incident radiation. The wavelength dependence is roughly proportional to the cube of λ. However, since no one has found an exact theoretical relationship for the wavelength dependence, we must resort to measuring μ/ρ for all of the commonly used wavelengths and tabulating them. An empirical relationship,

$$\frac{\mu}{\rho} = KZ^4 \lambda_{cm}^3, \qquad (1.12)$$

known as the Bragg–Pierce law, has been established, where Z is the atomic number and K is an empirical constant that is different on each side of each of the absorption edges shown in Figure 1.8.

The difference between I and I_0 for a fixed wavelength is dependent on the thickness of the absorber and on the *linear absorption coefficient* μ, which is a constant related to the absorbing material. Since all of the absorption processes shown in Figure 1.7 ultimately depend on the presence of electrons, clearly the ability of a material to absorb electromagnetic radiation is related to the density of electrons. In turn, the electron density of a material is determined by the types of atoms composing the material and the closeness of

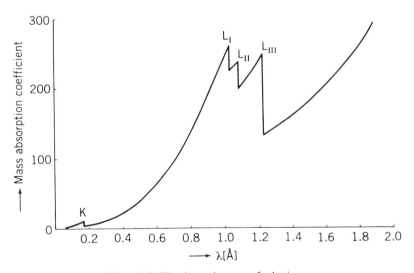

Figure 1.8. The absorption curve for barium.

their packing. The linear absorption coefficient of a material, therefore, depends on the types of atoms present and the density of the material. However, with the elimination of the functional dependence on density, which is determined by the type and strength of the chemical bonds in a material, a true constant for each type of element is obtained. Thus, μ/ρ is characteristic of each element at any specified wavelength. The absorption coefficients of the elements are listed in Appendix B.

Photoelectric absorption occurs at each of the energy levels of the atom, and the total photoelectric absorption τ(total) is determined by the sum of each of the individual absorptions. Thus,

$$\tau(\text{total}) = \tau(K) + \tau(L) + \tau(M) + \tau(N) + \cdots + \tau(n), \tag{1.13}$$

where $\tau(n)$ represents the outermost level of the atom containing electrons. It is apparent that all radiation produced as a result of electron transitions following ejection of orbital electrons must have a wavelength longer than that of the source which stimulated the excited state. Also, not all of the radiation produced is X-radiation; hence the photoelectric effect must be giving rise to X-radiation (λ) from the absorber and other photons.

The various contributions to the total absorption from the different energy levels is illustrated in Figure 1.9, which shows the absorption curve for barium. As is seen in the figure, the value of the mass attenuation coefficient increases steadily with wavelength in accord with the Bragg–Pierce law (Equation 1.12); also the curve has very sharp discontinuities, called absorption edges, indicated as K, L_I, L_{II}, L_{III}, etc. These absorption edges correspond to the binding energy of electrons in the appropriate levels. Where the absorbed wavelength is shorter than the wavelength of one of the edges, an electron from the corresponding level can be excited. For instance, in Figure 1.9 an absorbed wavelength of 0.3 Å is shorter than the K absorption edge of barium (0.332 Å) and hence photoelectric absorption in the K level can occur. For an absorbed wavelength of 0.4 Å, however, photoelectric absorption in the K level certainly cannot occur. Thus, in general terms, each time the wavelength increases to a value in excess of a certain absorption edge, one of the terms in Equation 1.13 drops out with a corresponding decrease in the value of the total absorption term. Mass attenuation coefficients are well documented for most of the X-ray region, and the data are readily available in tabular form [15, 16]. Where the specimen is made up of n elements, the total mass attenuation coefficient of the specimen μ_s is given by

$$\left(\frac{\mu}{\rho}\right)_s = \sum_{i=1}^{i=n} \left(\frac{\mu}{\rho}\right)_i w_i, \tag{1.14}$$

where w_i is the weight fraction of element i.

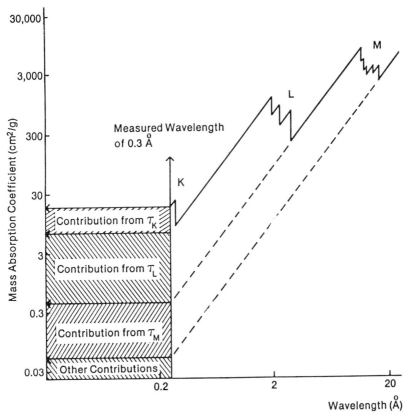

Figure 1.9. The absorption curve for barium, plotted on a log/log scale showing the contributions of the different levels. From R. Jenkins, *X-Ray Fluorescence Spectrometry*, p. 10, Fig. 1–4. Copyright © 1988, John Wiley & Sons, Inc. Reprinted by permission of the publisher.

1.7. SAFETY CONSIDERATIONS

It has been shown that X-rays are beams of energetic electromagnetic radiation that ionize matter with which they interact by ejecting electrons from their atoms. The extent of the ionization, absorption, and even molecular change of the material depends on the quantity (radiation flux and intensity) and the quality (the spectral distribution of the photon energy) of the radiation. Living organisms that are exposed to various doses of X-radiation can be injured by such exposures, and death may result if the exposure is particularly severe (see Figure 1.11). The amount of damage done to the body

CHARACTERISTICS OF X-RADIATION

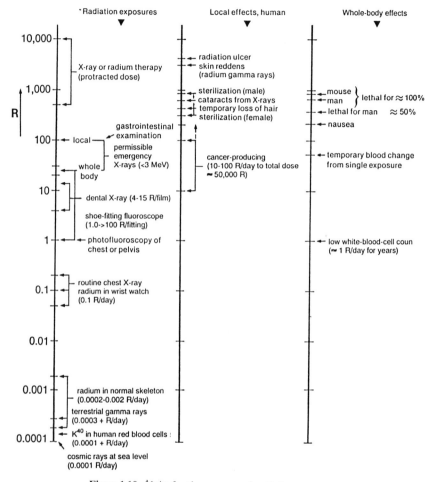

Figure 1.10. Units for the measure of radiation exposure.

by the radiation is often difficult to estimate, but Figure 1.10 gives typical effects of various levels of exposure to human beings. The extent of the damage depends on the energy of the radiation, the total dose received, the type of tissue exposed, the dose rate, and the volume of the body exposed. It is thus vitally imperative that all operators of X-ray instruments be knowledgeable in their use to protect themselves and others from injury [17–19].

The standard unit for measuring the quantity of ionizing radiation is the röntgen (R). The röntgen is the amount of radiation needed to create one electrostatic unit (2.093×10^9 ion pairs) in 1 cm^3 of dry air. This unit is based

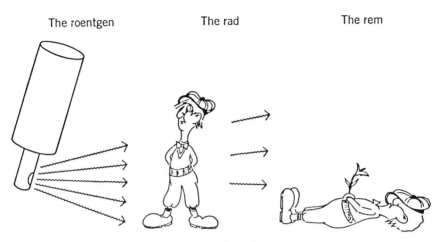

Figure 1.11. Effects of radiation exposure.

on the ionization energy of oxygen and nitrogen molecules and can be translated into the transfer of 5.24×10^{13} eV of energy per gram of air. For Cu radiation about 100 photons/cm^2 gives a dose rate of 1 mR/h, or a dose of 10^5 photons/cm^2 corresponds to 1 R. Because the röntgen is dependent on the energy of the radiation and the absorption of air, it does not directly relate to how human tissue absorbs radiation. Thus, another unit, the rad, is defined as the amount of radiation that will cause the absorption of 100 erg/g in tissue. A third unit is the rem (röntgen equivalent man), which is the absorbed dose corrected for the equivalent absorption of radiation in living tissue. Thus, the dose in rem = (dose in rad × RBE), where RBE is defined as the *relative biological effectiveness*. The RBE value for radiation typically employed in diffraction experiments is about unity. The gray (Gy) is the SI unit of absorbed dose: 1 Gy = 1 J/kg = 100 rad. Similarly, the sievert (Sv) is the SI unit of absorbed dose equivalent: 1 Sv = 1 J/kg = 100 rem.

REFERENCES

1. Le Galley, D. P. A type of Geiger–Muller counter suitable for the measurement of diffracted Mo K X-rays. *Rev. Sci. Instrum.* **6**, 279–283 (1935).
2. Parrish, W., and Gordon, S. G. Precise angular control of quartz-cutting by X-rays. *Am. Mineral.* **30**, 326–346 (1945).
3. Friedman, H. A Geiger counter spectrometer for industrial research. *Electronics* April (1945).

4. Kramers, H. A. On the theory of X-ray absorption and of the continuous X-ray spectrum. *Philos. Mag.* [6]. **46**, 836–871 (1923).
5. Moseley, H. G. J. The high frequency spectra of the elements. *Philos. Mag.* [6] **26**, 1024–1034 (1912); **27**, 703–714 (1913).
6. Dorgelo, H. B. Photographic photometry of line spectra. *Phys. Z.* **26**, 756–793 (1925).
7. Jenkins, R., Manne, R., Robin, J., and Senemaud, C. Nomenclature, symbols, units and their usage in spectrochemical analysis. VIII. Nomenclature system for X-ray spectroscopy. *Pure Appl. Chem.* **63**, 736–746 (1991).
8. Snyder, R. L. Analytical profile fitting of X-ray powder diffraction profiles in Rietveld analysis. In *The Rietveld Method* (R. A. Young, ed.), Chapter 7, pp. 111–131. Oxford Univ. Press, Oxford, 1993.
9. Bearden, J. A. X-ray wavelengths. *Rev. Mod. Phys.* **39**, 78–124 (1967).
10. *International Critical Tables for X-ray Crystallography*, Vol. IV: *Revised and Supplementary Tables*. Kynoch Press; Birmingham, England, 1974.
11. Cauchois, Y., and Senemaud, C. *International Tables of Selected Constants*, Vol. 18: *Wavelengths of X-ray Emission Lines and Absorption Edges*. Pergamon, Oxford, 1978.
12. Deslattes, R. D., Henins, A., and Kessler, E. G., Jr. Accuracy in X-ray wavelengths. In *Accuracy in Powder Diffraction*, NBS Spec. Publ. 567. U.S. Dept. of Commerce, National Bureau of Standards, Gaithersburg, MD, 1980.
13. Bearden, J. A., Henins, A., Marzolf, J. G., Sauder, W. C., and Thomsen, J. S. Precision determination of standard reference wavelengths for X-ray spectroscopy. *Phys. Rev. A* **135**, 899–910 (1964).
14. Härtwig, J., Hölzer, G., Wolf, J., and Förster, M. Remeasurement of the profile of characteristic Cu $K\alpha$ emission line with high precision and accuracy. *J. Appl. Crystallogr.* **26**, 539–548 (1993).
15. Leroux, J., and Thinh, T. P. *Revised Tables of X-ray Mass Attenuation Coefficients*. Corporation Scientifique Claisse Inc., Sainte-Foy, Quebec, Canada, 1984.
16. Wilson, A. J. C., ed. *International Tables for Crystallography*, Vol. C: *Mathematical, Physical and Chemical Tables*. Kluwer Academic Publishers, Dordrecht, The Netherlands, 1992.
17. Jenkins, R., and Haas, D. J. Hazards in the use of X-ray analytical instrumentation. *X-Ray Spectrom.* **2**, 135–141 (1973).
18. Jenkins, R., and Haas, D. J. Incidence, detection and monitoring from X-ray analytical instrumentation. *X-Ray Spectrom.* **4**, 33–42 (1975).
19. Upton, A. C., Health effects of low-level ionizing radiation. *Phys. Today*, August, pp. 34–39 (1991).

CHAPTER
2

THE CRYSTALLINE STATE

In this chapter all of the concepts from crystallography required for the applications of modern powder diffraction analysis will be introduced. The reader desiring a more comprehensive treatment is referred to a fundamental text on the subject (e.g., Phillips [1]).

2.1. INTRODUCTION TO THE CRYSTALLINE STATE

Solid substances form when the electrostatic attractions between atoms, ions, or molecules overcome thermal motion and cause the loss of translational freedom. The spatial arrangements that the species settle into depend on the nature of the bonding forces present but will always represent a configuration that minimizes the electrostatic interactions or lattice energy. Different types of chemical bonding cause solids to be organized in the following four different ways.

First, covalent species satisfy their primary bonding needs by forming strongly bonded molecules that are attracted to each other by the relatively weak van der Waals (dipolar) attractions. The covalent forces direct the construction of a very wide variety of molecular shapes. The orderly periodic packing of such shapes will produce the greatest variety of special packing arrangements.

A second type of solid is based on the packing of ions that do not form discrete molecules; their minimum energy spatial arrangements relate to the methods of packing together spheres of differing sizes.

The third type of solid is composed of atoms that display both ionic and covalent character in their bonding. Many materials like silicates and metal oxides form chains that extend through the entire crystal. These materials will also show a large variety of stable periodic crystalline arrangements.

A fourth type of solid, typified by metals, has valence electrons delocalized and packed together in simple arrangements. In the simplest cases the arrangements are exactly the same as will be found in a box filled with Ping-Pong balls.

Figure 2.1 shows four different crystalline arrangements of SiO_2. Figure 2.1a shows the arrangement of SiO_2 atoms known as low-temperature or

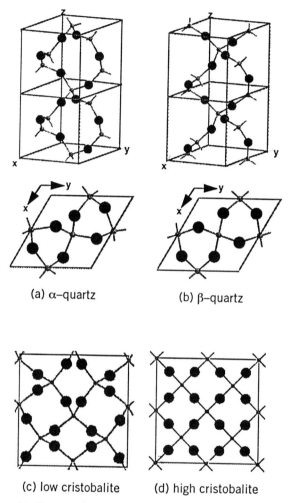

(a) α-quartz   (b) β-quartz

(c) low cristobalite   (d) high cristobalite

Figure 2.1. The crystal structures of the four crystalline phases of SiO_2.

α-quartz. Figure 2.1b shows another crystalline arrangement of SiO_2 atoms known as high-temperature or β-quartz, which is the minimum energy arrangement above 573 °C. In fact, SiO_2 formed under differing conditions can display a number of other crystalline arrangements such as low- and high-temperature cristobalite, shown in Figure 2.1c, d. When rapidly cooled from a melt, SiO_2 can also form an amorphous solid called silica. This category of solids includes packing arrangements that do not have any long-range periodicity. However, the atomic arrangements in these glasses still correspond to

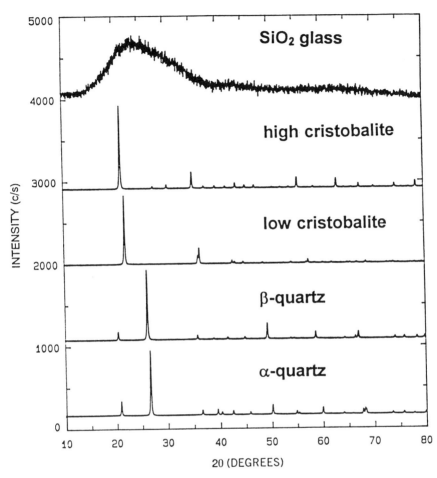

Figure 2.2. The X-ray powder diffraction patterns of α- and β-quartz, high- and low-temperature cristobalite, and amorphous silica.

a minimum packing energy, albeit one of higher energy than the thermodynamically stable crystalline phase, at a given pressure and temperature.

The most dramatic difference among crystalline and amorphous materials is in how they scatter an X-ray beam. Figure 2.2 shows the intensity of X-ray scattering at different angles from amorphous silica at the top, followed, in descending order, by high- and low-temperature cristobalite and β- and α-quartz. All of these phases have exactly the same chemical composition. The crystalline forms produce sharp lines due to diffraction, whereas the glass, like a liquid, produces a few broad peaks indicating only some short-range order in the atomic arrangement. Note that the diffraction patterns from each of the crystalline phases are diffferent from each other and may therefore be used in qualitative phase identification.

Because energy minimization is the only principle determining the arrangement of ions or molecules in space, an active area of science has grown in recent years in which the interionic and intermolecular forces are modeled with potential energy functions and a computer is used to calculate the energy of each possible spatial arrangement. When the atom–atom potential energy functions are well modeled, these simulations can not only predict the correct crystalline arrangement of atoms and molecules, they can also be used to predict many physical properties of the crystal and even describe the defect structure [2]. Science, however, remains an empirical endeavor, and so experimental crystal structure analysis through diffraction methods is still one of our most important tools.

In X-ray powder diffractometry one is generally dealing exclusively with crystalline (ordered) materials. The diffraction pattern can, moreover, be used to determine the degree of crystallinity, e.g., the mass fraction of the crystalline regions in otherwise amorphous substances.

2.2. CRYSTALLOGRAPHIC SYMMETRY

Nature appears to be full of symmetric patterns. Most animals have an approximate mirror plane through their vertical center lines. Snowflakes grow in dendritic patterns that repeat about the center every 60°. Many minerals form large beautiful crystals expressing various symmetries. Crystals form in accordance with the fundamental principle of energy minimization. Symmetry can be thought of as an invisible motion of an object. If an object is hidden from view and moved in a manner such that one cannot tell that it has been moved on reappearance, then the object is said to possess symmetry. For example, if a perfect snowflake is covered and then rotated about its center by 60°, after uncovering it an observer will not be able to tell that any movement has occurred. The rotation of an object around an imaginary axis is the simplest symmetry element to visualize and is referred to as *rotational symmetry*. To get the complete set of all possible symmetry elements that an object may possess, it is also necessary to allow for the possibility that on removing the object from view no movement at all was made. No motion, of course, produces an effect identical to rotating the object about any of its symmetry axes and must then also be a symmetry element. This element is called the identity operator by mathematicians but is more simply referred to as a 360° rotation by crystallographers.

CRYSTALLOGRAPHIC SYMMETRY   27

Figure 2.3. A proper and improper fourfold rotation.

Rotational symmetry is said to occur around a *proper axis* when the *handedness* of an object does not change. For example, think of two right hands arranged opposite each other on the circumference of a circle, i.e., separated by 180°. The *twofold axis* relating them is proper, in that the hands remain *right* during each rotation. A rotation that changes the "handedness" after each operation is called *improper*. An improper rotation is a rotation followed by the operation of inversion through the origin of a coordinate system at the center of symmetry. Inversion causes an object located at x, y, z to change coordinates to $-x, -y, -z$ written as $\bar{x}, \bar{y}, \bar{z}$. This, of course, turns a right hand into a left. Figure 2.3 shows a proper and improper fourfold rotation. The improper twofold axis is exactly the same as reflection in a mirror, and the symbol m is used for it rather than the equivalent $\bar{2}$. Table 2.1

Table 2.1. The Crystallographic Point Group Symmetry Elements

Degrees of Rotation ($360/n$)	Proper Axis (n)	Improper Axis ($\bar{n}$)
360°	1	$\bar{1}$
180°	2	$m(=\bar{2})$
120°	3	$\bar{3}$
90°	4	$\bar{4}$
60°	6	$\bar{6}$

shows the Hermann–Mauguin notation for the complete set of crystallographic symmetry elements.

An object, for example, a potato, may have a great number of possible symmetry elements that are not listed in Table 2.1. A simple exercise with a paring knife can put a fivefold rotation axis onto a potato. However, when such potatoes are poured into a box, the space-filling arrangement that they will take up will never show the fivefold symmetry of the individual potatoes. So, a spectroscopist will take account of a fivefold molecular axis in explaining the molecular properties of a material but, because a set of pentagons cannot be arranged periodically to fill space, this symmetry may never be expressed crystallographically. Thus, Table 2.1 shows the only symmetries that the periodic packing of objects may have. For example, the molecule benzene has a perfect molecular sixfold axis of rotation; however, when these molecules crystallize they align in a herringbone pattern that has only twofold and mirror crystallographic symmetry. In this case the sixfold molecular axis, which could be a crystallographic symmetry element, does not express itself in the packing arrangement.

In recent years, a new class of materials called *quasicrystals* has been discovered, composed of atoms that fill space but do so in an orderly but non-periodic manner. In these materials, classically forbidden symmetry elements like a fivefold rotation are permitted and have been observed.

2.2.1. Point Groups and Crystal Systems

Having examined the possible types of simple symmetry that an object which can fill space may possess, we may now ask the question: how many unique combinations of symmetry may these space-filling objects have? The simplest form of symmetry group is 1, or no symmetry at all. The addition of any other symmetry element to group 1 will, of course, raise the total symmetry and have to be classified as a new group. The addition of a center of symmetry (also called an inversion center) gives the group $\bar{1}$. Increasing the symmetry further gives objects containing only a twofold axis (group 2) or a mirror (group m). There is now an opportunity to combine two symmetries in an object to form a unique new group, $2/m$, which refers to objects having a twofold axis with a mirror perpendicular to it. Making the mirror plane parallel to the twofold axis would require that the two axes in the perpendicular plane also be at least twofolds, putting the object into the group 222, $mm2$, or mmm. Similar arguments lead to the conclusion that there are only 32 unique ways of combining symmetry elements in objects that can repeat in three dimensions to fill space. These 32 combinations are called *point groups* or *crystal classes* and are shown in Appendix D.

CRYSTALLOGRAPHIC SYMMETRY 29

Table 2.2. The Seven Crystal Systems

Crystal System	Axis System	Symmetry	Lattice
Cubic	$a = b = c, \quad \alpha = \beta = \gamma = 90°$	$m3m$	P, I, F
Tetragonal	$a = b \neq c, \quad \alpha = \beta = \gamma = 90°$	$4/mmm$	P, I
Hexagonal	$a = b \neq c, \quad \alpha = \beta = 90°, \quad \gamma = 120°$	$6/mmm$	P
Trigonal[a]	$a = b = c, \quad \alpha = \beta = \gamma \neq 90°$	$3m$	R
Orthorhombic	$a \neq b \neq c, \quad \alpha = \beta = \gamma = 90°$	mmm	P, C, I, F
Monoclinic	$a \neq b \neq c, \quad \alpha = \gamma = 90°, \quad \beta \neq 90°$	$2/m$	P, C
Triclinic	$a \neq b \neq c, \quad \alpha \neq \beta \neq \gamma \neq 90°$	1	P

[a] Or rhombohedral.

A comment should be made here on the type of coordinate systems required to specify these various symmetries. In groups 1 and $\bar{1}$ there need be no relationship between the three coordinate axes or the angles between them. This coordinate system is referred to as *triclinic* or *anorthic*. For groups 2, m and $2/m$, there need be no relation between the lengths of the three axes, but this symmetry requires that one axis meet the plane of the other two at right angles. This system, with one unique axis commonly labeled **b**, is called *monoclinic*. The groups 222, $mm2$, and mmm require that all three axes be orthogonal. This system is called *orthorhombic*. Increasing the symmetry further requires that two or more of the axes must be equal in length. In the *hexagonal* and *tetragonal* systems, the unique six- or fourfold axis is called **c** by convention. In the *cubic* system, all three orthogonal axes are equal. The last crystal system is called *trigonal*, which contains a unique 3 or $\bar{3}$ symmetry. A group of trigonal materials may be described with the unique *rhombohedral* cell shown later in Figure 2.5. The trigonal system differs somewhat from the others in that it can be represented on a hexagonal axis; thus, only six coordinate systems are essential to represent all symmetries. However, the simplest rhombohedral coordinate system has all three axes equal in length and the three interaxial angles also equal. The seven different coordinate systems required to represent all point group symmetries are called *crystal systems* and are shown in Table 2.2.

2.2.2. The Unit Cell and Bravais Lattices

Many crystalline substances have a habit or form that is retained no matter what the physical size of the crystal. For example, a potassium chloride crystal usually grows in the form of a cube. If a potassium chloride crystal were reduced in size until the smallest group of atoms repeating in space was found, it would be seen that a single K and Cl pair could be reproduced in space to make up the crystal. This smallest repeating unit is a point group and must possess one of the 32 allowed combinations of symmetries. It is much easier when thinking about three-dimensional periodicity to reduce the actual repeating chemical point group to a single representative point called a *lattice point*. Since the full crystal structure results from the repetition of the lattice point in space, solid state physicists commonly call it the *basis* of the structure; crystallographers more commonly refer to it as the *asymmetric unit* or *structural motif*. In potassium chloride a K and Cl pair is represented by such a lattice point; in more complex structures a lattice point might represent a whole or perhaps part of a protein. The three-dimensional array of points representing the full crystal structure is called a space lattice. If lines are drawn between adjacent lattice points a unit cell will be defined that can also be thought of as the unit that repeats in space to define the space lattice. The volume of the unit cell corresponds to the volume occupied by the atoms represented by the lattice point. Because the unit cell encloses a single lattice point, its volume corresponds to the spatial volume of, in the case of KCl, a K and a Cl atom. The edges of the unit cell are called *translation vectors* and are labeled **a**, **b**, and **c**. The interaxial angles are described by the Greek letter equivalent of the symbol not used to describe the translation vector: the angles between **c**/**b**, **c**/**a**, and **a**/**b** are referred to as α, β, and γ. The unit cell defines the directions and distances a lattice point must be translated in order to find another point in the space lattice.

The reader may be surprised to learn at this point that the unit cell we have described for KCl may not look like a cube. In fact, there are a large number of ways of connecting adjacent lattice points lying in different directions, defining a large number of possible unit cells. A few of the possible unit cells for the KCl space lattice are shown in Figure 2.4. In crystallography, one always uses symmetry as the criterion for choosing a unit cell or point group. The unit cell with maximum symmetry will be the unique cell that is chosen to describe any space lattice. Looking again at Figure 2.4 it will be seen that a unit cell drawn with an extra lattice point in the center of each of the six faces possesses cubic symmetry, where the angles between all of the principal faces are 90°. In fact, there are other crystal classes that also have angles between faces of 90°, but the cube is unique because of the combination of symmetry elements it possesses; the most characteristic being a threefold rotational axis along the body diagonal direction. It is the symmetry of the cell that defines the crystal class. The larger cubic unit cell for KCl, containing four lattice points has more symmetry and is therefore chosen.

The various types of three-dimensional unit cells and the lattices they describe are named after Auguste Bravais (1811–1863), an early French crystallographer who first described them mathematically. When the unit cell is defined by one equivalent point at each corner, it is called *primitive* and given

CRYSTALLOGRAPHIC SYMMETRY

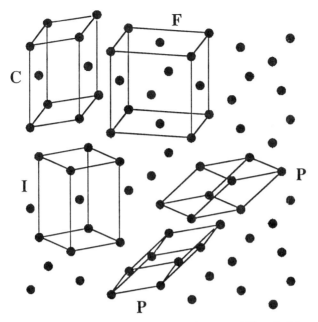

Figure 2.4. Possible primitive (*P*), face-centered (*C*), body-centered (*I*), and all-face-centered (*F*) unit cells for KCl.

the symbol *P*. If there is an additional equivalent lattice point in the middle of each cell it is called *body centered* and given the symbol *I*. When the cell has equivalent points at each corner and in the center of each face, it is called *face centered* and the symbol *F* is used. Each of these cell types are illustrated in Figure 2.4. The rather strange *rhombohedral* cell, in which the three interaxial angles are equal and acute, is given the symbol *R*. The last type of Bravais lattice occurs when only one face is centered. The symbol used to describe this type of cell depends on the name of the face that is centered: the face (or plane) parallel to the **a** and **b** directions of the lattice is called *C*; likewise, the face parallel to the **a** and **c** directions is called *B*. On considering the seven basic lattice symmetries and the various types of unit cells allowed to each symmetry, we find that there are 14 Bravais lattices, which are the only ways of filling space in a periodic manner. These are also indicated in Table 2.2 and shown in Figure 2.5.

2.2.3. Reduced Cells

Figure 2.4 shows a few of the possible unit cells that may be drawn in the KCl space lattice. In fact, a nearly infinite number of cells getting more and more

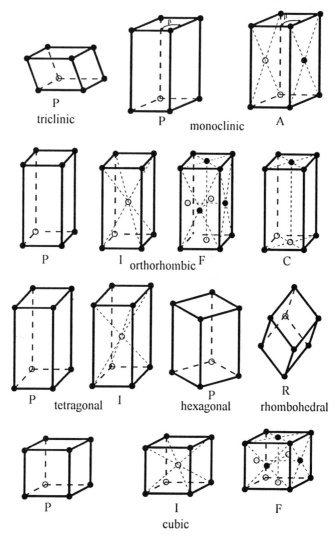

Figure 2.5. Unit cells of the 14 Bravais lattices: triclinic, P; monoclinic, P, A; orthorhombic, P, I, F, C; rhombohedral, R; hexagonal, P; tetragonal, P, I; cubic, P, I, F.

oblique may be drawn in any space lattice. All primitive unit cells in a space lattice will contain the same volume. All nonprimitive cells will have volumes equal to the number of lattice points in the cell times the primitive cell volume. Thus, there are relationships between all of the possible cells that may be drawn for a given lattice. It has been shown that the cell giving the maximum symmetry is the conventional cell, but this cell is not always easy to recognize. Chapter 3 will show that the process of diffraction permits us to determine the metric relationships between lattice points. However, because for any lattice a nearly infinite number of cells may be available from which to choose, recognizing this cell may at times be difficult. Another problem in choosing a unit cell to describe a space lattice is the arbitrary choice of cell edges and interaxial angles in crystal systems of low symmetry. For example, the triclinic system's unit cell is a completely arbitrary parallelepiped because it has no symmetry. Thus, any primitive unit cell in the lattice will describe the lattice, but separate workers who have chosen different unit cells will not easily recognize that they are describing the same lattice.

A number of arbitrary conventions have been proposed over the years to allow one both to identify hidden symmetry that a different cell setting would have and to produce the same unit cell for a given lattice. P. Niggli (see Mighell [3]) has described a mathematical procedure that will always produce the same reduced cell for any lattice. The reduced cell is defined as that cell whose axes are the three shortest noncoplanar translations in the lattice; consequently, there is only one such cell in any one lattice. It is, by convention, the standard choice for the triclinic cell. This Niggli cell is then a good place to start, once three potential translation vectors have been found that describe a lattice. Sublattices and superlattices may sometimes be obtained by halving the volume or centering the "correct" cell. The Niggli procedure will show that these cells also produce the same reduced cell. In fact, Niggli demonstrated that there are only 28 unique reduced cells. Mighell [3] has taken advantage of this fact in designing and creating the Crystal Data Database. This database contains the unit cell information for every material for which this information has been published. This extensive database is published by the International Centre for Diffraction Data [4] and may be used both to access the literature on crystallographic information for a particular phase and to identify materials that have unit cells similar to the one of interest.

In order to classify phases by their unit cells one must be able to transform a cell from one setting (i.e., one set of values for $\mathbf{a}, \mathbf{b}, \mathbf{c}, \alpha, \beta, \gamma$) to another setting. This transformation amounts to changing the coordinate system describing a space lattice. Cell transformations are done by creating the transformation tensor relating the two settings, which is well described in Chap. 5 of the *International Tables for Crystallography*, Vol. A [5], an invaluable reference work for a crystallographer. Program NBS*LATTICE [6], by Mighell and Himes, will not only automatically find the reduced cell from any arbitrary setting in a space lattice but will also attempt to identify the cell in the Crystal Data Database (computer version). Unit cell recognition is a very powerful tool in both phase identification and the identification of potential isostructural materials.

2.2.4. Space Groups

It has already been shown that there are 14 space lattices and 32 point groups. If the 32 point groups are arranged in the various patterns allowed by the 14 Bravais lattices, 230 unique three-dimensional patterns can be distinguished. These are called *space groups*. Every crystal structure can be classified into one of these 230 space groups. It has long been realized that the notation needed to describe these 230 space groups would need to be extremely complex. However, if it is recognized that space groups result from the combination of simple symmetry elements, proper and improper rotations and reflection, with the translational symmetries of the Bravais lattices, two new combinational symmetry elements can be defined; *glide planes* and *screw axes*. These non-simple-symmetry elements allow for a succinct space group notation.

The combination of a translational movement and a rotation changes the rotational axis into a screw axis as illustrated in Figure 2.6. The symbol for a screw axis is N_j, where N stands for the type of rotational axis (i.e., $360/N$ = the number of degrees in the rotation), and j indicates the fraction of the cell translated (i.e., j/N). So 2_1 means a rotation of $180°$ and a translation of $\frac{1}{2}$ of the unit cell along the direction of the rotation axis. A 3_2 means a rotation of $120°$ followed by a translation of $\frac{2}{3}$ of the cell. The combination of a translation with a mirror reflection is called a glide plane and is also illustrated in Figure 2.6. The location of the symbol for the glide plane in the space group symbol tells which mirror the reflection occurs across, while the actual symbol ($a, b, c, n,$ or d) tells the direction of the translation. After

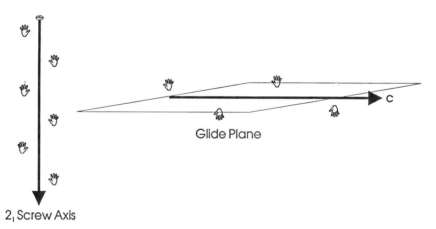

Figure 2.6. A 2_1 screw axis and a c glide plane.

reflection across a mirror, an *a* glide translates in the **a** direction, while *b* and *c* glides translate in the **b** and **c** directions, respectively. An *n* glide translates along the face diagonal of the unit cell, while the *d* glide moves along the body diagonal after reflection.

2.3. SPACE GROUP NOTATION

The orientation of the mirror plane and the translation for a glide plane introduces the subject of space group notation. Space group symbols are made of two parts: the first part is a capital letter designating the Bravais lattice type [*P*, *C* (*B* or *A*), *I*, *F*, or *R*]; the second part tells the type of symmetry the unit cell possesses (as discussed in the following subsections). The symbols used conventionally only specify the minimum symmetry to uniquely identify the space group (the irreducible representation). All implied symmetry is not mentioned in the symbol. This fact, along with possible arbitrariness in the choice of origin and definition of directions in each unit cell, can lead to confusion. Volume A of the *International Tables for Crystallography* [5] lists all possible views of a unit cell and all of the implied symmetry, along with a wealth of other information for each of the 230 space groups. The conventions used for the space group symbols in each of the crystal systems are discussed in the following subsections. The full collection of 230 space groups is given in Appendix D.

2.3.1. The Triclinic or Anorthic Crystal System

Because the only possible symmetries that can occur in the *triclinic crystal system* are 1 and $\bar{1}$ there are only two space groups: $P1$ and $P\bar{1}$. It is a general rule that when a center of symmetry is present it will be placed at the origin of the unit cell. Sometimes, for reasons of comparison, authors will use a *C* or an *I* triclinic centered cell; however, from symmetry considerations there is no reason to have anything other than a primitive lattice. Because the **a**, **b**, and **c** axes all equivalently lack symmetry, there is no need to position the 1 in the *P*1 symbol in any way to indicate a particular axis. However, the full symbol, describing the symmetry along each axis, would be *P*111, where the symbols following the Hermann–Mauguin notation refer to the symmetry of the **a**, **b**, and **c** directions.

2.3.2. The Monoclinic Crystal System

In order to increase the symmetry from triclinic it is necessary to add a twofold axis or a mirror. To do this requires the positioning of one of the axes so that it makes a 90° angle with the plane defined by the other two. This transformation results in the *monoclinic crystal system*. Because there can be only one symmetry plane and/or one twofold axis, the name will be purely arbitrary. The original choice of the unique axis was taken as **b** (with the perpendicular mirror plane being defined by the **a** and **c** axes). However, in the 1952 edition of the *International Tables for X-ray Crystallography* [7], it was decided to change and take **c** as unique (to be analogous with the tetragonal and hexagonal systems). This attempt at uniformity failed to do more than add confusion to one of the few arbitrary conventions in which there was nearly uniform agreement. The latest (1983) version of the *International Tables* shows the space groups with all possible choice of unique axes. The most commonly accepted convention still calls the unique axis **b**, as we shall do below.

The **a** and **c** axes in the plane normal to the unique axis have no symmetry and therefore their designation is arbitrary. In fact, any two directions in the plane perpendicular to the symmetry **b** axis will define a valid monoclinic unit cell. The various monoclinic space groups are composed of all possible combinations of the point group symmetries 2, m, $2/m$ and the corresponding translational elements 2_1, c, $2_1/m$, $2/c$ and $2_1/c$ with the Bravais lattices P and C. These combinations are shown in Table 2.3. Since the symmetry axis must be **b** and the mirror plane **ac**, then all symmetry notation in the symbols unambiguously must refer to these axes. It is not needed to specify the onefold axes along the **a** and **c** axes in a full symbol like $P12_11$. Note that in the table there are three blank entries for the groups $C2_1$ $C2_1/m$ and $C2_1/c$. This is because these combinations of symmetry are contained by implication in the groups $C2$, $C2/m$, and $C2/c$, respectively. Thus, there are only 13 unique monoclinic space groups and, if any more symmetry is added to a cell, the cell is forced into a higher symmetry crystal class.

Table 2.3. The 13 Monoclinic Space Groups[a]

Point Group	P	C
$2(C_2)$	$P2$	$C2$
	$P2_1$	
$m(C_s)$	Pm	Cm
	Pc	Cc
$2/m*(C_{2h})$	$P2/m$	$C2/m$
	$P2_1/m$	
	$P2/c$	$C2/c$
	$P2_1/c$	

[a]The asterisk indicates a Laue group.

Table 2.3 also lists the Schönflies symbols (in parentheses) for the monoclinic point groups. Schönflies notation is unambiguous for point groups and is commonly used by spectroscopists. However, because it does not account for translational symmetries, it is forced to just sequentially number each space group associated with a particular point group. The unambiguous Hermann–Mauguin notation is therefore preferred for describing space groups.

Powder diffraction analysis, unlike single crystal diffraction results, often does not supply enough information about the lattice to determine all of the symmetry elements unambiguously. The crystal system may always be determined; however, it is common to find that it may not be possible to establish the presence of a twofold axis or mirror. In these cases the undetermined symmetry element is written as an asterisk and the symbol is called the *aspect* rather than the space group symbol.

2.3.3. The Orthorhombic Crystal System

The starting point for describing symmetric arrangements was, in the triclinic case, a lattice containing no symmetry other than possibly a center of symmetry. This arrangement can be described by three translation vectors that are unequal and pointed in arbitrary directions. As soon as an attempt is made to increase symmetry by adding a mirror or a twofold axis (or both) it is found that one of the axes must meet the plane of the other two at right angles, giving a monoclinic cell. To further increase symmetry by adding yet another twofold axis or mirror, all three axes must be made to meet at 90° and this causes the remaining axis to also become orthogonal and twofold. Thus, the three point groups in the *orthorhombic crystal system* are 222, *mm*2 and *mmm*. The combinations of twofold axes and mirrors along three orthogonal axes produce 59 unique space groups.

In each orthorhombic space group symbol, for example, *Pbc*2, each symmetry specifier after the Bravais lattice symbol refers to the symmetry along the **a**, **b**, and **c** directions. However, because these three directions may be arbitrarily relabeled, the symmetry specifiers will of course change with the labeling. In fact, there are six different settings for each orthorhombic space group. The latest edition of the *International Tables for X-ray Crystallography* (Vol. A) shows each possible orientation of each cell and the space group symbol for that orientation. The older edition of the tables [7] shows the symbol for each permutation of axes.

2.3.4. The Tetragonal Crystal System

To add any more symmetry to the orthorhombic crystal system requires the use of a symmetry axis higher than twofold. The addition of a fourfold axis forces the two axes in the plane perpendicular to it to become of equal length. By convention, the unique symmetry axis is called **c**, and the **a** and **b** axes are equal and both are called **a**. The equivalence of **a** and **b** gives rise to a new situation for our notation. Rather than continuing with the non-Bravais part of our symbols referring to the **a**, **b**, and **c** axes the first symbol will refer to the symmetry along the **c** axis. The next symbol in the space group will refer to the symmetry along **a** and **b**. The third symbol, when present, refers to a new symmetry that can arise in *tetragonal space groups* along the face diagonal direction. There are 36 combinations of fourfold symmetry (not including a threefold axis, which pushes us into the cubic system).

2.3.5. The Hexagonal and Trigonal Crystal Systems

The combination of a threefold or sixfold axis with twofolds and mirrors produces a situation similar to the tetragonal system. The **a** and **b** axes perpendicular to the symmetry axis are equivalent. It must, however, be appreciated that the *trigonal crystal system* is not the same as the *hexagonal system* but a polar 3 or $\bar{3}$ symmetry axis makes a unique crystal class. A subset of the trigonal class can be described on a rhombohedral axis system, which is a unit cell in which the length of the three translation vectors is equal and they have the same angle, α, between them. Some argument is allowed here in that three of the trigonal or rhombohedral cells, when put together, can be described by a hexagonal cell. Thus, even though there is a unique trigonal symmetry, the system can be treated as hexagonal giving only six crystal systems rather than seven. There are 25 unique trigonal space groups.

The hexagonal system has a polar 6 or $\bar{6}$ symmetry axis along **c**, and the first symbol in the space group, following the Bravais letter, refers to it. The next symbol refers to the equivalent **a** and **b** axes. The last symbol refers to the face diagonal as in the tetragonal system. There are 27 unique hexagonal space groups.

2.3.6. The Cubic Crystal System

The *cubic* is the seventh crystal system and the one with the most symmetry. It is uniquely characterized by having a 3 or $\bar{3}$ axis along the four-body diagonal direction. Because all three translation vectors are equivalent, the first letter in the space group symbol (after the Bravais symbol) refers to this axis. The next symbol refers to the body diagonal direction [111] and is always a 3 (or $\bar{3}$). Note that the presence of a 3 as the second term after the Bravais symbol unambiguously tells us that the system is cubic and not trigonal (in which case the 3 would be in the first position). The last symbol refers to the face diagonal direction. There are 36 cubic space groups.

2.3.7. Equivalent Positions

The action of symmetry elements in a unit cell is to make one point in space equivalent to other points. An understanding of how atoms repeat by symmetry throughout a unit cell will greatly simplify all of the crystallographic computations that will be described in later chapters. For example, when a center of symmetry acts on a point at coordinates x, y, z it converts it to a point at $\bar{x}, \bar{y}, \bar{z}$.

$$x, y, z \stackrel{\bar{1}}{\Longleftrightarrow} \bar{x}, \bar{y}, \bar{z}.$$

Similarly, a twofold axis along **b** causes

$$x, y, z \stackrel{2}{\Longleftrightarrow} \bar{x}, y, \bar{z},$$

while a mirror in the **ac** plane converts x, y, z as

$$x, y, z \stackrel{m}{\Longleftrightarrow} x, \bar{y}, z.$$

The combination of a 2 along **b** and a mirror perpendicular to **b** causes

$$x, y, z \stackrel{2}{\Longleftrightarrow} \bar{x}, y, \bar{z}$$
$$m \Updownarrow \qquad \Updownarrow m$$
$$x, y, z \stackrel{}{\Longleftrightarrow} \bar{x}, \bar{y}, \bar{z}.$$

Note that the action of $2/m$ on x, y, z produces, among other coordinates, $\bar{x}, \bar{y}, \bar{z}$. This is the same relationship caused by the action of an inversion center. In fact, $2/m$ implies a center of symmetry. As will be seen later, the act of diffraction will add a center of symmetry to a diffraction pattern. For this reason, the 11 centrosymmetric point groups are called *Laue groups*. Any intersection of a mirror with a symmetry axis implies the presence of an inversion center.

Translational symmetry also applies to the locations at which equivalent atoms will be found in a unit cell. If the space group were $C2/m$ instead of $P2/m$, atoms would be found at the locations just derived and, in addition, there would be four more identical atoms located at those coordinates with the vector $[\frac{1}{2}, \frac{1}{2}, 0]$ added to each. A screw axis along **b** transforms

$$x, y, z \stackrel{2_1}{\Longleftrightarrow} \bar{x}, y + \tfrac{1}{2}, \bar{z}.$$

A c glide plane in the **ac** plane converts x, y, z as

$$x, y, z \stackrel{c}{\Longleftrightarrow} x, \bar{y}, z + \tfrac{1}{2}.$$

2.3.8. Special Positions and Site Multiplicity

For a phase in the space group $P2/m$, any atom located in the general position x, y, z will, by symmetry, also be found at coordinates $\bar{x}, y, \bar{z}$, $x, \bar{y}, z$ and $\bar{x}, \bar{y}, \bar{z}$. Thus, an atom in a general symmetry site must repeat four times in this space group. However, if an atom were to sit exactly on the twofold axis, it would not repeat itself on the action of rotation. This "special position" has a multiplicity of only 2. Atoms sitting in the mirror plane likewise have multiplicities of 2. An atom at the origin of the unit cell (which is conventionally chosen at a center of symmetry, if one exists) will not repeat itself by either reflection or rotation and therefore has a multiplicity of 1. Table 2.4 lists the complete set of *equivalent positions* for the space group $P2/m$.

A note is required here on the implication of putting the origin of the unit cell at the center of symmetry. In the case of $P2/m$, the center of symmetry is at the intersection of the twofold axis with the mirror and thus the centrosymmetric relationship x, y, z and $\bar{x}, \bar{y}, \bar{z}$ occurs in the list of equivalent positions. However, in a number of space groups the center of symmetry is not coincident with the intersection of the symmetry axis with the mirror plane. In $P2_1/c$ for

Table 2.4. The Equivalent Positions for the Various Symmetry Sites for the Space Group $P2/m$

Multiplicity	Wyckoff Symbol[a]	Symmetry	Equivalent Positions
4	o	1	x, y, z $\bar{x}, y, \bar{z}$ $x, \bar{y}, z$ $\bar{x}, \bar{y}, \bar{z}$
2	n	m	$x, \tfrac{1}{2}, z$ $\bar{x}, \tfrac{1}{2}, \bar{z}$
2	m	m	$x, 0, z$ $\bar{x}, 0, \bar{z}$
2	l	2	$\tfrac{1}{2}, y, \tfrac{1}{2}$ $\tfrac{1}{2}, \bar{y}, \tfrac{1}{2}$
2	k	2	$0, y, \tfrac{1}{2}$ $0, \bar{y}, \tfrac{1}{2}$
2	j	2	$\tfrac{1}{2}, y, 0$ $\tfrac{1}{2}, \bar{y}, 0$
2	i	2	$0, y, 0$ $0, \bar{y}, 0$
1	h	$2/m$	$\tfrac{1}{2}, \tfrac{1}{2}, \tfrac{1}{2}$
1	g	$2/m$	$\tfrac{1}{2}, 0, \tfrac{1}{2}$
1	f	$2/m$	$0, \tfrac{1}{2}, \tfrac{1}{2}$
1	e	$2/m$	$\tfrac{1}{2}, \tfrac{1}{2}, 0$
1	d	$2/m$	$\tfrac{1}{2}, 0, 0$
1	c	$2/m$	$0, 0, \tfrac{1}{2}$
1	b	$2/m$	$0, \tfrac{1}{2}, 0$
1	a	$2/m$	$0, 0, 0$

[a] Wyckoff notation describes each of the special symmetry positions in the unit cell.

SPACE GROUP THEORY 41

Table 2.5. The Equivalent Positions for the Various Symmetry Sites for the Space Group $P2_1/c$

Multiplicity	Wyckoff Symbol	Symmetry	Equivalent Positions			
4	e	1	x, y, z	$\bar{x}, y+\frac{1}{2}, \bar{z}+\frac{1}{2}$	$\bar{x}, \bar{y}, \bar{z}$	$x, \bar{y}+\frac{1}{2}, z+\frac{1}{2}$
2	d	$\bar{1}$	$\frac{1}{2}, 0, \frac{1}{2}$	$\frac{1}{2}, \frac{1}{2}, 0$		
2	c	$\bar{1}$	$0, 0, \frac{1}{2}$	$0, \frac{1}{2}, 0$		
2	b	$\bar{1}$	$\frac{1}{2}, 0, 0$	$\frac{1}{2}, \frac{1}{2}, \frac{1}{2}$		
2	a	$\bar{1}$	$0, 0, 0$	$0, \frac{1}{2}, \frac{1}{2}$		

example, the symmetry axis intersects the mirror at $0, \frac{1}{4}, \frac{1}{4}$. This causes the equivalent positions to be those listed in Table 2.5. The subtleties involved in the action of space group symmetry to produce the equivalent positions in the 230 space groups have been worked out in detail in Vol. A of the *International Tables for Crystallography*. Knowing what has been presented here and some practice with the use of this reference work will allow routine analysis of materials in complicated space groups. This reference should be on the library shelf of all workers interested in materials characterization.

2.4. SPACE GROUP THEORY

Crystal structure analysis is one of the most important applications of diffraction. A knowledge of the space group symmetry will sometimes allow a complete or partial analysis of the location of the atoms in the unit cell. As an example of such an analysis, let us now examine the structure of the compound KCl discussed previously. X-ray diffraction allows the determination of the space group to be $Fm3m$ with a cubic cell edge of 6.2917 Å. The measured density is 1.99 g/cm². What will our understanding of space groups allow us to determine about the crystal structure of KCl?

The density of a material will equal the weight of the contents of a unit cell divided by its volume. Thus,

$$\rho = \frac{WZ}{LV}, \quad (2.1)$$

where L is Avogadro's number; Z is the number of formula units per unit cell; W is the formula weight of 74.551 g/mol or $74.551/6.0221367 \times 10^{23} = 1.237949 \times 10^{-22}$ g; and V is the cell volume of 249.06 Å³, or 2.4906×10^{-22} cm³. Thus, using the observed density and solving for Z gives 4.0036, or

4 formula units per unit cell. Now, on looking up the space group in the *International Tables for Crystallography*, Vol. A, one finds that there are only two sites in such a unit cell that have the minimum possible multiplicity of 4. One is at the origin and the other in the cell center at $\frac{1}{2},\frac{1}{2},\frac{1}{2}$. Clearly, a K atom must go at one site and a Cl at the other. Thus, in this case the crystal structure can be completely determined from a knowledge only of the space group, density, and unit cell size.

Just as the density of a unit cell is determined by the mass, volume, and bonding of the atoms in it, the size of the unit cell will be determined by the volume and bonding of the atoms. Thus, if another atom of sufficiently similar bonding properties is to substitute for one of the original atoms, the volume of the unit cell will change. Vegard's law states this premise by noting that the volume of unit cells in a substitutional solid solution is linearly proportional to the fraction of sites substituted. Figure 2.7 shows this linear relationship between the lattice parameter C for CdS and CdSe. As sulfur substitutes for selenium in the compound, the lattice parameter uniformly expands. This fact permits a measurement of the lattice parameters of a substitutional solid solution to be used in determining the composition of the material. It should

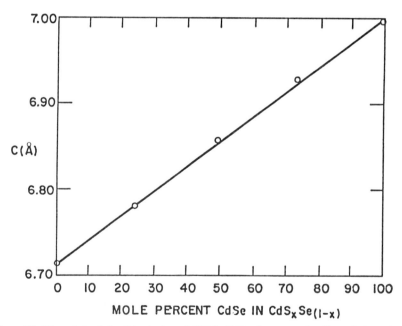

Figure 2.7. The substantial solid solution of CdS in CdSe, demonstrating Vegard's law of the linear dependence of the lattice parameter (C) on concentration.

2.5. CRYSTALLOGRAPHIC PLANES AND MILLER INDICES

The unit cell is a very useful concept and is used not only to characterize the symmetry of a crystal structure but also to specify crystallographic directions and even interatomic distances. To describe directions and distances, it is necessary to imagine sets of parallel planes intersecting the unit cell from various directions. Classical crystallography and modern materials science make extensive use of these imaginary planes. Any family of planes may be identified by integers called *Miller indices*. To determine these indices for the particular family of planes shown in Figure 2.8, a few simple rules must be established:

- Locations in a unit cell will always be specified in terms of fractions of the **a**, **b**, and **c** cell edges. These unitless *fractional coordinates* are formed by taking the absolute x, y, z coordinates and dividing by the corresponding cell edge: $x/\mathbf{a}$, $y/\mathbf{b}$, and $z/\mathbf{c}$.
- A valid family of planes will always have one member that passes through the origin of every unit cell.

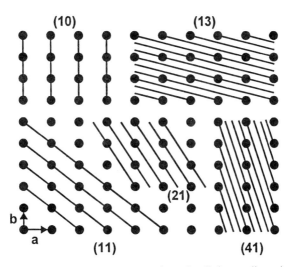

Figure 2.8. A family of planes intersecting unit cells in two dimensions.

- The plane to use in determining Miller indices is always the first plane in the family away from the origin. This plane may be found by moving in either direction from the plane passing through the origin. For example, in the lower right of Figure 2.8 there are the planes intersecting the **a** axis at $\pm\frac{1}{4}$ and the **b** axis at 1.
- The Miller indices named h, k, and l are the reciprocals of the fractional coordinates of the axial intercepts of the first plane away from the origin. Thus, the h index of the plane with an intercept of $\frac{1}{4}$ is simply 4. The k index for this plane is the reciprocal of the **b** axis intercept of 1. The indices of the plane going away from the origin in the opposite direction are -4 and -1.
- When a plane is parallel to an axis it is assumed that it will intersect the axis at infinity, and so its index is $1/\infty$ or zero. If it is assumed that in Figure 2.8 the **c** axis and the planes all rise vertically from the page, then the l index for the index determining plane is zero.

The lower right of Figure 2.8 shows a family of planes that will be characterized with the Miller indices $(4, 1, 0)$ or $(-4, -1, 0)$. These two sets of indices must of course be equivalent. Further, these planes can be used to define crystallographic directions. The normal to any plane is a mathematically unambiguous direction; thus, the indices [410] when written in square brackets will be used to designate that direction or zone. In some of the previous sections it was desired to refer to a unit cell's face and body diagonals. Now it will be clear that the face diagonal in the **ab** plane parallel to the **c** axis will be called the (110) plane, and the [110] direction will be perpendicular to this plane. Likewise, the plane that intercepts all three axes at 1 has the indices (111), and its normal [111] is the body diagonal of the unit cell.

Now that crystallographic planes and directions have been defined, it is possible to go one step further and characterize the Miller index determining plane by a vector that extends from the origin of the unit cell and meets the plane in question at 90°. This vector has the direction defined by the plane normal and has a magnitude equal to the well-defined perpendicular distance between planes. This vector is called $\mathbf{d}_{hkl}$ and, as will be seen in the next chapter, this quantity is directly determined by powder diffraction measurement.

REFERENCES

1. Phillips, F. C. *An Introduction to Crystallography*, 3rd ed. Wiley, New York, 1964.
2. Catlow, C. R. A. X-ray diffraction from powders and crystallites. *Appl. Synchrotron Radiat.* pp. 39–64 (1990).

3. Mighell, A. D. The reduced cell: Its use in the identification of crystalline materials. *J. Appl. Phys.* **9**, 491–498 (1976).
4. International Centre for Diffraction Data, 12 Campus Blvd., Newtown Square, PA 19073-3273, USA.
5. Hahn, T., ed. *International Tables for Crystallography*, Vol. A: *Space Group Symmetry*. Reidel, Dordrecht, The Netherlands, 1983.
6. Mighell, A. D., and Himes, V. L. Compound identification and characterization using lattice-formula matching. *Acta Crystallogr.* **A42**, 101–105 (1986).
7. Henry, N. F. M., and Lonsdale, K., eds. *International Tables for X-ray Crystallography*. Kynoch Press, Birmingham, England, 1952.

CHAPTER
3
DIFFRACTION THEORY

3.1. DIFFRACTION OF X-RAYS

It was shown in Section 1.5 that the oscillating electric field of a light wave will interact with the electrons in matter to cause coherent scattering. It has been known since the time of Christian Huygens (1629–1695) that each scattering point may be treated as a new source of spherical waves. Thus, waves scattering from two objects will expand in space around the objects until they interfere with each other. Their interaction will produce constructive interference at certain angles of view and destructive interference at other angles depending on the distance between the scatterers and the wavelengths of the radiation. When a periodic array of objects each scatter radiation coherently, the concerted constructive interference at specific angles is called *diffraction*. If the scatterers are lines ruled on a surface, separated by a fixed distance of 0.5 μm, then the interference effects will be seen for radiation with wavelengths in the visible light region of the spectrum and the scatterer is called an *optical grating*. Since such gratings have a fixed scatterer periodicity, each of the wavelengths in white light will require a different angle to achieve constructive interference. Thus, one observes all of the colors of the spectrum as the angle of view to a grating is changed. If the scatterers are atoms periodically arranged in a crystal, then the distances between them are on the order of angstroms, and the interference effects will be seen when the incident wave front has a wavelength in the X-ray region of the spectrum.

Unlike the simple one-dimensional periodicity of a conventional planar diffraction grating commonly used in visible–IR spectrometers, a crystal has a great many three-dimensional periodic relationships between the atoms that compose it. Thus, a crystal may diffract a monochromatic wave in a number of different directions in three-dimensional space. The angles of diffraction will only depend on the various periodic relationships between the atoms making up the crystal. The first description of the diffraction of X-rays by a crystal was developed by Max von Laue, who described the effect with three-vector dot-product equations. In 1913, William Henry Bragg and his son William Lawrence Bragg developed a much simpler way of understanding and predicting diffraction phenomena from a crystal, but to use it one must invoke a few artificial analogies. In Figure 3.1, three crystallographic planes are shown that

48 DIFFRACTION THEORY

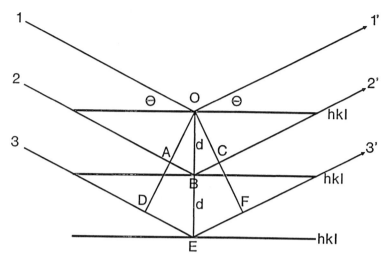

Figure 3.1. Bragg's law is easily seen to arise from an optical analogy to crystallographic planes reflecting X-rays.

may be described with Miller indices (*hkl*). The Bragg analogy views the incident X-ray beams as coming in from the left and reflecting from each of the planes in the family. If we assume that the initial waves are in phase with one another and that the waves reflect from each plane in the family, then the controlling equation can be quickly derived. In reflection, Snell's law requires that the angle of incidence θ and the angle of reflection must be equal. In Figure 3.1, the wave reflecting from the second plane must travel a distance ABC farther than the wave reflecting from the top plane. Similarly, the wave reflecting from the third plane must travel DEF farther. Thus, all waves reflecting from planes below the surface will be phase retarded with respect to the first wave, causing interference. Plane geometry shows that when the distance ABC is exactly equal to one wavelength (λ), the distance DEF will equal 2λ and the reflection from all planes at any depth in the crystal will emerge in phase, producing the constructive interference known as diffraction. The well-known Bragg equation follows readily from an analysis of the right triangle OAB. The distance AB may be obtained by taking the sine of θ:

$$AB = d_{hkl} \sin \theta.$$

When diffraction occurs ABC = λ; thus, the Bragg equation results:

$$\lambda = 2d_{hkl} \sin \theta. \tag{3.1}$$

In fact, all waves penetrating more deeply into the crystal will also have a phase lag of an integral multiple of wavelengths, so that all waves emerging at the Bragg angle will be in phase. In this development of the Bragg equation it was assumed that the incident X-rays are in phase and that they reflect from imaginary planes. In fact, Bragg's law is rigorously correct and may be derived with a bit more effort without these artificial assumptions.

3.2. THE RECIPROCAL LATTICE

The Bragg equation, with its analogy to reflection from crystallographic planes, simplified the purely mathematical description of von Laue in permitting a visual understanding of experimental diffraction effects and the predicting of diffraction geometries. However, the most useful method for describing and explaining diffraction phenomena was developed by P. P. Ewald and has the rather formidable name "reciprocal lattice" or "reciprocal space." Although the Ewald approach requires an additional concept, the reciprocal of d_{hkl}, it is the simplest and most powerful method for understanding diffraction effects. Its wide acceptance is an indication of its basic simplicity.

When a particular arrangement of an X-ray source, sample, and detector is proposed, it would be useful to be able to predict the various motions that will have to be applied to these three components in order to see particular diffraction effects. Let us consider the case of diffraction from the (200) planes of a large LiF crystal which has an identifiable (100) cleavage face. In order to use the Bragg equation to determine the angle at which the X-ray source and detector must be placed with respect to the crystal surface, one must first determine the value of d_{200}. This may be obtained by looking up the cubic lattice parameter of LiF in either the Powder Diffraction File [1] or the Crystal Data Database [2], finding that $\mathbf{a} = 4.0270$ Å. In the cubic system, d_{200} will simply be one-half of $\mathbf{a}$, or 2.0135 Å. Thus, from Bragg's law (Equation 3.1) one can calculate that the diffraction angle for the Cu $K\alpha_1$ wavelength of 1.54060 Å is 44.986° 2θ. Bragg's law therefore dictates that the (100) face of the LiF crystal should be placed so as to make an equal angle of $\theta = 11.03°$ with the incident X-ray beam and the detector. This is a simple example—but, in order to look at diffraction from the (246) planes, the visualization of the required orientation of the crystal becomes formidable.

The basic problem is that the Bragg planes in a space lattice are inherently three dimensional and restricting analysis to a two-dimensional plane within the lattice really does not reflect their three-dimensional nature. One possible solution is to remove a dimension from the problem by representing each two-dimensional plane as a vector: d_{hkl} is defined as the perpendicular distance from the origin of a unit cell to the first plane in the family (hkl), as illustrated

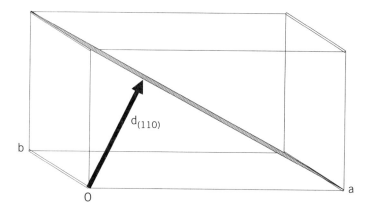

Figure 3.2. Here d_{110} is represented as a vector with magnitude equal to the value of d in angstroms and in the direction from the origin of the unit cell meeting the plane at a right angle.

for the (110) plane in Figure 3.2. A diagram may be made by representing each d_{hkl} vector as a point, at its tip, in the appropriate crystallographic coordinate system. Figure 3.3 shows the **ab** plane of such a diagram. Even for the limited number of vectors shown in Figure 3.3 it is clear that the density of points, representing planes, increases to infinity as the origin is approached. Thus, rather than simplifying our real space picture, this vector space construction has definitely made it more complex.

The full three-dimensional vector representation of the planes in a unit cell will be a sheaf of vectors projecting out of a sphere in all directions, getting thicker and thicker as the center is approached. Examination of Figure 3.3 shows that the vectors are approaching the origin according to the reciprocals of the d_{hkl} values. Ewald, on noting this relationship, proposed that instead of plotting the d_{hkl} vectors, the reciprocal of these vectors should be plotted. The reciprocal d vector is defined as

$$\mathbf{d}^*_{hkl} \equiv \frac{1}{d_{hkl}}. \qquad (3.2)$$

Figure 3.3 can now be reconstructed plotting $\mathbf{d}^*_{hkl}$ vectors instead of d_{hkl} values. Figure 3.4 shows this construction. The units are in reciprocal angstroms and the space is therefore a *reciprocal space*. Notice that the points in this space repeat at perfectly periodic intervals defining a space lattice called a *reciprocal lattice*. The repeating translation vectors in this lattice are called **a***, **b***, and **c***. The interaxial (or reciprocal) angles are α^*, β^*, and γ^*, where the

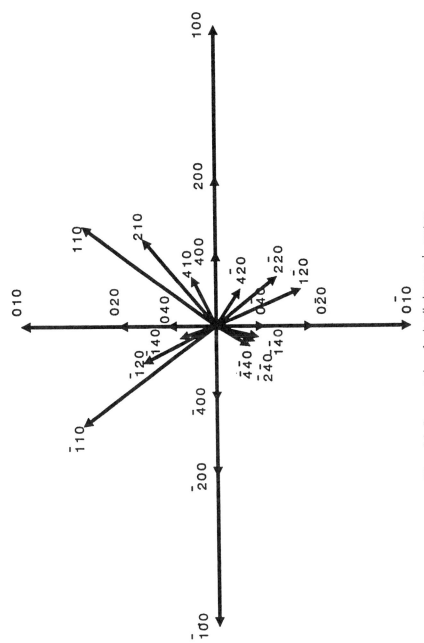

Figure 3.3. Representation of unit cell planes as d_{hkl} vectors.

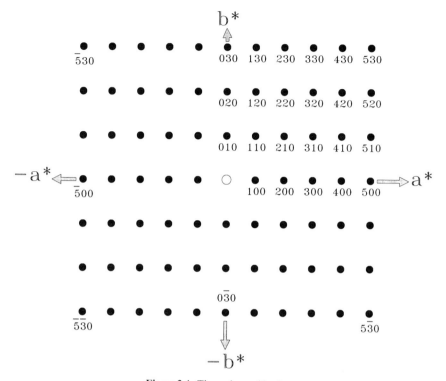

Figure 3.4. The reciprocal lattice.

reciprocal of an angle is defined as the complement, or 180° minus the real-space angle.

The concept of the reciprocal lattice makes the visualization of the Bragg planes extremely easy. To establish the index of any point in the reciprocal lattice, one simply has to count out the number of repeat units in the $\mathbf{a}^*$, $\mathbf{b}^*$, and $\mathbf{c}^*$ directions. Figure 3.4 shows only the ($hk0$) plane of the reciprocal lattice; however, the concept is of a full three-dimensional lattice that extends in all directions in reciprocal space. If the innermost points in this lattice are connected, a three-dimensional shape of the reciprocal unit cell will be seen that is directly related to the shape of the real-space unit cell. Thus, the symmetry of the real-space lattice propagates into the reciprocal lattice. The reciprocal lattice has all of the properties of the real-space lattice described in Chapter 2. Any vector in this lattice represents a set of Bragg planes and can be resolved into its components:

$$\mathbf{d}^*_{hkl} = h\mathbf{a}^* + k\mathbf{b}^* + l\mathbf{c}^*. \tag{3.3}$$

THE RECIPROCAL LATTICE   53

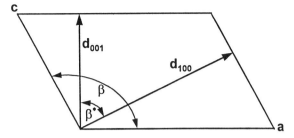

Figure 3.5. The **ac** plane of a monoclinic unit cell showing the relationship between the translation vectors and d_{hkl} vectors.

An important point to note here is that the integers in Equation 3.3 are, in fact, equal to the Miller indices of the (hkl) plane. The relationship between d and $\mathbf{d}^*$ is simply a reciprocal as defined in Equation 3.2. However, in crystal systems where the angle between translation vectors is not 90°, the d and $\mathbf{d}^*$ vectors are not collinear with the translation vectors. For example, Figure 3.5 shows the **ac** plane of a monoclinic unit cell in which the d_{100} vector meets the (100) plane at right angles according to its definition. However, this causes the **a** translation vector to be unequal in magnitude and direction to d_{100}. This, of course, implies that the relationship between $\mathbf{d}_{100}^*$ and $\mathbf{a}^*$ will also involve the sine of the interaxial angle. The relationship between the real and reciprocal translation vectors are given in Table 3.1. In most cases it is a trivial matter to compute the reciprocal lattice dimensions; however, for the difficult triclinic cell the following relations are required:

$$\cos\alpha^* = \frac{\cos\beta\cos\gamma - \cos\alpha}{\sin\beta\sin\gamma}, \tag{3.4}$$

$$\cos\beta^* = \frac{\cos\alpha\cos\gamma - \cos\beta}{\sin\alpha\sin\gamma}, \tag{3.5}$$

$$\cos\gamma^* = \frac{\cos\alpha\cos\beta - \cos\gamma}{\sin\alpha\sin\beta}, \tag{3.6}$$

$$V^* = \frac{1}{V} = \mathbf{a}^*\mathbf{b}^*\mathbf{c}^*(1 - \cos^2\alpha^* - \cos^2\beta^* - \cos^2\gamma^* + 2\cos\alpha^*\cos\beta^*\cos\gamma^*)^{1/2} \tag{3.7}$$

DIFFRACTION THEORY

Table 3.1. Direct and Reciprocal Space Relationships

System	a*	b*	c*
Orthogonal	$a^* = \dfrac{1}{a}$	$b^* = \dfrac{1}{b}$	$c^* = \dfrac{1}{c}$
Hexagonal	$a^* = \dfrac{1}{a \sin \gamma}$	$b^* = \dfrac{1}{b \sin \gamma}$	$c^* = \dfrac{1}{c}$
Monoclinic	$a^* = \dfrac{1}{a \sin \beta}$	$b^* = \dfrac{1}{b}$	$c^* = \dfrac{1}{c \sin \beta}$
Triclinic	$a^* = \dfrac{bc \sin \alpha}{V}$	$b^* = \dfrac{ac \sin \beta}{V}$	$c^* = \dfrac{ab \sin \gamma}{V}$

3.3. THE EWALD SPHERE OF REFLECTION

In Figure 3.6 a cross section through an imaginary sphere with a radius of $1/\lambda$ is shown around a real crystal. The reciprocal lattice associated with the crystal's real-space lattice is viewed as being tangent to this sphere at the point where an X-ray beam entering from the left and passing through the crystal would exit the sphere on the right. Thus, this *Ewald sphere* contains all of the components that are needed to visualize the diffraction process geometrically, instead of mathematically using the Bragg equation. It is clear that a rotation of the crystal (and its associated real-space lattice) will also rotate the reciprocal lattice because the reciprocal lattice is defined in terms of the real-space lattice. Figure 3.7 shows this arrangement at a specific time in the rotation of the crystal, when the (230) point is brought into contact with the sphere, where, by definition,

$$CO = \frac{1}{\lambda},$$

and

$$OA = \frac{d^*_{(230)}}{2}.$$

Hence,

$$\sin \theta = \frac{OA}{CO} = \frac{d^*_{(230)}/2}{1/\lambda},$$

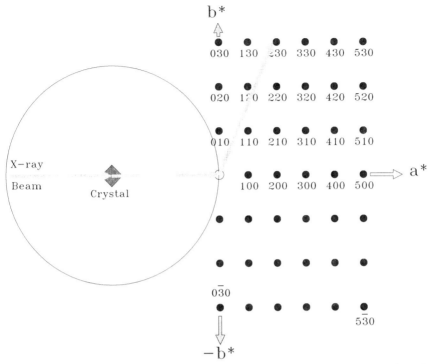

Figure 3.6. The Ewald sphere of reflection with a crystal in the center and its associated reciprocal lattice tangent to the sphere, at the point where the X-ray beam exits.

or

$$\lambda = \frac{2\sin\theta}{\mathbf{d}^*_{(230)}}.$$

But from the definition in Equation 3.2 we have

$$d_{(230)} \equiv \frac{1}{\mathbf{d}^*_{(230)}};$$

therefore,

$$\lambda = 2d_{(230)}\sin\theta.$$

Thus, it is clear that the reciprocal lattice and sphere of reflection concepts incorporate Bragg's law. As each lattice point, representing a **d***-value, touches

DIFFRACTION THEORY

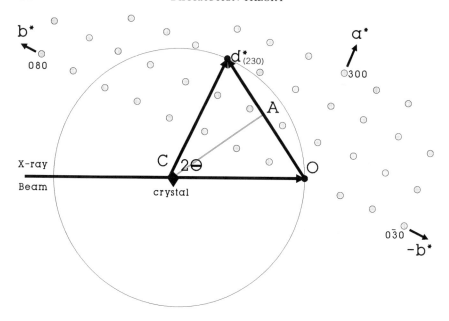

Figure 3.7. The Ewald sphere of reflection with the crystal rotated so that the (230) reciprocal lattice point touches it, permitting it to diffract.

the sphere of reflection, the condition for diffraction is met and diffraction occurs. The diffracted intensity is directed from the crystal, C, through the lattice point touching the sphere. In terms of Bragg notation, the real-space lattice plane, represented by **d***, "reflects" the incident beam. In the notation of the Ewald sphere, whenever the crystal is rotated to permit a point in the reciprocal lattice to touch the sphere, the condition for diffraction will be met and the diffracted beam will emerge in the direction connecting the center of the sphere to the reciprocal lattice point in contact with the sphere. The Ewald construction allows the easy analysis of otherwise complex diffraction geometries, showing what motion must be applied to a crystal in order to produce a diffracted beam in a particular direction.

The Ewald sphere construction is particularly useful in explaining diffraction phenomena in any type of geometry. The principal advantage is that it avoids the need to do calculations to explain the phenomena and instead allows us to visualize an effect using a pictorial, mental model. In addition, it permits the simple analysis of otherwise complex relationships among the crystallographic axes and planes that will be described in Section 3.5.

3.4. ORIGIN OF THE DIFFRACTION PATTERN

3.4.1. Single Crystal Diffraction

The various techniques for diffracting X-rays from single crystal or powders may simply be thought of as methods for visualizing the reciprocal lattice with various distortions introduced by a particular experimental geometry. For example, in the case of electron diffraction, the wavelength is so short that the radius of the Ewald sphere ($1/\lambda$) is very much larger than in the case of Cu $K\alpha$ radiation. Thus, when a small crystallite is bathed in the electron beam of a transmission electron microscope, all of the reciprocal lattice points in the plane normal to the electron beam will simultaneously be touching the sphere of reflection and hence be in diffracting position. If a film is placed behind the crystallite, an undistorted image of that plane of the reciprocal lattice will be recorded, similar to that shown in Figure 3.8. In fact, the image of the ($0kl$) plane of the reciprocal lattice of a 0.2 mm crystal of ammonium oxylate (shown in Figure 3.8) was taken using Mo $K\alpha$ radiation and a *precession camera*. The precession camera is specially designed to bring the points of the reciprocal lattice onto the surface of the sphere of reflection so as to produce undistorted images (i.e., it is a camera designed to photograph an imaginary concept!).

In fact, the image of the reciprocal lattice shown in Figure 3.8 does not include all of the points—only those that have sufficient intensity to be recorded on the film. The intensities of the reciprocal lattice points on the film are related to the types and locations of the atoms in the unit cell and will be discussed later. This recorded pattern of spots is referred to as the intensity-weighted reciprocal lattice. Note that the hkl values for any point are established by simply counting out the number of the layers in the $\mathbf{a}^*$ and $\mathbf{b}^*$ directions. Thus, the problem of *indexing* a single crystal diffraction pattern is trivial. This is not the case for powder diffraction patterns.

Today, most single-crystal diffraction is done on automated single-crystal diffractometers. These devices have four individual degrees of freedom, allowing the positioning of each reciprocal lattice point on the sphere of reflection and then scanning an electronic detector past it to record its intensity. In this manner, a few thousand data points can be measured per day. For very large unit cell materials, two-dimensional array detectors have been developed to permit rapid measurement of hundreds of thousands of diffraction intensities. The rapid recording of large quantities of diffraction intensities is currently the subject of intense development. An example of this thrust is the two-dimensional array of charge-coupled devices (CCD), which can record X-rays over a region of about 30 cm^2 with a resolution on the order of 100 μm.

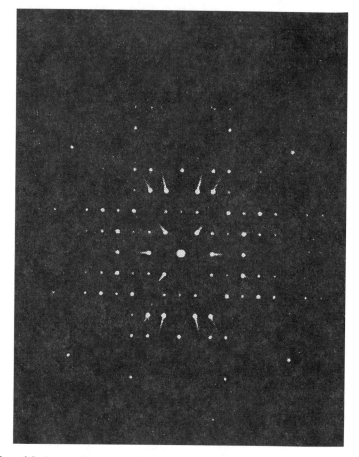

Figure 3.8. A precession camera photograph of the (0kl) zone of ammonium oxylate.

3.4.2. The Powder Diffraction Pattern

It has been shown that a single crystal with a particular set of atomic planes oriented toward the X-ray beam will diffract X-rays at an angle θ determined by the distance between the planes. However, most materials are not single crystals but are composed of billions of tiny crystallites. This type of material is referred to as a powder or a polycrystalline aggregate. Most materials in the world around us—ceramics, polymers, and metals—are polycrystalline because they were fabricated from powders. In any polycrystalline material there will be a great number of crystallites in all possible orientations. When a powder with randomly oriented crystallites is placed in an X-ray beam, all possible interatomic planes will be seen by the beam. However, diffraction from each type of plane will only occur at its characteristic diffraction angle θ. Thus, if we change the experimental angle 2θ, all of the possible diffraction peaks that can be produced from the differently oriented crystallites in the powder will be produced.

In order to understand the diffraction pattern produced by a powder, consider the following points:

- There is a d^*_{hkl} vector associated with each point in the reciprocal lattice with its origin on the Ewald sphere at the point where the direct X-ray beam exits.
- Each crystallite located in the center of the Ewald sphere has its own reciprocal lattice with an orientation determined by the orientation of the crystallite with respect to the X-ray beam.

For example, the vector associated with the (100) reflection of each randomly oriented crystallite will form a sphere of vectors radiating outward from the origin of the reciprocal lattice. The number of (100) vectors in this sphere is equal to the number of crystallites in the X-ray beam. Of all of these

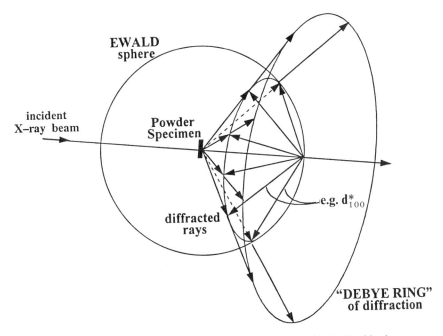

Figure 3.9. The intersection of d^*_{100} vectors from a powder with the Ewald sphere.

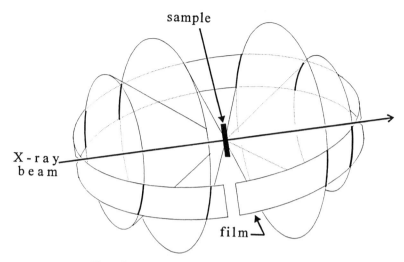

Figure 3.10. The origin of Debye diffraction rings.

d^*_{100} vectors, the only ones in position to diffract are those in contact with the Ewald sphere. Figure 3.9 shows that these vectors will intersect the Ewald sphere creating an annulus. Diffraction intensity will originate at the crystallites in the center of the sphere and project outward through this annulus producing a cone of diffracted intensity. Thus, each point in the reciprocal lattice as imaged in Figure 3.8 will produce a cone of diffraction subtending its Bragg angle. Figure 3.10 shows a series of these concentric cones, each associated with a single (hkl). These cones will intersect a film placed normal to the beam to produce a set of concentric circles. These circles are known as *Debye rings*. Figure 3.10 indicates a film strip intersecting the various powder rings in a manner called a Debye–Scherrer diffraction photograph (see Figure 7.2). Figure 3.11 shows an actual Debye–Scherrer film that was produced from the experiment shown in Figure 3.10, where the sample was KCl. Each line, in the pattern shown in Figure 3.11, corresponds to diffraction from a particular set of interatomic planes whose spacing, d_{hkl}, may be calculated from the Bragg equation (3.1) because the wavelength for a given experiment is always known.

3.5. THE LOCATION OF DIFFRACTION PEAKS

There are an infinite number of sets of planes that may be thought of as intersecting a unit cell. These planes are characterized by their (hkl) values and the d_{hkl} interplanar spacings. The d_{hkl} values are a geometric function of the size

Figure 3.11. The Debye–Scherrer diffraction pattern of KCl. (Courtesy of H. Hitchcock and J. Hurly, NASA.)

and shape of the unit cell. The relationship between d_{hkl} and the real unit cell is cumbersome and usually stated in a different form for each crystal system. However, the functional relation between the square of the reciprocal lattice vectors and the size and shape of the reciprocal unit cell is easily derived, has a simple form, and applies equally to all crystal systems. The $\mathbf{d}^{*2}$ equation, coupled with the relations given in Table 3.1, will allow simple computations relating d_{hkl} to the lattice parameters in any crystal system and will permit an easy understanding of the indexing techniques used to assign d_{hkl}'s to the d-values measured by powder diffraction techniques.

To derive the expression for $\mathbf{d}^{*2}$, one must only remember that since $\mathbf{d}^{*}_{hkl}$ is a vector it is necessary to take the dot product:

$$\mathbf{d}^{*2}_{hkl} = \mathbf{d}^{*}_{hkl} \cdot \mathbf{d}^{*}_{hkl}. \tag{3.8}$$

Substituting the vector components given in Equation 3.3,

$$\mathbf{d}^{*2}_{hkl} = (h\mathbf{a}^* + k\mathbf{b}^* + l\mathbf{c}^*) \cdot (h\mathbf{a}^* + k\mathbf{b}^* + l\mathbf{c}^*), \tag{3.9}$$

and squaring yields

$$\mathbf{d}^{*2}_{hkl} = h^2\mathbf{a}^{*2} + k^2\mathbf{b}^{*2} + l^2\mathbf{c}^{*2} + 2hk\mathbf{a}^*\mathbf{b}^* \cos \gamma^*$$
$$+ 2hl\mathbf{a}^*\mathbf{c}^* \cos \beta^* + 2kl\mathbf{b}^*\mathbf{c}^* \cos \alpha^*. \tag{3.10}$$

This expression relates the square of the inverse of d_{hkl} to the size and shape of the reciprocal unit cell for any plane in any crystal system. It is useful to point out that in orthogonal crystal systems the final three terms of the equation include a cos 90° term and therefore go to zero. For many applications, then, Equation 3.10 reduces to a particularly simple form. For example, in the cubic system it becomes

$$\mathbf{d}^{*2}_{hkl} = (h^2 + k^2 + l^2)\mathbf{a}^{*2}; \tag{3.11}$$

for the tetragonal system,

$$\mathbf{d}^{*2}_{hkl} = (h^2 + k^2)\mathbf{a}^{*2} + l^2\mathbf{c}^{*2}; \tag{3.12}$$

for the hexagonal system,

$$\mathbf{d}^{*2}_{hkl} = (h^2 + hk + k^2)\mathbf{a}^{*2} + l^2\mathbf{c}^{*2}; \tag{3.13}$$

and for the orthorhombic system,

$$d_{hkl}^{*2} = h^2 \mathbf{a}^{*2} + k^2 \mathbf{b}^{*2} + l^2 \mathbf{c}^{*2}. \tag{3.14}$$

Equation 3.10 permits the computation of all of the possible d_{hkl} values for any unit cell. Each of these d_{hkl} values fits into Braggs's law (Equation 3.1) and permits the computation of the angle at which diffraction may occur from a particular set of planes in a crystal.

For example, quartz specimens are commonly used in the alignment of diffractometers. α-Quartz is well known to crystallize in hexagonal space group $P3_221$ with lattice parameters $\mathbf{a}$ = 4.9133 Å and $\mathbf{c}$ = 5.4053 Å. The most intense diffraction peak in the quartz powder pattern occurs from the (101) reflection. In aligning a diffractometer at a synchrotron facility at which the incident beam monochromator is producing a beam of λ = 1.328921 Å, at what angle should the researcher look to find the (101) reflection from a polycrystalline quartz alignment block?

To answer this question one must first compute the reciprocal cell dimensions from the expression in Table 3.1:

$$\mathbf{a}^* = \frac{1}{4.9133 \sin(120)}$$

$$\mathbf{a}^* = 0.23501 \text{ Å}^{-1}$$

$$\mathbf{c}^* = \frac{1}{5.4053} = 0.18506 \text{ Å}^{-1}.$$

Next Equation 3.13 is used to determine $\mathbf{d}_{101}^{*2}$:

$$\mathbf{d}_{101}^{*2} = (1^2 + (1 \times 0) + 0^2)0.23501^2 + 1^2 \times 0.18506^2 = 0.08948 \text{ Å}^{-1}$$

Since d_{hkl} in any crystal system is simply the square root of the reciprocal of $\mathbf{d}_{hkl}^{*2}$, then d_{101} is 3.3434 Å. Once d is known, Bragg's law permits the computation of the diffraction angle 2θ:

$$\lambda = 2 d_{hkl} \sin \theta$$
$$1.32890 = 2 \times 3.3434 \sin \theta$$
$$2\theta = 22.926°.$$

The last piece of information required in order to compute a full diffraction pattern for a material is the intensity of each diffraction peak.

3.6. INTENSITY OF DIFFRACTION PEAKS

All of the diffraction theory discussed until now has looked at the metric aspects of a diffraction pattern, that is, the positions of the possible diffraction maxima and how they are related to the size and shape of the unit cell. In order to understand what determines the intensity of a diffraction peak, it is necessary to examine a number of independent phenomena. First, it is necessary to consider how much intensity a single electron will coherently scatter. Then, one must extend this scattering concept to include interference effects that will occur due to the electrons being distributed in space around atoms and the fact that atoms are not stationary in a lattice but vibrate in an anisotropic manner. Lastly, the interference effects caused by the scattering from atoms in different regions of the unit cell must be considered. At the conclusion of this analysis, all of the equations required to compute the intensities of diffraction maxima will be in place so that one may accurately calculate a powder pattern.

3.6.1. Electron Scattering

In Section 1.5, X-rays were described as electromagnetic radiation that, from a fixed point in space, will be seen as an oscillating electric field. The field will also cause an electron to oscillate (i.e., accelerate and decelerate) and therefore reradiate the energy as a spherical wave. An electron is limited in its ability to radiate energy by the dictates of electrodynamics. In 1906, J. J. Thompson showed that the intensity scattered from an electron is

$$I = \frac{I_0}{r^2}\left[\frac{e^2}{m_e c^2}\right]^2 \frac{1 + \cos^2(2\theta)}{2}, \qquad (3.15)$$

where I_0 is the intensity of the incident beam; e, the charge on the electron; m_e, the mass of the electron; c, the speed of light; and r, the distance from the scattering electron to the detector (its appearance as r^2 in the denominator expresses the familiar inverse square law). The final term involving the cosine function results from the fact that the incident X-ray beam is unpolarized and the process of scattering polarizes it. This term is called the polarization factor. Equation 3.15 addresses only coherently scattered radiation; all other absorption mechanisms are ignored. The term $e^2/m_e c^2$ is the classical radius (r_e) of the electron and is derived from the fundamental electrodynamic theory of J. C. Maxwell. It is only the electromagnetic r_e term that differentiates electron from neutron scattering.

3.6.2. The Atomic Scattering Factor

In order to quantitatively model the scattering from a crystal one must allow for any interference effects that can modify the inherent scattering that each of the electrons surrounding an atom will produce. The fact that an atom has a size on the order of the wavelengths used for diffraction experiments gives rise to an interference effect due to scattering from different regions of the electron cloud. Figure 3.12 illustrates this effect in a rather simplified way, showing X-ray beams X and X' coherently scattered from electrons at A and B around an atom. Spherical scattered waves will emerge from each scattering source. When these scattered waves are viewed at an angle of 0.0°, waves Y and Y' are exactly in phase, but when viewed from any nonzero angle, it is seen that wave Z' has to travel BC–AD farther than wave Z. This gives rise to a phase shift and destructive interference which will depend on the angle of view. Because atomic dimensions are on the same order as the wavelength of the scattered X-rays, the BC–AD path difference will always be less than one wavelength and the resulting interference will always be partially destructive rather than constructive. This phenomenon is described by the quantity f_0, which is called the *atomic scattering factor*. The function f_0 is normalized in units of the amount of scattering occurring from a single electron as given in

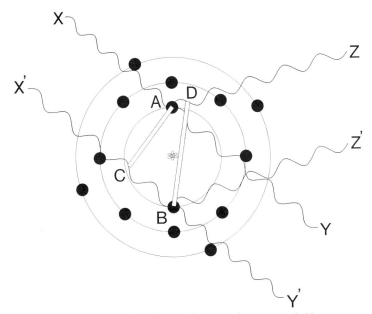

Figure 3.12. The scattering of X-rays from a real atom extended in space.

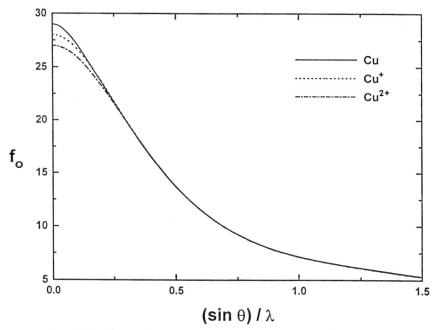

Figure 3.13. The atomic scattering curve for neutral copper, Cu^{1+} and Cu^{2+}.

the Thompson equation (3.15). At zero degrees, f_0 will be equal to the number of electrons surrounding any atom or ion, as may be seen in Figure 3.13 for the three valence states of copper. Since the phase difference depends on the wavelength and the angle of view, the function f_0 is specified as a function of $(\sin \theta)/\lambda$.

The actual shape of the f_0 function must be calculated by integrating the scattering over the electron distribution around an atom. The wave functions for all atoms except hydrogen must be obtained by quantum computational approximation methods and are compiled in the *International Tables for Crystallography*, Vol. C [3]. These compilations do a good job of representing the scattering from spherical atoms, while nonspherical bonding electron density introduces only small deviations between the model and observation. In fact, the careful analysis of both X-ray and neutron diffraction data obtained at temperatures near absolute zero has permitted crystallographers to determine the exact charge on atoms by looking at the departure of the observed electron scattering density from spherical symmetry. Experience in using computed f_0 values has shown that the error is less than the typical errors from other sources and therefore may be considered negligible in powder diffraction.

Atomic scattering factors are usually given in two forms: either in tables as a function of $(\sin\theta)/\lambda$, or as the coefficients of a polynomial fit to curves of the type shown in Figure 3.13.

3.6.3. Anomalous Scattering

There are two other factors that influence the intensity of the scattering from an atom, and it is convenient to consider these as modifying the atomic scattering factor: the first is a phenomenon called *anomalous scattering* or *anomalous dispersion*; the second, treated in Section 3.6.4, below, is due to the fact that atoms vibrate about their average lattice positions. In normal Thompson or coherent scattering, the electron acts as an oscillator under the stimulation of the oscillating electric field of the incident radiation. However, when the electron is bound to a nucleus in an atom, there will be quantum restrictions on the energies it can scatter. A simplistic picture views the electron oscillation in response to an electric field as a true movement of the electron back and forth away from the nucleus. When the frequency of the radiation gets high enough to cause the electron on its next oscillation away from the nucleus to no longer feel the restoring attraction, ionization occurs. This is a wave description of the photoelectric effect. Using this model one may think of the removal of an electron as a resonance phenomenon and the various absorption edges as natural frequencies of vibration.

The number of quantum states available to an electron bound to an atom rises rapidly as its energy approaches the ionization limit. As the energy of radiation incident on the electron approaches the ionization potential, the oscillating electron finds itself instantaneously raised high enough in energy to be able to exist in one of the quantum allowed energy states near the ionization edge. Thus, one may think of a true quantum transition occurring to a state with energy near the absorption edge. The collapse of this excited state, with the transition of the electron back to its ground state orbital, will be accompanied by the emission of a photon of the absorbed energy. However, the lifetime of the excited state will cause the phase of this "fluoresced photon" to lag behind that of a normally scattered photon, producing an interference effect that shows up as an anomalous change in the scattered intensity.

It is most common to describe waves as vectors in the complex plane of an Euler coordinate system and this notation will be employed in each of the remaining sections of this chapter. In this notation, the phase lag is viewed as a rotation of the vector in the complex plane and thus the factor required to correct the normal scattering factor f_0 for anomalous scattering will have a real $(\Delta f')$ and an imaginary $(\Delta f'')$ term. The effective scattering from an atom will be

$$|f|^2 = (f_0 + \Delta f')^2 + (\Delta f'')^2. \quad (3.16)$$

Values for $\Delta f'$ and $\Delta f''$ may be found tabulated in the *International Tables for Crystallography* [3]. In calculating the intensity of scattered X-rays these correction factors only need be employed when the wavelength of the scattered X-rays is near an absorption edge.

3.6.4. Thermal Motion

It was shown in Section 3.6.2 that the size of the atom causes some destructive interference from scattering by electrons displaced at distances on the order of a few percent of the wavelength of the X-rays. If the atom in question is vibrating about its lattice site, then its effective size is larger and the interference effects are, in turn, larger. The interference in a stationary atom causes the atomic scattering factor to fall off somewhat exponentially as a function of $(\sin\theta)/\lambda$, as shown in Figure 3.13. In order to describe the enhancement of this fall off in f_0, caused by thermal motion, it is necessary to first define a parameter, B, which is related to the vibrational amplitude of the atom,

$$B = 8\pi^2 U^2. \quad (3.17)$$

Here B is called the Debye–Waller temperature factor. It is directly related to U^2, the mean-square amplitude of vibration of an atom. The amount and direction of atomic vibration will depend on the temperature (i.e., the amount of kT, thermal energy available), the atomic mass, and the direction and strength of the force constants holding the atom bonded in its location. The effect of increasing B on f_0 is described by

$$f = f_0 \exp\left[-\frac{B\sin^2\theta}{\lambda^2}\right]. \quad (3.18)$$

Figure 3.14 shows the effect of increasing thermal motion on the atomic scattering factor for various values of B. If the surroundings of the atom have cubic symmetry [4], the temperature factor B is the same for all directions of vibration of an atom and is therefore called the isotropic temperature factor. In fact, most atoms in solids will have special directions in which they can vibrate with higher amplitudes. While the *proper anisotropic motion* of an atom may always be described by three vibrational amplitudes, in general six terms of a second-rank tensor are required to transform the proper motion into the crystallographic coordinate system. Note that the $(\sin^2\theta)/\lambda^2$ term in Equation 3.18 is simply $\mathbf{d}^{*2}/4$. Since $\mathbf{d}^{*2}$ can be defined in terms of its vector components,

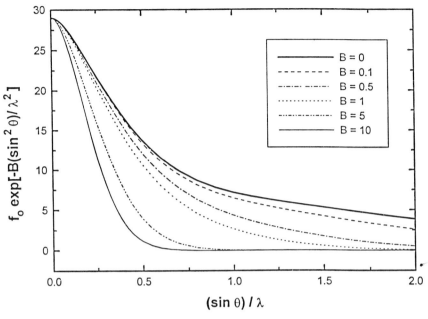

Figure 3.14. The effect of atomic thermal motion on the copper scattering factor.

as shown in Equation 3.10, B can be broken into six B_{ij} anisotropic terms:

$$f = f_0 \exp\left[-\frac{B\mathbf{d}^{*2}}{4}\right], \tag{3.19}$$

$$f = f_0 \exp -\tfrac{1}{4}(B_{11}h^2\mathbf{a}^{*2} + B_{22}k^2\mathbf{b}^{*2} + B_{33}l^2\mathbf{c}^{*2} \\ + 2B_{12}hk\mathbf{a}^*\mathbf{b}^* + 2B_{13}hl\mathbf{a}^*\mathbf{c}^* + 2B_{23}kl\mathbf{b}^*\mathbf{c}^*). \tag{3.20}$$

In Equation 3.20, the cosines of the reciprocal interaxial angles are incorporated into the values of the B_{ij} cross terms. Another common form of the anisotropic temperature factor is made by replacing the B_{ij}'s in Equation 3.20 with the tensor components of the mean-square amplitude of vibration:

$$f = f_0 \exp -2\pi^2(U_{11}h^2\mathbf{a}^{*2} + U_{22}k^2\mathbf{b}^{*2} + U_{33}l^2\mathbf{c}^{*2} \\ + 2U_{12}hk\mathbf{a}^*\mathbf{b}^* + 2U_{13}hl\mathbf{a}^*\mathbf{c}^* + 2U_{23}kl\mathbf{b}^*\mathbf{c}^*). \tag{3.21}$$

Equation 3.21 expresses the atomic vibrational motion in terms of anisotropic mean-square amplitudes of vibration. The most common form of anisotropic temperature factor used is one in which the reciprocal lattice dimensions, all constants, and the cosine terms are incorporated into terms called β_{ij}'s:

$$f = f_0 \exp - (\beta_{11}h^2 + \beta_{22}k^2 + \beta_{33}l^2 + 2\beta_{12}hk + 2\beta_{13}hl + 2\beta_{23}kl). \quad (3.22)$$

A word of warning is needed here. No single convention for expressing anisotropic temperature factors has been adopted, and some authors of crystal structure papers further modify Equation 3.22 to incorporate the constant 2's in the last three terms into the β_{ij}'s. One must carefully examine an author's table of temperature parameters before using them in a computation.

3.6.5. Scattering of X-rays by a Crystal: The Structure Factor

It has been shown that a single electron scatters coherently according to the Thompson equation. Multiple electrons distributed around an atom will scatter as described by the atomic scattering factor as modified by thermal motion and anomalous dispersion. In order to predict the intensity of a diffraction line arising from a crystal, the interference effects between scattering atoms in the unit cell must next be considered. In order to understand these interference effects, it is useful to consider the one-dimensional array of copper atoms (shown in Figure 3.15) being irradiated by Cu $K\alpha_1$ X-rays with $\lambda = 1.5406$ Å. Any unit cell may be chosen to describe a periodic arrangement of atoms, so the one shown in Figure 3.15, beginning at atom A and ending at atom C, although unconventionally nonprimitive, is perfectly acceptable. When X-ray beams 1 and 3 fall onto atoms A and C, they will be coherently scattered in all directions. The distance between the (100) planes is twice the copper–copper interatomic distance, or 5.112 Å, and thus the angle at which constructive interference occurs will be given by Equation 3.1 as

$$1.5406 = 2 \times 5.112 \sin \theta,$$

or $\theta = 8.667°$. The diffracted beams from atoms A and C are shown as 1' and 3' in Figure 3.15. In order to establish the amplitude of the scattered waves from the (100) planes, Figure 3.13 may be consulted to evaluate the atomic scattering factor of a neutral copper atom at a $\sin \theta/\lambda$ of 0.1 to be 27.5. Thus, atoms A and C at the origin of the two unit cells shown in Figure 3.15 each scatter 27.5 times the amount of scattering of one electron. If only atoms A and C were present then the net scattering would be 2(27.5) or 55 electron units and the intensity of the (100) "reflection" would be related to this number. However, there is another atom in the unit cell at position B, halfway between A and C. The triangle YCY' must be equal to one wavelength at the Bragg

INTENSITY OF DIFFRACTION PEAKS 71

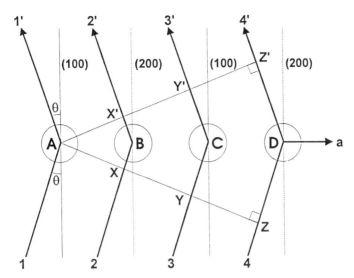

Figure 3.15. Interference between four copper atoms arranged along a straight line.

angle, and thus the triangle XBX' must, by geometry, be exactly one-half λ causing the scattered waves labeled 2' and 4' to be fully out of phase with waves 1' and 3'. Since the scattering amplitude of waves 2' and 4' are also determined by the copper scattering factor, their amplitudes will be identical to that of wave 1' and 3', and so all of the scattering from the (100) planes will be completely extinguished, leaving an intensity of exactly zero. The highly idealized one-dimensional crystal shown in Figure 3.15 was described with a centered unit cell—an additional lattice point being located in the center of the one-dimensional unit cell. Thus, it is the space group of this unit cell which dictates that the (100) reflection must have zero intensity. Such an absence is referred to as a systematic extinction condition.

Next, consider the same arrangement of atoms used in Figure 3.15 but for the (200) reflection. The diffraction angle is now determined by the distance between the (200) planes and occurs at

$$1.5406 = 2 \times 2.556 \sin \theta,$$

or 17.54° θ. The Bragg condition dictates that triangle XBX' is now equal to 1.0λ and so, by geometry, triangle YCY' and ZDZ' must be 2λ and 3λ and all waves 1', 2', 3', and 4' will now be exactly in phase. To find the intensity of the (200) reflection from the unit cell, Figure 3.13 must again be consulted to determine that a neutral copper atom scatters 26 units at $(\sin \theta)/\lambda$ of 0.2. Thus, for the (200), all four copper atoms will scatter in phase, each with an amplitude of 26 units. A more convenient way of finding the intensity is by determining the scattering of the two copper atoms in each unit cell and multiplying by the number of unit cells that make up the crystal.

The artificially simple example shown in Figure 3.15 illustrates all of the principles required to compute the intensity from any set of planes in any unit cell. For example, orthoclase ($KAlSi_3O_8$), a monoclinic feldspar, has 13 atoms in each of the four asymmetric units in a $C2/m$ unit cell. It should be clear that a real crystal structure like orthoclase will require a more powerful mathematical formalism to allow for the many interference effects that will occur among the 52 atoms in its unit cell. To find the overall resultant scattering from any general unit cell, it is necessary to sum the amplitudes of all of the waves scattering from each atom present, allowing for the phase relationships.

As mentioned in Section 3.6.3, above, it is conventional to describe waves as vectors in a complex Euler coordinate system, where the amplitude of the scattered wave is represented by the vector length and the polar coordinate angle of the vector represents the wave's phase. Figure 3.16 illustrates this case for two atoms a and b. The mathematical description of a vector in an Euler coordinate system is

$$\text{vector} = A\exp(i\phi), \quad (3.23)$$

where A is the vector amplitude, or f for an atomic scattering vector, and ϕ is the phase angle. The net scattering from all of the atoms in a unit cell will be the

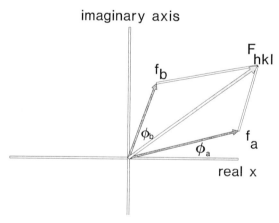

Figure 3.16. Atomic scattering from two atoms represented as scattering vectors ($\mathbf{f}_a$ and $\mathbf{f}_b$) in a complex plane. The resultant scattering from the two is the vector sum $\mathbf{F}_{hkl}$.

INTENSITY OF DIFFRACTION PEAKS 73

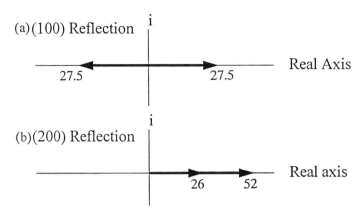

Figure 3.17. Vector description of the net scattering from the line of copper atoms: (a) for the (100) reflection; (b) for the (200) reflection.

vector sum of the individual scattering vectors. This vector sum is known as the structure factor, F_{hkl}, and the intensity will be related to its square. For the situation described in Figure 3.15 for the (100) reflection, the scattering from atom A is a vector with $\phi = 0.0°$ while the vector from atom B has $\phi = 180°$, as shown in Figure 3.17a. The vector sum is of course zero, indicating that the (100) reflection will have no intensity. For the (200) reflection, both of the phase angles are $0.0°$ and so the resultant net scattering from the unit cell (i.e., F_{200}) is the sum of the scattering from atoms A and B, as shown in Figure 3.17b. Note that the structure factor is also a scattered wave that has a phase of its own.

A convenient mathematical way to describe the vectors and vector sums under discussion is to define the phase angle of the scattering from each atom in the unit cell relative to a plane wave through the origin of the unit cell as

$$\phi = 2\pi i(hx + ky + lz), \tag{3.24}$$

where x, y, and z are the fractional coordinates of the atom. The structure factor then becomes

$$F_{hkl} = \sum_{j=1}^{m} f_j \exp[2\pi i(hx_j + ky_j + lz_j)], \tag{3.25}$$

where the sum is taken over the m atoms in the unit cell. In order to understand why the phase angle is chosen with this form, the reader need only note that the function exp $n\pi i$ will be equal to $+1$ when n is even and -1 when n is odd. In this standard notation, the (100) example from Figure 3.15 is now viewed as

$$\mathbf{F}_{100} = f_A \exp[2\pi i(1 \times 0.0)] + f_B \exp[2\pi i(1 \times 0.5)] = f_A - f_B = 0.$$

For the (200) reflection the structure factor is

$$\mathbf{F}_{200} = f_A \exp[2\pi i(2 \times 0.0)] + f_B \exp[2\pi i(2 \times 0.5)] = f_A + f_B = 2f_{Cu}.$$

Whenever complex algebra is invoked to describe a physical phenomenon, it is clear that the physically observable quantity may not contain the $\sqrt{-1}$ term, which has no place in our reality. The square of i produces the value -1, which is physically acceptable, and so "expectation values" are always the square of complex quantities. The observed intensity for any Bragg reflection (hkl) is then proportional to $\mathbf{F}_{hkl}^2$. To compute an intensity, $\mathbf{F}_{hkl}^2$ must first be evaluated and then any other correction terms specific to the physics of the experiment must be applied.

Systematic Extinctions. Translational symmetry relates atoms in one part of a unit cell to identical atoms in another region. This effect generates certain phase relationships between the scattering from these symmetry-related atoms, causing certain classes of reflections to have exactly zero intensity. For example, a centered Bravais lattice permits us to divide the atoms in a unit cell into groups associated with each lattice point. Since a lattice point represents some grouping of atoms, each lattice point in a centered cell must represent exactly the same group. Thus, in a body-centered cell, for each atom located at x, y, z there will be an identical atom located at $x + \frac{1}{2}, y + \frac{1}{2}, z + \frac{1}{2}$. Then $\mathbf{F}_{hkl}$ can be factored into a sum over the atoms represented by each lattice point:

$$\mathbf{F}_{hkl} = \left(\sum_{j=1}^{m/2} f_j \exp[2\pi i(hx_j + ky_j + lz_j)] \right)$$
$$+ \left(\sum_{n=1}^{m/2} f_n \exp\left[2\pi i \left(hx_j + \frac{h}{2} + ky_j + \frac{k}{2} + lz_j + \frac{l}{2} \right) \right] \right). \quad (3.26)$$

If the sum of $h + k + l$ is even, the second term in Equation 3.26 will contain an exponent with an integer, n, in it. An integral number of 2π's will have no effect on the value of this term, and in this case the equation reduces to

$$\mathbf{F}_{hkl} = \sum_{j=1}^{m/2} 2f_j \exp[2\pi i(hx_j + ky_j + lz_j)]. \quad (3.27)$$

Table 3.2. Example of Conditions for the Extinction of Reflections Due to Translational Symmetries

Symmetry	Extinction Conditions
P	none
C	$hkl; h + k =$ odd
I	$hkl; h + k + l =$ odd
F	$hkl; h, k, l$ mixed even and odd
$2_1 \parallel b$	$0k0: k =$ odd
$c \perp b$	$h0l: l =$ odd

However, if the sum of $h + k + l$ is odd, then the second term in Equation 3.26 will contain an exponent with a $2\pi(n/2)$ term; here n is any integer representing a full rotation of the scattering vector. This causes each term in the second sum to be negative. Thus, for each atomic scattering vector in the first sum, there is an equal and opposite scattering vector in the second sum. The result is that the structure factor (and hence the intensity for all reflections with the sum of $h + k + l =$ odd) is exactly equal to zero. This condition is called a *systematic extinction*. A similar analysis will show different extinction conditions for the other Bravais lattices, as summarized in Table 3.2. Screw axes and glide planes will also introduce systematic extinctions into various classes of reflections. Reflections that have zero intensity due to the atomic scattering vectors happening to cancel each other out and not due to a systematic symmetry condition, are called *accidentally absent*.

3.7. THE CALCULATED DIFFRACTION PATTERN

The foregoing sections have detailed a number of the points that need to be taken into account in order to compute the intensity of a diffraction line. The structure factor clearly shows that the intensities of diffraction lines depend on the atomic locations and, of course, site occupancies and thermal motion. In principle, these fundamental atomic parameters may be extracted from Equation 3.25, but to do so will require an exact modeling of the intensity. The case of diffraction is a fortunate one because the physics of this process is very well understood and hence modelable. The most important aspects of this modeling have already been presented. However, a few more considerations will be required to carry out a quantitative calculation. The additional corrections fall into two categories: those that vary from one reflection to another or depend on diffraction angle, and those that affect only the absolute value of the intensity and so contribute the same proportional amount to each (hkl).

3.7.1. Factors Affecting the Relative Intensity of Bragg Reflections

a. Multiplicity of Bragg Planes. The number of equivalent planes cutting a unit cell in a particular hkl family is called the plane *multiplicity factor* M_{hkl} and will directly affect the intensity. If the shape of the crystallites and the mounting technique do not induce any preferred orientations of the crystallites in a specimen, or if a solid polycrystalline specimen is not textured, then there will be as many crystallites oriented to diffract from the (100) plane as from the ($\bar{1}$00). If the system is cubic, then the (010), (0$\bar{1}$0), (001), and (00$\bar{1}$) have identical d values and structure factors, so six different crystallite orientations will all contribute equally to the intensity observed at the diffraction angle of the (100) reflection, making $M_{100} = 6$. If the crystal system is tetragonal, M_{100} will be 4 and M_{001} will be 2.

Each cubic face has a diagonal (110) and an equivalent ($\bar{1}$10) plane. With six faces, there are 12 crystallographic orientations [(110), ($\bar{1}$10), (1$\bar{1}$0), ($\bar{1}\bar{1}$0), (101), ($\bar{1}$01), (10$\bar{1}$), ($\bar{1}$0$\bar{1}$), (011), (0$\bar{1}$1), (01$\bar{1}$), (0$\bar{1}\bar{1}$)], which will all contribute to the intensity of the (110) reflection. Thus, the (110) reflection will have 12/6 more crystallites contributing to its intensity than the (100) and would be twice as intense as (100), if all other factors were equal. The plane multiplicity factors for the various classes of reflections in the different crystal systems are shown in Table 3.3.

b. The Lorentz Factor. The Lorentz factor is a measure of the amount of time that a point of the reciprocal lattice remains on the sphere of reflection during the measuring process. Since diffraction takes place over a narrow range of angles (see Section 3.9.2), the points have small but finite size and are inversely related to the crystallite size. Thus, if a detector scans through 2θ at a constant angular velocity, then the amount of time each reciprocal lattice point remains in diffracting position on the Ewald sphere becomes a function of diffraction angle. The high-angle (longest) reciprocal lattice vectors cut the Ewald sphere in a manner approaching a tangent. This causes the high-angle reflections to remain in diffracting position longer, as the vector slides along the back of the sphere, and increase their intensity relative to the low-angle reflections. For a powder diffractometer this effect can be compensated, and all reflections put onto the same intensity scale, by including the term $1/(\sin^2\theta \cos\theta)$ in the expression for calculating diffraction intensities. This factor also includes the geometric effect that the diffractometer only collects intensity from a fixed slit length rather than from a whole Debye ring. Usually, the Lorentz factor is combined with the polarization term from the Thompson

Table 3.3. Plane Multiplicity Factors, M_{hkl}

System	hkl	hhl	hh0	0kk	hhh	hk0	h0l	0kl	h00	0k0	00l
Cubic	48[a]	24	12	(12)	8	24[a]	(24[a])	(24[a])	6	(6)	(6)
Tetragonal	16[a]	8	4	(8)	(8)	8[a]	8	(8)	4	(4)	2
Hexagonal	24[a]	12[a]	6	(12)	(12)	12[a]	(12[a])	12[a]	6	(6)	2
Orthorhombic	8	(8)	(8)	(8)	(8)	4	(4)	(4)	2	(2)	(2)
Monoclinic	4	(4)	(4)	(4)	(4)	(4)	(2)	(4)	2	(2)	(2)
Triclinic	2	(2)	(2)	(2)	(2)	(2)	(2)	(2)	(2)	(2)	(2)

[a]When all permutations of indices do not produce equivalent planes, M must be reduced by half.

equation and called the *Lp correction*. Note that a different diffraction geometry may have a different Lorentz factor associated with it.

c. Extinction. Another effect that must be treated to permit the computation of a powder diffraction pattern is one that affects only the most intense lines from crystallites with a high degree of perfection. The theory that has been presented in this chapter is referred to as kinematic theory. To exactly describe all of the effects that may occur between the field of the incident electromagnetic X-ray wave and the periodic field within a crystal, one must invoke a full electrodynamic description. The first such description was published by C. G. Darwin in 1914 [5]. This was followed in 1916 by P. P. Ewald, who in his elegant treatment also developed the concept of the reciprocal lattice. A subsequent treatment in 1922 by Darwin explained the (by then already observed) phenomenon he called *extinction*.

The observed reduction in intensity from intense reflections is due to an interference effect from secondary reflections of the diffracted beam from the undersides of the atomic planes. This is illustrated in Figure 3.18. To understand this effect, one must realize that there is a phase shift of $\pi/2$ each time a beam is "reflected" from a plane. Thus, in Figure 3.18, an incident X-ray beam penetrates to successively deeper planes in a perfect crystal. On reflection (with a phase shift of 90°) the beam strikes the underside of the planes and reflects again (now with a phase shift of 180°) into the crystal the direction of the incident beam. The diffracted intensity that should have emerged from the crystal has been directed back into the crystal fully out of phase with the rest of the incident beam, destructively interfering with it. This interference causes an extinction of some of the power that would otherwise be in the direct diffracted

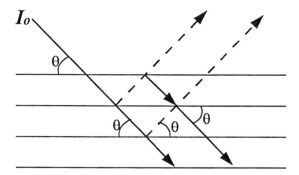

Figure 3.18. The primary extinction effect reducing diffracted power from strong reflections of a perfect crystal by reflecting diffracting power back into the crystal out of phase with the incident beam.

beam and causes a lowering of the observed intensity of strong reflections from very perfect crystals.

There is another dynamical effect in perfect crystallites called *secondary extinction*. This results when the strongest reflections diffract most of the intensity in the incident beam out of the crystal before the beam can penetrate to any significant depth. Therefore, the lower lying planes never have a chance to diffract the amount of intensity they are actually capable of reflecting. This also causes the most intense reflections to have a lower relative intensity than weaker reflections.

Zachariasen [6] derived a correction for extinction effects that requires the knowledge of the domain size, a parameter typically not known. Researchers usually try to eliminate extinction effects by inducing strains into a specimen, for example, by thermally shocking it between room temperature and the temperature of liquid nitrogen. In an "ideally imperfect" crystal, there are very small misorientations between mosaic blocks, perhaps induced by crystal imperfections, and the effect of extinction is not large. In practice, powders making up the specimen for examination by X-ray powder diffractometry contain crystallites lying between the two extremes of "perfect"and "ideally imperfect." Nevertheless, it is important that extinction effects be minimized. Therefore, powders should be ground as fine as possible, since this decreases the size of the individual crystallites and induces strains in the mosaic blocks. However, even at a particle size of $1\,\mu$m some materials may still show extinction in their most intense reflections, which can lead to reduction of the observed intensity by as much as 25% [7].

d. Absorption. Clearly the absolute intensity of a diffraction peak will be proportional to the number of unit cells diffracting. This in turn will be related to the footprint the incident beam makes on the sample combined with its depth of penetration. Bragg–Brentano geometry with a briquette-shaped homogeneous specimen produces the fortunate case where the irradiated volume of the specimen is constant at all diffraction angles. This is the result of the exact offsetting of the effect of the incident beam shadow shrinking as 2θ increases, by the beam penetrating deeper, producing a constant volume of irradiation independent of diffraction angle. Thus, to allow for the effect of specimen absorption on the diffracted intensities, the intensity equation need only contain a constant factor $1/\mu$. For other geometries, such as a cylindrical sample, the absorption term will have a trigonometric component and become a function of 2θ.

As will be described in Section 7.5.4, low sample absorption in a Bragg–Brentano diffractometer specimen and high sample absorption in cylindrical and transmission geometries can cause serious profile asymmetry and peak location displacement.

e. Microabsorption. There is another absorption effect that can disturb the relative intensities of diffraction peaks in a polyphase mixture. Microabsorption occurs when large crystallites of phase α lie above or below crystallites of phase β. Since the incident beam may spend a disproportionate amount of time inside an α-crystallite, it will not be absorbed as if it passed through a medium having the average mass attenuation coefficient of the mixture. Instead, if it spends more time in an α-crystallite, it will act as if it had been absorbed by a material with a mass absorption closer to that of pure α. This phenomenon was first described by Brindley [8] and is of most concern in quantitative phase analysis applications. This, like extinction, will be reduced in importance as crystallite size decreases.

f. Monochromator Polarization. As described above, the process of diffraction causes the polarization of the diffracted beam. Thus, if an additional diffraction step is employed to monochromatize the incident or diffracted beams, an additional correction to the intensity will be required. For a diffracted beam monochromator, the polarization factor in Equation 3.15 becomes

$$\frac{1 + \cos^2(2\theta)\cos^2(2\theta_m)}{2},$$

where θ_m is the Bragg angle of the monochromator crystal. This angle is fixed when the monochromator is aligned, so the factor contributes equally to all reflections. A word of warning is required here concerning the popular diffracted-beam pyrolytic graphite crystals. These crystals have such a high mosaic spread that they might better be considered as a bandpass filter. As such, they do not polarize the beam to the degree implied in the correction factor.

3.7.2. The Intensity Equation

Summarizing the above and adding a few constants that apply to the case of an ideally imperfect crystal, we may write the integrated intensity diffracted from a phase α in a flat briquette-shaped sample, measured by a diffractometer with fixed receiving slit, neglecting air absorption, as

$$I_{(hkl)\alpha} = \frac{K_e K_{(hkl)\alpha} v_\alpha}{\mu_s}. \tag{3.28}$$

Here K_e is a constant for a particular experimental system:

$$K_e = \frac{I_0 \lambda^3}{64\pi r}\left(\frac{e^2}{m_e c^2}\right)^2, \tag{3.29}$$

where

- I_0 = incident-beam intensity;
- r = distance from the specimen to the detector;
- λ = wavelength of the X-radiation;
- $(e^2/m_e c^2)^2$ is the square of the classical electron radius;
- μ_s = linear attenuation coefficient of the specimen;
- v_α = volume fraction of phase α in specimen.

Also $K_{(hkl)\alpha}$ is a constant for each diffraction reflection hkl from the crystal structure of phase α:

$$K_{(hkl)\alpha} = \frac{M_{hkl}}{V_\alpha^2} |F_{(hkl)\alpha}|^2 \left(\frac{1 + \cos^2(2\theta)\cos^2(2\theta_m)}{\sin^2 \theta \cos \theta} \right)_{hkl}, \qquad (3.30)$$

where

- M_{hkl} = multiplicity for reflection hkl of phase α;
- V_α = volume of the unit cell of phase α;
- the fraction in parentheses, $(\cdots)_{hkl}$, equals the Lorentz and polarization corrections for the diffractometer $(Lp)_{hkl}$, including a correction for the diffracted beam monochromator;
- $2\theta_m$ = the diffraction angle of the monochromator;
- $F_{(hkl)\alpha}$ = the structure factor for reflection hkl and includes anomalous scattering and temperature effects.

It is a triumph of our understanding of physics that computations of integrated diffraction intensities, using Equation 3.28, produce the observed values. This triumph has been heavily exploited in determining the crystal structure of materials. This ability, first used by the Braggs in 1913, has established the basis of modern solid state science. Bragg's law (Equation 3.1) is used to test that a particular model for a crystal structure is correct, and the lattice parameters are used (with Equation 3.10) to calculate the positions of possible diffraction lines. Then Equation 3.28 is used to compute the intensities of these lines. Even slight errors in the crystal structure model will be seen as discrepancies between the observed and calculated intensities. When each of the intensities can be independently measured, as in the case of single-crystal diffractometry, they can be used in a least squares procedure to refine the atomic coordinates. When the reflections are difficult to separate as in the powder pattern of a low-symmetry material, the least squares procedure is performed against the whole powder pattern. In the materials laboratory, the most common application of the computed powder pattern is to look for effects like preferred orientation of the crystallites, solid solution effects, and any structural modifications that may have taken place. The most commonly used program for this purpose is POWD by Smith et al. [9].

3.8. CALCULATION OF THE POWDER DIFFRACTION PATTERN OF KCl

The computation of a powder diffraction pattern may involve millions of discrete steps and is therefore typically carried out with a computer program. In order to understand any large computation, a simple example should be carried out manually. Potassium chloride will provide such an example. The experimental powder pattern of KCl has already been seen in Figure 3.11 as a Debye–Scherrer photograph and is shown again as a diffractometer trace in Figure 3.19. The experimental d and I^{rel} values may be found in the Powder Diffraction File as pattern number 4-587, which is also shown in Figure 3.19.

The computation of a powder pattern is simplified by the fact that for a given material, measured under fixed instrumental conditions, the K_e term in Equation 3.28 is the same for all diffraction lines. As long as relative intensities $(I/I_{max}$ or $I^{rel})$ are to be computed, all constant terms will cancel out and therefore need not be computed. So, for a phase pure material only the $K_{(hkl)\alpha}$ term is required. An additional simplification occurs due to the systematic extinction conditions shown in Table 3.2. The only reflections which may have intensity are shown in Table 3.4.

Table 3.4. Reflections Allowed for Various Bravais Lattices

(hkl)	$h^2 + k^2 + l^2$	Primitive	BCC	FCC
(100)	1	Yes	No	No
(110)	2	Yes	Yes	No
(111)	3	Yes	No	Yes
(200)	4	Yes	Yes	Yes
(210)	5	Yes	No	No
(211)	6	Yes	Yes	No
None	7	None	None	None
(220)	8	Yes	Yes	Yes
(221)	9	Yes	No	No
(301)	10	Yes	Yes	No
(311)	11	Yes	No	No
(222)	12	Yes	Yes	Yes

POWDER DIFFRACTION PATTERN OF KCL

As shown in Section 2.4, KCl crystallizes in the face-centered cubic space group $Fm3m$ (No. 225), with four KCl units per unit cell (i.e., $Z = 4$) and a lattice parameter of 6.2917 Å. The location of the K ion at the origin of the unit cell and the Cl ion at $\frac{1}{2}, \frac{1}{2}, \frac{1}{2}$ allows for an algebraic manipulation of the structure factor to give, for mixed (i.e., even and odd) indices $F^2 = 0$. For unmixed indices the following relationships hold: where $(h + k + l)$ is even,

$$F^2 = 16(f_K + f_{Cl})^2,$$

and where $(h + k + l)$ is odd,

$$F^2 = 16(f_K - f_{Cl})^2.$$

Table 3.5 gives a typical scheme of the steps involved in calculating the relative intensities from KCl. The lattice parameter may be used in Equation 3.10 to calculate the d_{hkl}, which in turn may be used in Equation 3.1, for a wavelength of 1.5406 Å, to compute the 2θ values given in column 4. The Lp factor in column 7 is computed from

$$\left(\frac{1 + \cos^2(2\theta)}{\sin^2 \theta \cos \theta}\right)_{hkl}$$

The atomic scattering factors, f_K and f_{Cl}, are obtained from published tables [10]. From the published crystal structure, the isotropic Debye–Waller temperature factor, B, for potassium is found to be 1.50 and for chlorine, 1.05. The exponential argument of Equation 3.18, i.e., $[-(B\sin^2 \theta)/\lambda^2]$, is listed as B_K and B_{Cl}, and the temperature-corrected scattering factors as f_K^T and f_{Cl}^T. Following the calculation of the structure factor and the multiplicity, intensity values are calculated according to Equation 3.28 and then normalized to the strongest line to give the relative intensities (I_{calc}^{rel}). It will be seen that good agreement is obtained between calculated and observed intensities.

The agreement between the results of such a computation and the observed powder pattern verifies the validity of the model expressed in Equation 3.28. It is occasionally argued that the calculated powder pattern is more "accurate" than the observed pattern and so experimental patterns should be replaced in the Powder Diffraction File (PDF) by calculated patterns. But, in light of the various distortions that occur in real materials due to stacking faults, anisotropic crystallite size and strain broadening, preferred orientation, etc., this argument is generally rejected in favor of experimental data. However, when only the crystal structure of a material has been reported, its calculated pattern is welcome in the PDF.

Table 3.5. Calculated Pattern for Potassium Chloride

Line	hkl	$h^2 + k^2 + l^2$	2θ	$\sin\theta$	$\sin\theta/\lambda$	Lp
1	111	3	24.49	0.2121	0.1377	41.60
2	200	4	28.35	0.2449	0.1590	30.52
3	220	8	40.52	0.3463	0.2248	14.02
4	311	11	47.92	0.4061	0.2636	9.62
5	222	12	50.19	0.4241	0.2753	8.65
6	400	16	58.65	0.4897	0.3179	6.08
7	331	19	64.51	0.5337	0.3464	4.92
8	420	20	66.40	0.5476	0.3554	4.63
9	422	24	73.71	0.5998	0.3893	3.75

Line	f_K^0	B_K	f_K^T	f_{Cl}^0	B_{Cl}	f_{Cl}^T
1	15.66	0.028	15.22	14.22	0.020	13.94
2	15.02	0.038	14.46	13.41	0.027	13.06
3	13.01	0.078	12.06	11.25	0.053	10.67
4	11.90	0.104	10.72	10.27	0.073	9.55
5	11.59	0.114	10.34	10.01	0.080	8.29
6	10.56	0.152	9.07	9.22	0.106	7.76
7	10.02	0.180	8.37	8.80	0.126	7.76
8	9.81	0.189	8.12	8.69	0.133	7.61
9	9.23	0.227	7.35	8.28	0.159	7.06

Line	F	$\|F^2\|$	M_{hkl}	I_{calc}	I_{calc}^{rel}	I_{obs}^{rel}
1	1.28	26	8	0.87	0	0
2	27.52	12118	6	221.92	100	100
3	22.73	8266	12	139.11	63	59
4	1.17	22	24	0.51	0	0
5	19.59	6139	8	42.51	19	23
6	17.37	4825	6	17.59	8	8
7	0.61	6	24	0.71	0	0
8	15.73	3957	24	43.93	20	20
9	14.41	3324	24	29.90	13	13

Crystallographic conditions may sometimes occur that cause an "allowed" reflection not to be observed. As an example, it was stated earlier that for face-centered cubic lattices the odd values of $(h + k + l)$ are equal to the square of the differences of the two atomic scattering factors. If the example of NaCl, KCl, and RbCl are considered, as illustrated in Figure 3.19, an interesting fact is observed. The atomic scattering factor of the chloride ion is 18 electrons (at zero degrees 2θ), and that of the sodium ion 10 electrons. Thus, the intensities of odd (hkl) values are proportional to 8^2, or 64. Since the potassium ion is isoelectronic with the chlorine ion, the difference in the atomic scattering factors is zero so all odd (hkl) values are absent. The atomic scattering factor of the rubidium ion is 36 electrons thus, the intensities of odd (hkl) values is proportional in this instance to $(36-18)^2$ or 324. Examination of the three patterns reveals that while the even (hkl) values are all present, there are no odd values present in the case of KCl. It will also be noted that the odd (hkl) values for RbCl are much stronger than the equivalent values for NaCl due to the larger squared difference term in the case of RbCl.

3.9. ANISOTROPIC DISTORTIONS OF THE DIFFRACTION PATTERN

In Section 3.7.1 a number of items were described that affect the relative intensity of one line with respect to another. There are three additional things that can affect the diffraction lines of particular crystallographic zones. Figure 3.20 illustrates how atoms bond together in a unique number, type, and arrangement to form unit cells. If the unit cells are stressed uniformly (i.e., macrostress; see below), the diffraction lines will be shifted; if they are stressed nonuniformly (i.e., microstress), the diffraction lines will be broadened. A *crystallite* comprises a number of cells systematically grouped together to form a coherently diffracting domain. If the cells are not identical but show a variation in atomic position destroying long-range order, the material is called *amorphous*. Where the individual cells are highly ordered, the specimen is referred to as *crystalline*. The specimen itself is made up of a large number of crystallites. The ideal arrangement of individual crystallites in the specimen for diffraction purposes is that they be completely random. Where the crystallites take up some particular orientation, the specimen is referred to as suffering from *preferred* orientation.

3.9.1. Preferred Orientation

Preferred orientation of crystallites in bulk, polycrystalline materials is a vital subject to many materials industries. A number of industrial materials are based on tensor physical properties of the crystallites. For example, barium hexaferrite ceramic magnets, used commonly as seals on refrigerator doors, are polycrystalline materials in which only the $(00l)$ crystallite direction has a magnetic moment. Thus, the fabrication and quality control procedures must involve manipulating and measuring the degree of preferred orientation in the ceramic. Extruded wires show a characteristic preferred orientation, as do most pressed powder materials. The most common way of evaluating the type and extent of preferred orientation is to measure the pole figure for a particular crystallographic direction. The pole figure is simply the intensity

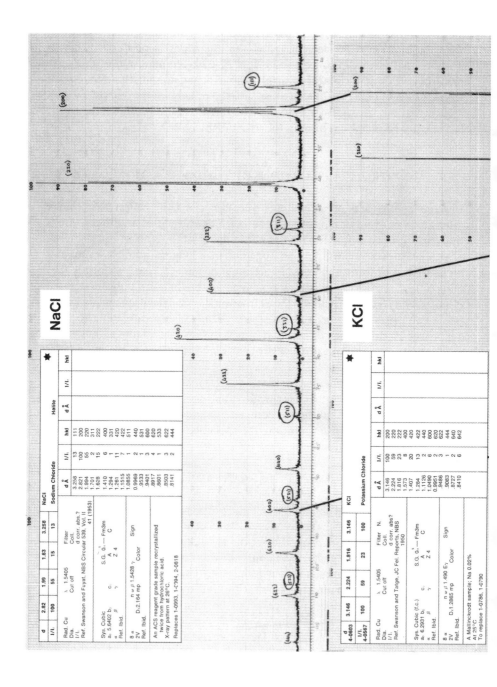

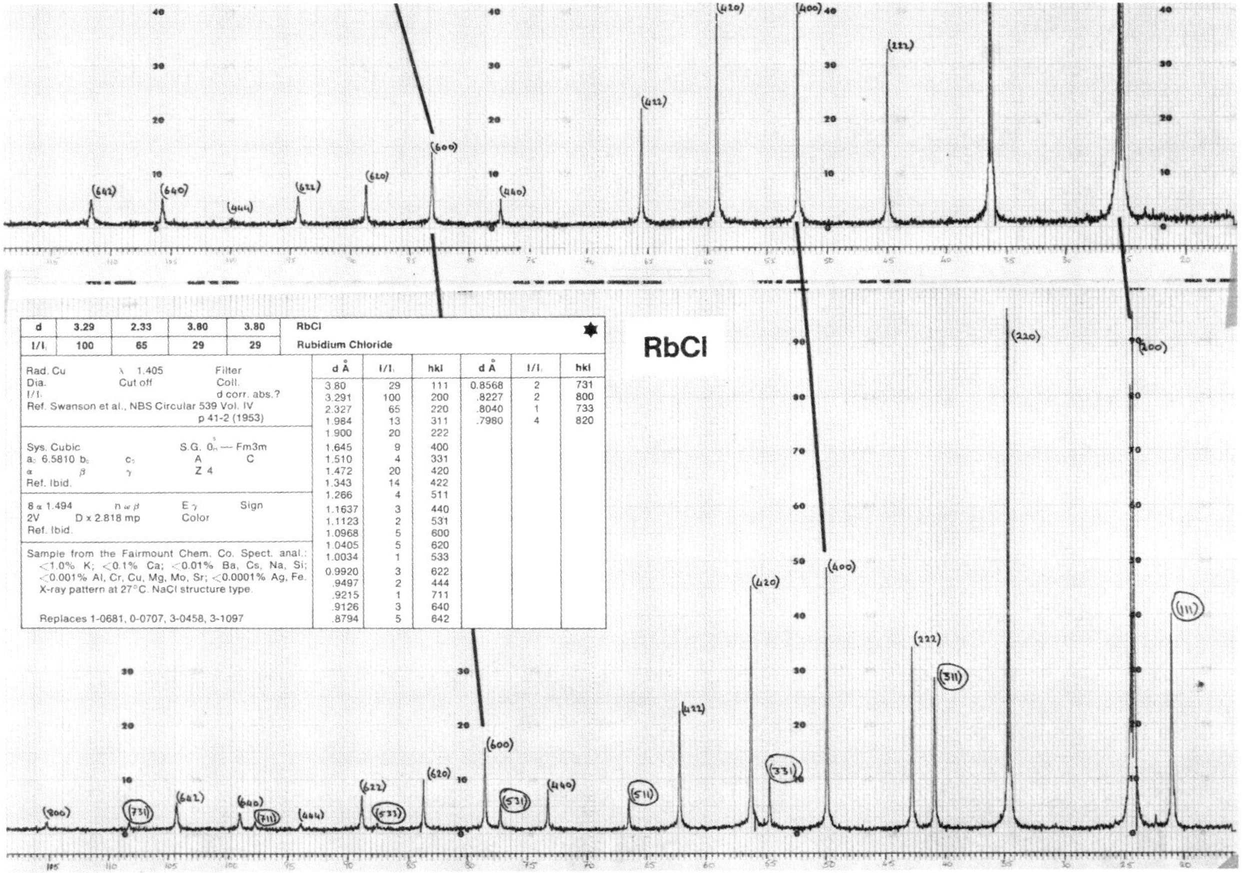

d	3.29	2.33	3.80	3.80	RbCl
I/I₁	100	65	29	29	Rubidium Chloride

Rad. Cu   λ 1.405   Filter
Dia.   Cut off   Coll.
I/I₁   d corr. abs.?
Ref. Swanson et al., NBS Circular 539 Vol. IV p 41-2 (1953)

Sys. Cubic   S.G. O_h^5 — Fm3m
a₀ 6.5810 b₀ c₀ A C
α β γ Z 4
Ref. Ibid.

ε α 1.494 n ω β ε γ Sign
2V D x 2.818 mp Color
Ref. Ibid.

Sample from the Fairmount Chem. Co. Spect. anal.: <1.0% K; <0.1% Ca; <0.01% Ba, Cs, Na, Si; <0.001% Al, Cr, Cu, Mg, Mo, Sr; <0.0001% Ag, Fe. X-ray pattern at 27°C. NaCl structure type.

Replaces 1-0681, 0-0707, 3-0458, 3-1097

d Å	I/I₁	hkl	d Å	I/I₁	hkl
3.80	29	111	0.8568	2	731
3.291	100	200	.8227	2	800
2.327	65	220	.8040	1	733
1.984	13	311	.7980	4	820
1.900	20	222			
1.645	9	400			
1.510	4	331			
1.472	20	420			
1.343	14	422			
1.266	4	511			
1.1637	3	440			
1.1123	2	531			
1.0968	5	600			
1.0405	5	620			
1.0034	1	533			
0.9920	3	622			
.9497	2	444			
.9215	1	711			
.9126	3	640			
.8794	5	642			

Figure 3.19. Diffractograms of NaCl, KCl, and RbCl.

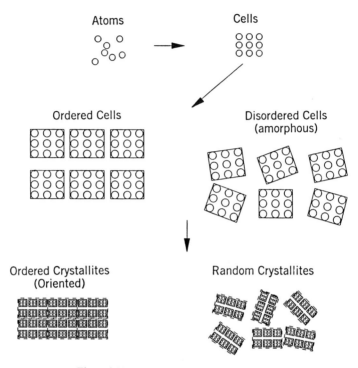

Figure 3.20. The structural makeup of solids.

of a particular Bragg diffraction line plotted as a function of the three-dimensional orientation of the specimen. It is determined on a pole figure diffractometer, which is essentially the same as a single-crystal diffractometer, being able to rotate the specimen through all orientations while monitoring the diffraction intensity of a reflection. The results are displayed as a pole figure, which is a two-dimensional stereographic projection. Bunge [11] has written a comprehensive treatment of preferred orientation evaluation techniques.

A nonconventional method should be mentioned that avoids the requirement of an expensive automated pole figure device while often providing the required information. All of the preferred orientation on crystallites lying within a materials' surface can be ascertained by comparing the observed powder pattern diffraction intensities to the "ideal" intensities obtained by calculating the powder pattern as described above. The trick is to find a way to compare these two patterns, which are on different scales. Snyder and Carr [12] devised a scale-independent function that allows this comparison and permits direct computation of a tensor properties enhancement or disenhancement in a particular crystallographic direction.

3.9.2. Crystallite Size

In Figure 3.1 it was shown that when diffraction occurs the path length distance between adjacent planes, ABC, must exactly equal 1λ. Geometry requires that when this condition is met, the path length distance DEF between the three sets of planes must be 2λ and all similar triangles from lower lying planes must also be an integral multiple of λ. If the angle of incidence θ is set so that the path length distance ABC becomes 0.5λ, then DEF will be 1.0λ, etc. Thus, each pair of planes exactly cancels each other's scattering. This not only means that diffraction is not observed, but in fact all incident intensity will be extinguished by what may be called *Bragg extinction*. Next consider the situation when the incident X-ray beam is 10% away from the Bragg angle such that $ABC = 1.1\lambda$. Geometry then causes triangle DEF to be 2.2λ, and the phase shift from the scattered waves from the sixth plane down from the surface will be 5.5λ, which will therefore be exactly out of phase (0.5λ) with the waves scattered from the first plane. The scattering from the second and seventh planes will also be exactly out of phase. When all the planes from all of the unit cells are considered, it will be seen that no net scattering will occur; that is, there will be another unit cell at some depth within the crystal that will exactly cancel the scattering from any other cell except exactly at the Bragg angle where all scattering is in phase.

If θ is set closer to the Bragg angle so that the distance ABC is equal to 1.001λ, then the scattering from the first plane will be canceled by the scattering from the plane 501 layers deep in the crystal, with a phase shift of 500.5λ. Similarly, when ABC is 1.00001 the scattering will be canceled by a plane 50,001 layers deep in the crystal. Thus, it is clear that Bragg reflections should occur only exactly at the Bragg diffraction angle, producing a sharp peak. However, if the crystal is only 1000 Å in size, then the planes needed to cancel scattering from, for example, the (100) plane, with an ABC distance of 1.0001λ (i.e., the 5001st plane), are not present. Thus, the diffraction peak begins to show intensity at a lower θ and ends at a higher θ than the Bragg angle. This is the source of "particle size broadening" of diffraction lines. The observed broadening can be used to determine the crystallite sizes of less than 1 μm in materials. Crystallites larger than 1 μm typically have a sufficient number of planes to allow the diffraction peak to display its inherent Darwin width [5] (i.e., the width dictated by the uncertainty principle), additionally broadened by instrumental effects, with no contribution from size broadening (see Section 11.3.4).

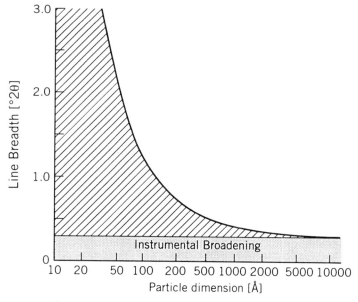

Figure 3.21. Line width as a function of particle dimension.

The crystallite size broadening (β_τ) of a peak can usually be related to the mean crystallite dimension (τ) via the Scherrer equation [13]:

$$\tau = \frac{K\lambda}{\beta_\tau \cos\theta}, \qquad (3.31)$$

where β_τ is the line broadening due to the effect of small crystallites. Here β_τ is given by $(B - b)$, B being the breadth of the observed diffraction line at its half-intensity maximum, and b the instrumental broadening or breadth of a peak from a specimen that exhibits no broadening beyond inherent instrumental peak width. Note that β_τ must be given in radians; K is the so-called *shape factor*, which usually takes a value of about 0.9 [14]. Figure 3.21 gives a plot of the line breadth as a function of the particle dimension. Here it will be seen that below 10,000 Å (1 μm) the observed line breadth becomes significantly broader than the inherent instrumental breadth. Where this is the case, the assumption can no longer be made that the peak height of a diffraction line is proportional to its integrated area.

The additional broadening in diffraction peaks beyond the inherent peak widths due to instrumental effects can be used to measure crystallite sizes as low as 10 Å. However, a second cause of broadening, namely, the effects of stress, can complicate the picture.

3.9.3. Residual Stress and Strain

Strain in a material can produce two types of effects in diffraction patterns. If the stress is uniformly compressive or tensile, it is called a *macrostress* and the distances within the unit cell will either become smaller or larger, respectively. This will be observed as a shift in the location of the diffraction peaks. These macrostresses are measured by an analysis of the lattice parameters (described later in Section 9.6.4). Figure 3.22 shows the diffraction from AgBr in a photographic film. The differential expansion between the film substrate and the AgBr causes a *macrostrain* that changes the lattice parameter and shifts the peak in the dry specimen from that of the wet specimen.

Microstrains are produced by a distribution of both tensile and compressive forces, and the resulting diffraction profiles will be broadened about the original position. Both the size and strain broadening effects generally produce a symmetric broadening. The observed asymmetry in diffraction profiles is usually due to instrumental effects (see Section 7.5) or to depth gradients in composition or strain. *Microstress* in crystallites can come from a number of sources: dislocations (the most important source), vacancies, defects, shear planes, thermal expansions and contractions, etc. (see Warren [4]). Whatever the cause of the residual stress in a crystallite, the effect will cause a distribution of d-values about the normal, unstrained, or macrostrained d_{hkl}-value. Figure 3.23 shows a somewhat simplified view of the effect of the application of nonuniform stress to a series of rows of unit cells (for simplicity, just one row of cells is shown in the figure). The overall effect of the stress is to bend the rows of cells into an arc. In order that the cells remain in contact, the cells are distorted to a wedge shape. While the average d-spacing remains constant, the average d-spacing at the bottom of the wedge is less than the average value, and the average d-spacing at the top of the wedge is greater than the average. Thus, there is now a range of d-values with an equivalent range of 2θ values and a broadening of the diffraction line.

Figure 3.24 [15] shows an actual example of the relief of stress by annealing. The lower figure shows one of the strong lines of a chill-cast brass sample. Because the cooling was rapid, the brass solidified quickly, causing a buildup of strain within the sample. As the specimen is annealed at increasingly higher temperatures, the stress is relieved, as indicated by a sharpening of the diffraction line.

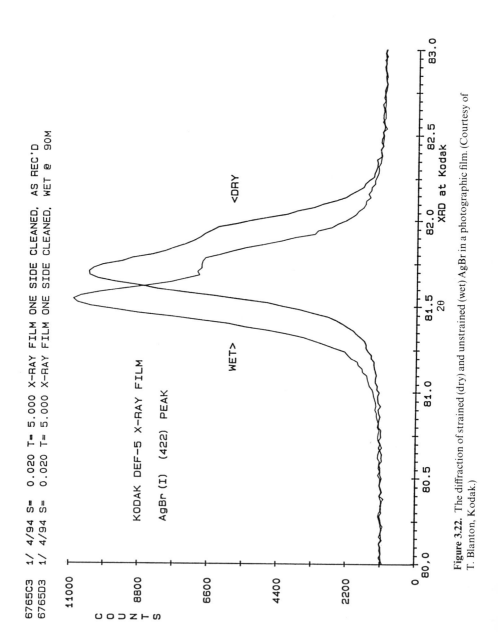

Figure 3.22. The diffraction of strained (dry) and unstrained (wet) AgBr in a photographic film. (Courtesy of T. Blanton, Kodak.)

ANISOTROPIC DISTORTIONS OF THE DIFFRACTION PATTERN 93

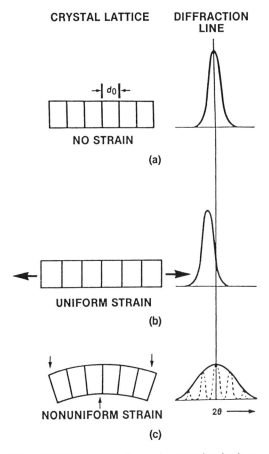

Figure 3.23. Strain expanding and contracting d-values.

The broadening in a peak due to stress has been shown to be related to the residual strain ε by

$$\beta_\varepsilon = 4\varepsilon \tan \theta. \qquad (3.32)$$

As with β_τ, β_ε (in radians) is the additional broadening of an observed diffraction peak—beyond the inherent instrumental breadth—caused by stress. The fact that stress-induced diffraction peak broadening follows a $\tan \theta$ function whereas crystallite size broadening has a $1/\cos \theta$ dependence allows us to separate these effects. An excellent up-to-date treatment of strain analysis may be found in Noyan and Cohen [16].

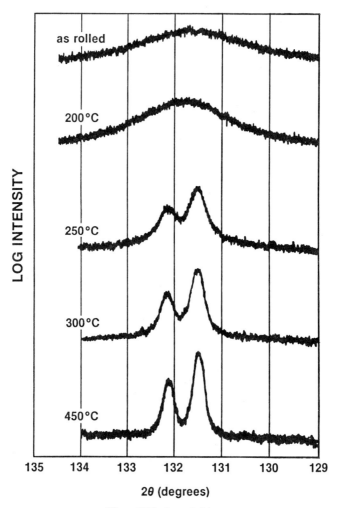

Figure 3.24. Annealed brass.

REFERENCES

1. *Powder Diffraction File.* International Centre for Diffraction Data, 12 Campus Blvd., Newtown Square, PA 19073-3273.
2. *Crystal Data File.* International Centre for Diffraction Data, 12 Campus Blvd., Newtown Square, PA 19073-3273.
3. Wilson, A. J. C., ed. *International Tables for Crystallography,* Vol. C: *Mathematical, Physical and Chemical Tables.* Kluwer Academic Publishers, Dordrecht, The Netherlands, 1992.

4. Warren, B. E. *X-Ray Diffraction*, p. 191. Addison-Wesley, Reading, MA, 1969.
5. Darwin, C. G. The theory of X-ray reflexion. *Philos. Mag.* [6] **27**, 315, 675 (1914).
6. Zachariasen, W. H. *Theory of X-Ray Diffraction in Crystals*. Dover, New York, 1945.
7. Cline, J. P., and Snyder, R. L. The effects of extinction on X-ray powder diffraction intensities. *Adv. X-Ray Anal.* **30**, 447–456 (1987).
8. Brindley, G. W. The effect of grain or particle size on X-ray reflections of mixed powders and alloys, considered in relation to the quantitative determination of crystalline substances by X-ray methods. *Philo. Mag.* [7] **36**, 347 (1945).
9. Smith, D. K., Nichols, M. C., and Zalinsky, M. E. POWD 10: A FORTRAN IV programs for calculating X-ray powder diffraction patterns. The Pennsylvania State University; University Park, PA, 1982. (Program available by anonymous FTP to XRAY.ALFRED.EDU.)
10. Wilson, A. J. C., ed. *International Tables for Crystallography*. Vol. C: *Mathematical, Physical and Chemical Tables*. Kluwer Academic Publishers, Dordrecht, The Netherlands, 1992.
11. Bunge, H.-J. *Texture Analysis in Materials Science*. Butterworth, London, 1982.
12. Snyder, R. L., and Carr, W. L. Method for determining the preferred orientation of crystallites normal to the surface. In: *Interfaces of Glass and Ceramics* (V. D. Frechette, ed.), pp. 85–99. Plenum, New York, 1974.
13. Scherrer, P. Estimation of the size and internal structure of colloidal particles by means of röntgen. *Nachr. Ges. Wiss. Göttingen, Math.-Phys. Kl.* **2**, 96–100 (1918).
14. Klug, H. P., and Alexander, L. E. *X-Ray Diffraction Procedures*, 2nd ed., pp. 656 et seq. Wiley (Interscience), New York, 1974.
15. Cullity, B. D. *Elements of X-Ray Diffraction*, 2nd ed., Chapter 9, pp. 285–292. Addison-Wesley, Reading, MA, 1978.
16. Noyan, I. C., and Cohen, J. B. *Residual Stress; Measurement by Diffraction and Interpretation*. Springer-Verlag; New York, 1987.

CHAPTER
4
SOURCES FOR THE GENERATION OF X-RADIATION

This chapter will consider mainly sealed X-ray tube sources, as well as giving a brief mention to rotating anode tubes. Use of neutron source radiation is beyond the scope of this text. Though some brief mention will be made of synchrotron source radiation, for a more detailed description of the synchrotron source the reader is referred to other publications dealing specifically with this topic, e.g., refs. 1 and 2.

4.1. COMPONENTS OF THE X-RAY SOURCE

A typical laboratory source of X-rays for powder diffraction measurements has three major components: the line-voltage supply, the high-voltage generator, and the X-ray tube. The primary requirements for an X-ray source are that it must have the following features.

1. High photon output
2. High specific intensity
3. Selectable levels of kilovolts, and milliamperes
4. Stable output

Figure 4.1 shows how these four components interact in a typical $\theta:2\theta$ vertical diffractometer (or goniometer) configuration. In this instance, the tube is mounted in a shockproof shield that is bolted to the tabletop of the high-voltage generator. In the case of $\theta:\theta$ arrangement, the tube shield is driven around a vertical circular plane by the goniometer. In horizontal diffractometer configurations, the X-ray tube is mounted in a special tube shield that clamps horizontally either to the tabletop of the high-voltage generator or directly to a horizontal goniometer. These various diffractometer configurations are described in detail in Section 7.6. While the main thrust of this chapter is the high-voltage generator and the X-ray tube, the characteristics of the line-voltage supply cannot be ignored.

SOURCES FOR THE GENERATION OF X-RADIATION

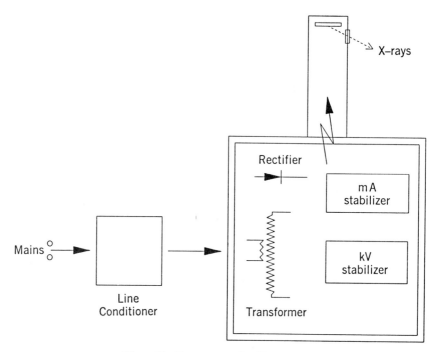

Figure 4.1. Components of an X-ray source.

4.2. THE LINE-VOLTAGE SUPPLY

The line-voltage supply is typically 110 or 220 V, single-phase ac or 220 and 400 V three-phase ac. In most parts of the world the line-voltage supply is reasonably constant, although some variations may occur during those periods of the day when use is heaviest. Figure 4.2 shows examples of the common types of line-voltage supply variation that are encountered. Variations generally fall into one of three main categories:

1. A slow (minutes/hours) variation in the actual voltage level
2. A cyclical variation in the amplitude of the waveform
3. Superimposed short-term (milliseconds) burst of higher voltage spikes

The second of the variations is illustrated at the lower portion of Figure 4.2. It will be seen that a periodic variation (*ringing*) of the cycle occurs, which is beyond the nominal peak and valley of the ac waveform. Most high-voltage generators will adequately cope with both of these problems provided that the variations are within ±10% of the specified voltage. The third type of

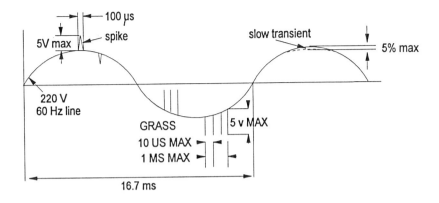

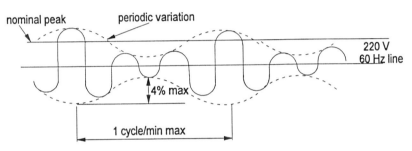

Figure 4.2. Waveshape aberrations for an ac source.

variation is, however, much more of a problem. Although the high-voltage generator does contain stabilization circuits, the response time of the circuits is finite. If the spike duration is shorter than the response time of the stabilizing circuit, the burst of excess voltage will pass to the X-ray tube, creating an equivalent *spike* in the photon output of the tube (see also Section 4.3.2). The presence of excessive numbers of spikes nearly always presents counting and display problems, and in extreme circumstances there may be a need to employ a special isolation transformer. With increasing use of solid state electronic devices, and need for better and better source stability, it is always advisable to carefully check the *cleanliness* of the line-voltage supply before a new instrument is installed.

4.3. THE HIGH-VOLTAGE GENERATOR

The purpose of the high-voltage generator is to transform the line-voltage supply to around 10,000–50,000 V, typically in steps of 5000 V. The generator employed may be of the constant-potential, half-wave rectified, or full-wave rectified type. In recent years, most manufacturers have moved to generators of the *high-frequency* type. The major advantages offered by this newer technology are lower cost, smaller size and weight, coupled with greater conversion efficiencies (therefore less heat loss). The transformer supplies filament current (i) and high voltage (V_0) to the X-ray tube. The filament current is derived from a step-down transformer that reduces the stabilized 220 V ac line to 5–10 V ac to heat the X-ray tube filament. The high voltage is derived by use of a high-potential transformer and a rectifier. While older (pre-1970) high-voltage generators used electron tube rectifiers (which were themselves potential sources of unwanted X-radiation and thus had to be carefully shielded), more modern units employ solid state rectifiers. All of these devices require high electrical insulation, and to this end most of the components are mounted in a high dielectric oil-filled tank.

The intensity $I(\lambda)$ of a characteristic wavelength from the X-ray tube anode is given by the expression

$$I(\lambda) = Ki[V_0 - V_e]^n, \quad (4.1)$$

where K is a constant, and V_e is the critical excitation potential of the wavelength concerned. The exponent n generally takes a value of about 1.6 for values of V_0 up to twice V_e, after which it approaches unity [3]. Since excitation of the characteristic line only occurs when the value of V_e is exceeded, emission of characteristic radiation will only occur during a definite part of the wave cycle.

Figure 4.3 shows wave forms for half-wave, full-wave, and constant-potential sources. The upper figure shows one cycle for a half-wave system. The cycle starts at zero volts (V_0), reaching a maximum (V_m) at one-quarter of the cycle; dropping back to zero volts at half the cycle; dropping to a minimum ($-V_m$) at three-quarters of the cycle; and finally increasing back to zero volts at the completion of the cycle. Note that if the excitation potential for a given characteristic line (e.g., Cu Kα) is V_e, the value of V_e is only exceeded for a specific period of the cycle. This effective part of the cycle is called the *duty cycle* of the generator. In the case of a half-wave rectified generator, this time is only about 30% of the cycle (depending upon the values of V_e and V). By rectifying the source to the full-wave condition shown in the center diagram, the duty cycle can be doubled. The third case shown is the constant potential source. In this instance *smoothing* is applied such that the maximum value of V_m is maintained for the whole cycle. Thus, in principle, the duty cycle of the constant potential generator should approach 100%.

Even with so-called *constant-potential* high-voltage generators, there is some variation of the value of V_m, this effect being called *ripple* (illustrated in

THE HIGH-VOLTAGE GENERATOR

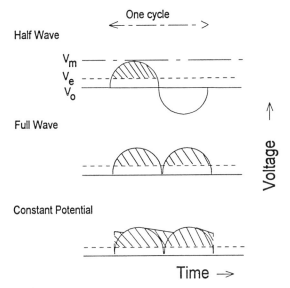

Figure 4.3. Half-wave, full-wave, and constant-potential waveforms.

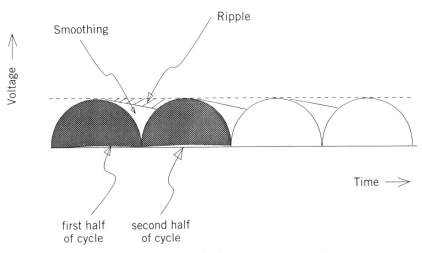

Figure 4.4. The effect of ripple on the duty cycle of the constant-potential high-voltage generator.

Figure 4.4). Practical considerations limit the effectiveness of the smoothing process, resulting in a small falloff in the value of V_m going from the one half cycle to the next. The overall effect is to reduce the practical duty cycle to about 90% under optimum excitation conditions. From the foregoing it follows also that the output of a constant-potential generator should be about three times that of a half-wave rectified unit. Since the duty cycles of both half wave and full wave are much lower than that of a constant-potential generator, when using a half- or full-wave generator one is able to exceed the recommended loading of an X-ray tube by a significant amount without damaging the tube by melting the target. However, because of the much higher efficiency, most powder diffractometry today is performed with constant-potential generators, and tube loadings are always specified for these generators.

4.3.1. Selection of Operating Conditions

The output from an X-ray tube powered by a high-voltage generator is described in terms of the radiation *flux*. The flux is essentially the density of X-ray photons per unit area per second. The takeoff angle is the angle between the plane of the tube target and an incident slit of an experiment. Since most experiments are performed at a constant takeoff angle and with carefully chosen slits and collimators, the degree of flux is determined by the voltage and current to the X-ray tube, as described by Equation 4.1. Since the requirement in most powder diffraction experiments is to employ the maximum available flux from the X-ray tube, the optimum settings of any variables are important. The actual choice of tube conditions will generally be determined by a number of factors including the following:

1. Maximum power rating (i.e., mA × kV) of tube
2. Type of generator employed
3. Optimum kilovolt level
4. Takeoff angle of X-ray tube
5. The choice of monochromatization conditions
6. Desired lifetime of the tube

While Equation 4.1 does relate the characteristic line intensity to V_0 and i, it does not describe the actual selection of the values. Since the power rating of the tube is restricted by its *specific loading* (see Section 4.4.2), the product of V_0 and i must not exceed the maximum rating of the tube. This maximum rating may be anywhere between 1.5 and 3.5 kW depending on the design of the tube. The optimum choice of voltage and current can be derived from an *isowatt curve*, a plot of the operating voltage vs. the total X-ray intensity from the tube. The isowatt curve shows a peak whose position depends upon the takeoff

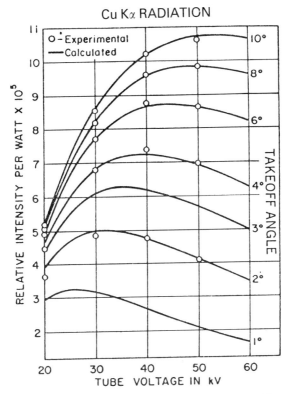

Figure 4.5. Experimental and calculated isowatt curves for a copper target X-ray tube. Reprinted from Short [5, p. 569], with permission.

angle of the tube. As an example, Figure 4.5 shows calculated and experimental isowatt curves for a copper target X-ray tube at various takeoff angles [4]. As can be seen in the figure, the curve peaks at about 45 kV for a 6° takeoff angle and 35 kV for a 3° takeoff angle. The peak in the isowatt curve occurs because the penetration of the electrons into the tube target increases with the tube potential. While more X-rays are produced at the higher potential, the absorption of the characteristic Cu K radiation increases.

In addition to working at the optimum tube voltage for a specific takeoff angle, it is also important that one remains within the *power curve* of a given tube. Figure 4.6 shows a typical power curve. The power curve defines the acceptable limits for the operation of a tube. Too high a current at low voltage may cause too high a filament temperature, leading in turn to the stripping of tungsten from the tube filament. Such a process rapidly increases the contamination level of a tube, as well as shortening its life. Conversely, too high a

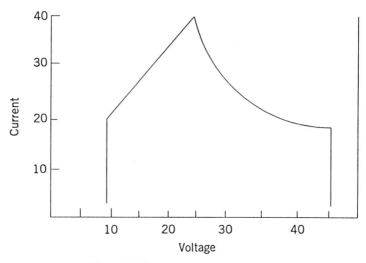

Figure 4.6. Power curve for an X-ray tube.

voltage at low current may exceed the X-ray specifications of the tube. In short, a tube should never be turned on until the user is familiar with the appropriate power curve. As was previously stated, if the generator employed is a half- or full-wave system, it may be safe to exceed the rated loading of the tube.

4.3.2. Source Stability

An important property of any X-ray source is its drift characteristics. *Drift* can be defined as a variation in the output of the source. Equation 4.2 was derived by differentiating Equation 4.1, and it will be seen from the derived equation that, should the tube current or voltage vary, there will be an effect on the intensity output:

$$\frac{\Delta I}{I} = \frac{\Delta i}{i} + \frac{1.6 \Delta V}{V - V_e}. \tag{4.2}$$

While the drift caused by variations in the tube current varies linearly with the tube current drift, the effect of variations in the tube voltage is much more complicated. As a rough approximation, when a copper target tube is used at 45 kV, the intensity variation caused by high-voltage drift is roughly twice the magnitude of the high-voltage drift.

There are several forms of drift, and the more important of these are summarized in Table 4.1. Drift characteristics are generally classified in terms

Table 4.1. Drift Characteristics of an X-ray Source

Type	Time	Magnitude (%)	Source
Ultralong	Months/years	1–20	Aging of tube
Long	Days/weeks	0.2–0.5	Thermal, focal spot wander
Short	Minutes/hours	<0.1	Stabilization circuitry
Ultrashort	Milliseconds	0.2–10	Transients

of their duration. Short-term drift is due mainly to limitations of the high voltage and current stabilization circuitry. It occurs over a period of minutes to hours—generally within the time cycle of a given diffraction measurement. The magnitude of short-term drift is typically around 0.1%. Long-term drift has a time cycle of days to weeks and is generally about two to five times the magnitude of short-term drift. The main source of long-term drift is due to thermal changes and to *wander* of the focal spot of the tube. In quantitative measurements, the effects of long-term drift can be removed by use of a reference standard that is measured within the same short-term period as the experimental measurement. Ultra-long-term drift is due mainly to aging of the X-ray tube. Its effect can be large, but—as in the case of long-term drift—its influence on a specific measurement can be calibrated out by means of a reference standard. By far the most difficult drift characteristic to handle is ultra-short-term drift. It arises mainly from transient output pulses from the line-voltage supply, which have a duration less than that of the high-voltage generator stabilization circuits. Its effect can only be removed by use of a *clean* input supply to the generator or by means of an isolation transformer between the line-voltage supply and the high-voltage generator input.

It is common practice to design high-voltage generators with very good short-term drift characteristics. The effects of long- and ultra-long-term drift are then calibrated out by use of monitor or reference standards. The actual short-term stability of a high-voltage generator is a specified design characteristic of the generator and, as such, may vary with the cost and age of the generator. Typical values for Δi and ΔV in a modern high-voltage generator are around 0.02%. Putting these values into Equation 4.2 gives a value for $\Delta I/I$ of something less than 0.1%.

4.4. THE SEALED X-RAY TUBE

For many years almost all X-ray powder diffraction measurements were carried out using a sealed or demountable fixed-anode X-ray tube. The sealed tube offered the advantages of good stability, reasonably high photon yield, and reliability. One disadvantage of the sealed tube, as it was employed in the 1950s and 1960s, is that contamination of the tube, due mainly to deposition of tungsten from the filament on the surface of the anode, as well as some degree of anode pitting, decreased the usefulness of the tube, even though it still operated. For this reason, some laboratories preferred to use demountable fixed-anode tubes, which could be disassembled and cleaned. As the power loading of X-ray tubes increased over the years, the tubes would fail long before contamination became a serious problem. Thus the popularity of demountable tubes waned in the 1970s.

Today, the sealed X-ray tube is the most popular source for probably 90% of all diffraction installations. However, other sources have gained in popularity over the last two decades. As an example, improvements in the reliability of the high-power rotating-anode tube have made it a very attractive alternative for some laboratories. More recently, the growth in the availability and application of synchrotron sources has provided diffractionists with a very bright and continuous wavelength source, allowing a new generation of experiments heretofore impossible with sealed tubes.

4.4.1. Typical X-ray Tube Configuration

A typical X-ray tube configuration is illustrated in Figure 4.7. An X-ray tube consists essentially of a tungsten filament (a) and an anode (d). Passage of a current through the tungsten spiral causes it to glow and emit electrons. These electrons are accelerated toward the anode by means of a high potential, usually on the order of 30–60 kV. The conversion of electrons to X-rays is a very inefficient process (i.e., < 1%), with most of the electron energy being converted to heat. For this reason, the anode must be water cooled and, in order to make best use of the cooling water, it is filtered (g) and directed in a jet onto the back of the anode, diametrically opposite the shadow of the filament. The jetting of the cooling water onto the back of the anode is important because the force of the water breaks up any water-vapor barrier that may form between the cooling water and the anode block. Should such a vapor barrier form, it may cause loss of physical contact with the cooling medium with consequent breakdown of the cooling system. The whole upper part of the X-ray tube is constructed from a single block of copper in order to ensure good cooling. This block is sealed to the glass envelope, which acts mainly as an electrical insulator for the leads (b) that carry filament current at high voltage. It is usual to keep the cathode at a high negative potential and to hold the anode assembly at ground. The filament is located in a slightly more negative *Wehnelt cup* (c) to ensure a finely focused beam of electrons. In so-called *fine-focus tubes*, this filament assembly is carefully shaped to give an

THE SEALED X-RAY TUBE

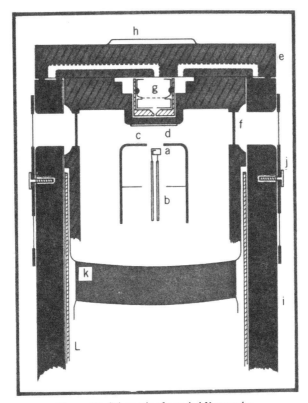

Figure 4.7. Schematic of a sealed X-ray tube.

extremely fine beam of electrons (see Figure 4.8). The X-ray tube is evacuated to about 10^{-6} mmHg and sealed.

Figure 4.9b shows a photograph of an actual X-ray tube that has been cut away in order to reveal the features identified in Figure 4.9a. Seen at the center of the photo are the focusing cylinder and cathode, behind which can be seen one of the X-ray tube windows. Two other (partially cutaway) windows can be seen, one on either side of the tube. The anode of the tube is at the top of the picture and the cooling jet/filter assembly can be clearly seen at the upper side of the anode block. The lower portion of the picture shows the glass body of the tube through which the high-voltage and filament current connectors pass. The very latest X-ray tubes [5] are now fitted with ceramic, rather than glass, bodies. The ceramic body tube has better shockproofing and insulation properties. The conventional X-ray diffraction tube is normally fitted with four windows. As shown in Figure 4.10, the filament is of a rectangular shape

(a)   (b)   (c)

Figure 4.8. X-ray tube filament assemblies: (a) broad focus; (b) normal focus; (c) fine focus. Reprinted from R. Jenkins and J. L. de Vries, *An Introduction to Powder Diffractometry*, p. 19, Fig. 23. Copyright © 1977, N. V. Philips, Eindhoven, The Netherlands.

and, when viewed at the two long faces, will give two line foci; when viewed at the two short faces, it will give spot foci. The spot foci are usually employed for camera work, where a single cylindrical collimator is generally used.

Although the copper anode tube is by far the most popular in X-ray powder diffractometry, copper is by no means the only target material employed. Table 4.2 lists characteristics of some of the common anode materials. Chromium $K\alpha$ has a wavelength of 2.29 Å and is useful for the measurement of X-ray patterns from materials with large unit cells, especially organic materials. Since chromium is not as good a heat conductor as copper, the specific loading obtainable with chromium target tubes is about 20% less than is achievable with copper. In addition 50% of the Cr $K\alpha$ is absorbed by air across the diffractometer beam path, compared to only 10% for Cu $K\alpha$. This means that the usable intensities from a system employing Cr $K\alpha$ radiation will be only about 40% of those obtainable with an equivalent system employing Cu $K\alpha$. Iron $K\alpha$ radiation has a wavelength about midway between Cu $K\alpha$ and Cr $K\alpha$, and is especially useful in the examination of materials containing a lot of iron. Copper $K\alpha$ is very effective at fluorescing Fe $K\alpha$ radiation from the sample, and the signal-to-noise ratio obtained with Cu $K\alpha$ on iron-containing specimens is often rather poor—especially in film work where there is no diffracted beam monochromator to remove specimen fluorescence. Finally, Mo $K\alpha$ radiation has a rather short wavelength and is good for dealing with small unit cells, especially for samples showing high absorption for Cu $K\alpha$ radiation. For this reason, Mo $K\alpha$ is often a popular choice for the study of metals and metal alloy systems. One disadvantage with the use of Mo $K\alpha$ radiation is that the optimum voltage for the excitation of Mo $K\alpha$ radiation is about 80 kV. Since most high-voltage generators do not operate above about 60 kV, the intensity output from a molybdenum target tube is frequently not optimum.

4.4.2. Specific Loading

The conventional modern X-ray tube holds the Wehnelt cup around the tungsten filament at a potential of a few hundred volts more negative than the cathode so that the electrons are repelled and focused onto the target. The focal spot is actually a line about 1 × 10 mm in dimension. Intensity is defined as the photon flux passing a unit area in unit time. Thus, focusing the X-rays into a smaller area increases the intensity. Various modifications of design parameters produce fine-focus, normal-focus, and broad-focus tubes. The limitation on the intensity that may be produced is generally determined by the efficiency of the cooling system. *Microfocus* tubes use the focusing cup to squeeze the electron beam down to about 0.1 × 1 mm, which can produce a 50 μm spot focus. These units are used for experiments requiring extremely small, intense beams for high-resolution work. Microfocus tubes usually have replaceable targets.

The maximum rating of the X-ray tube depends mainly upon the ability of the anode to dissipate heat; thus the *specific loading* (in W/mm^2) of the anode is an important parameter. Table 4.3 lists data from three different focal area copper anode tubes and shows the relationship between focal area and specific loading. Fine-focus[1] tubes are usually employed for powder diffractometry depending upon the resolution–intensity compromise required. These have typical loadings approaching 350 W/mm^2. Calculations [4] show that, when used under optimum conditions, the fine-focus tube gives 1.5 to 2 times more useful Cu $K\alpha$ intensity than that of the broad- and normal-focus tubes.

Table 4.4 gives a rough indication of the conversion efficiencies in the powering of a sealed X-ray tube. While the specific loading for a fine-focus tube

---

[1]There may be some confusion between the terms *fine-focus* and *long fine-focus*. Until the mid-1970s the typical cathode areas employed were about 0.5 × 8 mm. These tubes were the original *fine-focus* tubes. After that time the cathode area was increased to 0.5 × 12 mm. In order to differentiate this tube from the previous *fine-focus* tube, the new tube was referred to as the *long fine-focus* tube. In recent years, the old 0.5 × 8 mm cathode tubes became less popular and most workers now use the 0.5 × 12 mm cathode tubes. These tubes are now referred to as *fine-focus* tubes, even though, in actual fact, they are really *long fine-focus tubes*.

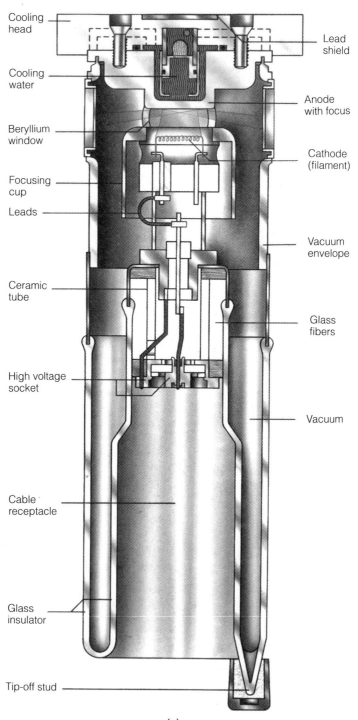

(a)

THE SEALED X-RAY TUBE 111

(b)

Figure 4.9. Cutaway pictures of a sealed X-ray tube: (a) schematic; (b) photograph.

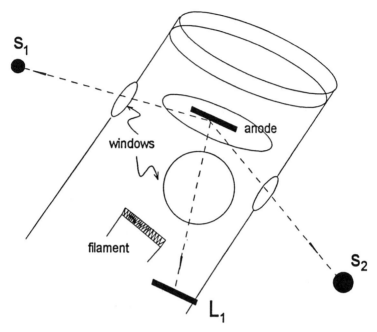

Figure 4.10. Line (L) and spot (S) foci.

Table 4.2. Characteristics of Common Anode Materials

Anode Material	Atomic Number	$K\alpha$* (Å)	Critical Excitation Potential (keV)	Optimum Voltage (kV)
Cr	24	2.291	5.99	40
Fe	26	1.937	7.11	40
Cu	29	1.542	8.98	45
Mo	42	0.710	20.00	80

is around 300 W/mm², because the conversion of electrons to X-rays is only about 1% efficient, and because only about 10% of the X-ray output is *useful*, the actual rating efficiency for Cu $K\alpha$ is only around 0.3 W/mm². This number can be compared to the value for a rotating anode of 3 W/mm² and for synchrotron 25 W/mm². Note, however, that while the increase for the synchrotron looks enormous (as indeed it is), heating of the beryllium window and monochromator limit the practical loading to about 10 W/mm².

THE SEALED X-RAY TUBE

Table 4.3. Typical X-ray Tube Loadings

Tube Type	Dimensions (mm)	Loading (kW)	Specific Loading (W/mm^2)
Fine focus	0.5 × 12	2.0	333
Normal focus	1.0 × 12	2.5	208
Broad focus	2.0 × 12	3.0	125
Rotating anode	0.5 × 10	15.0	3000

Table 4.4. Conversion Efficiencies in the Powering of a Sealed X-ray Tube

Source	Conversion	Specific Loading (W/mm^2)
Electrons	Loading = 1800 W Area = 6 mm^2	300
All X-rays	Electrons → X-rays = 1%	3
Useful X-rays	Cu $K\alpha$ = 10%	0.3

It is not generally possible to choose just any combination of mA and kV levels for a given X-ray tube because the filament temperature must stay within certain limits to ensure optimum tube life, contamination rates, etc. High-mA/low-kV conditions, for example, lead to high filament temperatures resulting in, among other things, a high rate of tungsten deposition on the anode, with subsequent spectral contamination. The acceptable ranges of kV and mA are usually defined by power-rating curves supplied with the X-ray tube.

4.4.3. Care of the X-ray Tube

X-ray tubes are generally rather expensive items, and every attempt should be made to extract as much *life* as possible from a given tube. New X-ray tubes, as well as X-ray tubes which have been unused for a period of months, require a *running-in* period before they are used at full loading. When X-ray tubes are made, one of the more difficult problems is to remove all traces of air from the tube, which would otherwise oxidize the tube filament, causing the tube to fail. Air is typically removed by a combination of high vacuum and baking at high temperature while applying a high current to a special oxidizable filament called a *getter*. Unfortunately, there is invariably some residual air left in the tube. Provided that the tube is in fairly constant use, a space charge is built up on the inside glass walls of the tube, and this static charge causes any air molecules to adhere to the walls, where they do no harm. If the tube is left unused for a long period of time, the space charge dissipates, releasing the air into the body of the tube. If a high current is applied to the normal filament of the tube, it may immediately oxidize, and the tube fails. When a tube is run in, the high voltage and tube current are first applied at the lowest level, then gradually raised to full loading over a period of about 3 h. This process re-establishes the space charge, and the tube filament is preserved.

Probably the most common cause of tube breakdown is the failure of the cooling system. As was explained in Section 4.4.1, the jetting action of the coolant water onto the back of the X-ray tube anode ensures the breakup of any vapor barrier between the anode block and the coolant water. Any blockage of the jet aperture can cause a disruption of the jetting system. Such blockage can arise from the presence of particles of dirt in the cooling water. For this reason, it is important to keep the internal X-ray tube water filter clean as well as to ensure that the general cooling water supply is free of contaminants.

In laboratories where exact reproducibility of measurements is important, X-ray tube degradation must be monitored. The best procedure for observing this is by means of routine running of a standard specimen, such as a cut and polished quartz block, a hot-pressed silicon wafer, a corundum plate or flat crystallized glass ceramic such as Corning Ware. A weekly plot of the integrated intensity of the selected diffraction line provides both a good record of tube degradation and a monitor of diffractometer alignment.

4.5. EFFECTIVE LINE WIDTH

The effective intensity of the source is dependent both on the receiving slit aperture and the focal spot characteristics of the X-ray tube. The takeoff angle of the X-ray tube can be important in the selection of tube conditions owing to the finite depth within the anode at which the characteristic X-radiation is produced. Also, because the takeoff angle determines the effective width of the X-ray beam, it influences the peak and background responses obtained with the diffractometer. Figure 4.11 shows that the effective width W of the filament shadow on the anode is determined by the angle of view α (i.e., the takeoff angle) and the true width of the filament shadow B. From the diagram, it is seen that

$$W = B \sin \alpha. \quad (4.3)$$

Equation 4.3 indicates that the smaller the value of B, the narrower W will be. Figure 4.12 shows plots of the beam width as a function of the takeoff angle for

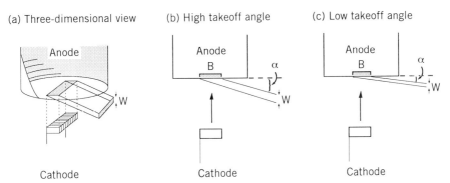

Figure 4.11. Effect of the takeoff angle on line width and intensity.

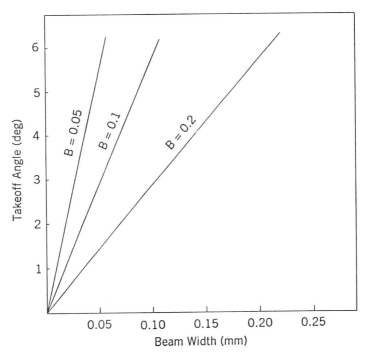

Figure 4.12. Correlation between beam width and takeoff angle for different types of X-ray tube.

broad-, normal-, and fine-focus X-ray tubes. Table 4.5 gives data for the beam width at the X-ray tube for the same types of tube. A small value of α gives a narrow beam and a reasonably high intensity, but this intensity will be very sensitive to the surface roughness of the anode. The optimum size of the receiving slit is typically two to three times larger than the projected beam

SOURCES FOR THE GENERATION OF X-RADIATION

Table 4.5. Relationship Between Takeoff Angle and Effective Beam Width

Type of Tube	Cathode Width (mm)	Beam Width (mm)
Fine focus	0.5	0.05
Normal focus	1.0	0.1
Broad focus	2.0	0.2

width W because of the broadening of the beam due to axial divergence. As an example, for a fine-focus copper target tube with focal spot dimensions of 12×0.5 mm, at a takeoff angle of $6°$ and with Soller collimators giving $\pm 2.5°$ beam divergence, and a goniometer circle of 173 mm, the beam width at the receiving slit is about 0.2 mm.

4.6. SPECTRAL CONTAMINATION

One of the important requirements of an X-ray source is that it should be spectrally pure. While it is true that the monochromatization device employed may significantly *clean up* the diffracted beam, a combination of, for example, a poorly aligned diffracted-beam monochromator and a spectrally impure source can result in many weak additional lines in a diffraction pattern (see also Section 6.3.1). Since these weak lines are diffracted unwanted wavelengths, they can be difficult to identify. Common X-ray tube contaminants are listed in Table 4.6. It will be seen that the two most critical are tungsten from the tube filament and copper from the anode cooling block assembly. The latter of these causes significant problems when a non-copper-anode X-ray tube is employed. Both copper and tungsten contamination will increase with

Table 4.6. Sources of Common X-ray Tube Contaminants

Element	Specific Source	Effect
Cu	Anode block	Increases with time
W	Filament	Increases with time
Fe	Window seal	Generally very small
Ca	Window	Generally very small

the operational life of the X-ray tube, and a contamination check should be made every 6 months or so to monitor the degree of the contamination. Contamination checks are best made by taking a well-known specimen (α-quartz, corundum, etc.) and running a scan *under the normal operating conditions*, then checking the diffractogram for additional peaks (see also Section 6.3.1). It is important to emphasize that the diffracted beam monochromator must be removed from the instrument before this measurement is made. Thus, the normal operating conditions are those used for this periodic measurement.

4.6.1. X-ray Tube Life

The useful lifetime of an X-ray tube is dependent on many factors. One very critical factor is the specific loading at which the tube has been run. Even though a normal or broad-focus tube may last for several tens of thousands of hours, it is uncommon to find fine-focus tubes lasting more than about 10,000 h. While the practice of *underrunning* (which essentially means operating an X-ray tube at a loading significantly less than its maximum rated value) is common, there is little evidence to support the claim that the practice *significantly* extends tube life. As has been previously discussed, a tube may finally fail due to mechanical shock or some other reason; however, the *useful* life of a tube may be reduced even though the tube still operates. An example of the reduction in useful tube life is the increase in spectral contamination already discussed in Section 4.6. Clearly, such a problem is more important when one is attempting to detect low concentrations of contaminant phases. Beyond the contamination issue there are two other common causes of tube life reduction: decrease of spectral output and loss of correct focusing.

The decrease in intensity of spectral output occurs for a number of reasons, the most important of which is the deposition of cathode material (typically tungsten) on the anode surface and on the insides of the beryllium windows. The characteristic radiation from the X-ray tube is attenuated by the tungsten, with a subsequent loss in intensity. A second reason for loss of spectral output is due to *pitting* of the anode surface. Pitting of the anode occurs due to the intense electron flux on the cooled surface of the anode block. Decrease in spectral output is generally a gradual process that takes place over a period of many months, and the short-term effect on diffractometer performance is generally unnoticeable. However, over a period of several years of use the characteristic line intensity from a sealed tube may drop by as much as 50%. For this reason, a reference standard should be used to measure the absolute efficiency of a tube on a weekly or monthly basis.

A second reason for the reduction of useful tube life is loss of focusing. As was described in Section 4.4.1, a focusing cup is used to roughly focus the

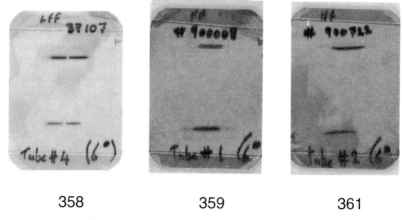

Figure 4.13. Pinhole pictures of X-ray tube filaments.

filament electrons onto the surface of the anode. If, for any reason, the position of the filament should change relative to the active surface of the focusing cup, the emission from the tube may change. As an example, Figure 4.13 shows three different pinhole pictures of filaments from different X-ray tubes. The center picture (359) shows a normal filament as evidenced by a full, straight image. The filament picture on the right (361) shows some curvature and was taken from a tube where the anode surface was not flat. While this tube is usable, the distortion of the filament image would be carried through the optical path of the diffractometer to the receiving slit (see Section 7.6.1), causing significant additional broadening of the diffracted line profile. The filament at the left (358) shows significant sagging, so much so that the center portion of the filament has dropped below the Wenheld cylinder and so does not emit any radiation within the boundary defined by the X-ray tube window. Thus, while the filament is still continuous, X-ray emission from the tube appears as a broken line. The effect of such irregular emission on the focusing of the diffractometer is somewhat unpredictable. What is certain, however, is that the total emission of X-rays is significantly less than it should be.

4.7. THE ROTATING ANODE X-RAY TUBE

Another way to increase the intensity of an X-ray tube is to increase the amperage and avoid melting the target by using a much larger anode mass, rotating the anode at high speed. Figure 4.14 compares the anode configurations of the *fixed* and *rotating* anode X-ray tubes. In the case of the fixed

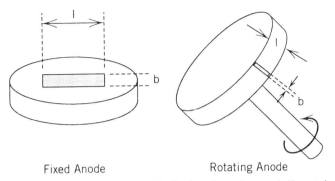

Fixed Anode   Rotating Anode

Figure 4.14. Anode configurations of the fixed- and rotating-anode X-ray tubes.

anode, the cathode shadow is a line $l \times b$ (where l is the length and b is the breadth), or about $10 \times 1 = 10 \, \text{mm}^2$. In the case of the rotating anode, the total irradiated surface is $(\pi d/b)$, where d is the diameter of the anode disk. For an anode disk diameter of 20 mm, the total irradiation area is $\pi \times (d/2)^2 = 1505 \, \text{mm}^2$. Two advantages accrue for the rotating anode: first, the irradiation area is larger by about a factor of 60; second, the rotation continuously brings cooler metal into the path of the focused electron beam. Because of the more efficient cooling, rotating anode systems can be run at loadings around an order of magnitude higher than an equivalent fixed anode system. Thus, rotating anode tubes produce very intense X-ray beams. However, due to the mechanical difficulties of a high-speed motor drive that must feed through into the vacuum, there are difficulties in routine continuous operation. Fortunately, in recent years rotating anode units have become much more reliable, due mainly to the use of *ferro-fluidic* seals and turbomolecular pumps.

Another less-common method for generating X-rays that is in current use is to charge a very large bank of capacitors and to dump the charge to a target. Such devices date back to the early 1940s when Tsukermann was able to measure Laue patterns of aluminum single crystals using a pulsed X-ray generator and millisecond exposures [6]. Johnson and co-workers at the Lawrence Livermore Laboratory [7] spent several years improving the performance and reliability of a relatively low-voltage (50 kV) pulse X-ray source that ran at 50 kA, giving a 30 ns X-ray pulse. One more recent flash X-ray device was designed to run at 300 kV peak, giving a 30 ns pulse [8]. Despite the fact that this duration of the flash is very short, advantage is taken of this incredibly small time window to perform very sophisticated fixed and dynamic experiments.

REFERENCES

1. Cox, D. E., Hastings, J. B., Cardoso, L. P., and Finger, L. W. Synchrotron X-ray powder diffraction at X13A: A dedicated powder diffractometer at the National Synchrotron Light Source. *Mater. Sci. Forum.* **9**, 1–20 (1986).
2. *Journal of Synchrotron Radiation.* International Union of Crystallography, 1994 to date.
3. Schreiner, W. N., Jenkins, R., and Dismore, P. F. XRD instrument sensitivity results from a round robin study. *Adv. X-Ray Anal.* **35**, 333–340 (1992).
4. N. V. Philips Gloeilampenfabrieken, Eindhoven, The Netherlands.
5. Short, M. A. X-ray intensities from copper-target diffraction tubes. *Adv. X-Ray Anal.* **20**, 565–574 (1977).
6. Tsukermann, V. A., and Avdeenko, V. *Zh. Tekh. Fiz.* **12**, 185 (1942).
7. Johnson, Q., Mitchell, A., Keeler, R. N., and Evans, L. X-ray diffraction evidence for crystalline order and isotropic compression during the shock wave process. *UCRL Rep.* **73140**, April 13 (1971).
8. Charbonnier, F. M. Proposed flash X-ray system for X-ray diffraction with submicrosecond exposure. *Adv. X-Ray Anal.* **15**, 446–461 (1972).

CHAPTER
5
DETECTORS AND DETECTION ELECTRONICS

5.1. X-RAY DETECTORS

The nature of the X-ray detectors and electronics used with them is an important part of understanding the approaches to and limitations of X-ray applications. An X-ray detector consists of two basic parts, the *transducer* and the *pulse formation circuit*, with its counting electronics. The function of the transducer is to convert the energies of the individual X-ray photons to an electric current. The pulse formation circuit converts the current into voltage pulses that are counted and/or integrated by the counting equipment, allowing various types of visual indication of X-ray intensity. In practice, the term *transducer* is not commonly employed by the X-ray community, and the term *detector* (or *counter*) is employed somewhat arbitrarily to describe the transducer with or without counting circuitry. Detectors used in conventional X-ray powder diffractometers are generally one of four types: gas proportional counters, scintillation counters, Si(Li) detector diodes, and the intrinsic germanium detector. Of these detector systems, the scintillation counter is by far the most commonly employed. In addition to these, there are also some more specialized detectors that are used for unique applications. All X-ray detectors, however, have one thing in common, in that they all depend upon the ability of X-rays to ionize matter.

While its use is diminishing, one should certainly not overlook the use of photographic film, since this is one of the oldest methods of detecting X-rays. The only two considerations that need to be mentioned concern the type of film and its linear range. Due to the low demand for X-ray film sensitive to the relatively low energies used in diffraction, suppliers and availability vary. When the silver halide grains in a film are being exposed so that they will become developable (i.e., reduced to pure silver on the plastic substrate), the grain size is important. Each particle in the emulsion struck by three to five X-ray photons becomes sensitized. Once sensitized, the entire particle will reduce to silver in the development process. Thus, if a region of the film is exposed to an intense X-ray flux, the same particles will be struck more than once. This causes a loss of information analogous to the dead time in electronic counters. The film darkening is only proportional to the intensity of the exposing X-rays over what is called the *linear range* of the film. To allow for this effect when measuring intense sources one must use a film of smaller grain size to extend the linear range or reduce the X-ray intensity with filters in front of the film or reduce the incident beam flux.

5.2. DESIRED PROPERTIES OF AN X-RAY DETECTOR

Of the several specific properties that are sought in an X-ray detector, the most important are *quantum-counting efficiency, linearity, energy proportionality,* and *resolution* [1]. Each of these properties will be discussed with reference to Figure 5.1. The terms used are defined as follows:

- E = energy of an X-ray photon (eV);
- V = average size of a voltage pulse (V);
- δV = spread in output voltage for a given E (%);
- I_0 = photon flux incident on the transducer (photons/s);
- I = photon flux passing through the transducer (photons/s);
- R = pulse rate produced by the detector (c/s, or counts/s).

5.2.1. Quantum-Counting Efficiency

In Figure 5.1a X-ray photons are incident on the absorber at a flux of I_0 photons/s; I is the flux of photons that pass through the absorber to the

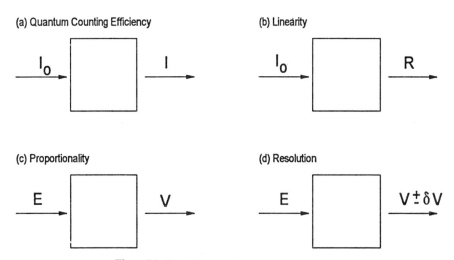

Figure 5.1. Desired properties of an X-ray detector.

detector. *Quantum-counting efficiency* usefully describes the efficiency of the detector in collecting radiation incident upon it. A detector exhibiting ideal quantum-counting efficiency is one in which $I \sim I_0$. Ideally, in the case of X-ray diffraction, one would prefer that the detector be efficient in collecting the diffracted characteristic photons. At the same time, it should be inefficient in collecting the generally unwanted diffracted and scattered short-wavelength bremsstrahlung, since this greatly contributes to the background, especially at low angles. The gas proportional detector to be discussed in Section 5.3.1 is effective in this regard, since the gas in the detector has a much lower absorption for high-energy, short-wavelength radiation. In the case of scintillation and Si(Li) detectors, some short-wavelength discrimination is designed into the detector by careful selection of the thickness of the X-ray transducer. In the case of the scintillation counter, short-wavelength discrimination is achieved by careful choice of the phosphor; in the case of Si(Li), it is determined by the actual thickness of the silicon disk. It must be appreciated that most detectors that have been designed for application in X-ray diffraction have probably been optimized for the measurement of Cu $K\alpha$ radiation, since this is generally the radiation of choice. Where diffraction measurements are being made with a shorter wavelength (e.g., Mo $K\alpha$, $\lambda = 0.71$ Å) or a longer wavelength (e.g., Cr $K\alpha$, $\lambda = 2.29$ Å), some loss in quantum counting efficiency will probably result.

5.2.2. Linearity

A second important property of an X-ray detector is its *linearity*. While the process of conversion of individual X-ray photons to voltage pulses is extremely fast, it is finite—perhaps on the order of a few microseconds. If the photon flux is very high, a photon that arrives while the detector is still processing the previous X-ray photon conversion may be lost. The time required for the detector to collect a photon, convert it to a pulse, and count the pulse is related to the *dead time* (τ) of the detector. The fraction of photons that may go unconverted is described by the *linearity* of the detector. Figure 5.1 shows a single X-ray photon of energy E entering the detector producing a pulse V. A number of photons are incident upon the detector at a rate of I photons/s, producing voltage pulses at a rate of R pulses/s. The detector is said to be linear when there is a direct proportionality between R and I. Where this is not true, the dead time of the counter is referenced.

Since the counter is dead for a certain time τ, the measured count rate R_m will always be lower than the true count rate R_t. The following relationship holds approximately true:

$$R_t = \frac{R_m}{1 - R_m \tau}. \qquad (5.1)$$

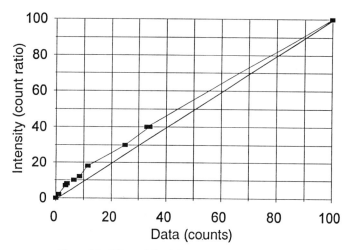

Figure 5.2. Effects of dead time on relative intensities.

Dead time can arise from a variety of different sources, and the combination of these different effects can be very complicated. However, in simplified terms, dead-time phenomena lead to two different types of dead time: *paralyzable* (nonextending) dead time and *nonparalyzable* (extending) dead time. This nomenclature is used because the effect of paralyzable dead time is to completely saturate the detector, causing it to cease counting. Nonparalyzable dead time causes an increasing loss in counts at increasing count rate, but never reaches the stage where the counter saturates. The process described above and illustrated in Figure 5.2 [2] is paralyzable dead time, and its magnitude will be dependent on the rate at which photons arrive. A further complication, however, is the width of the voltage pulse produced by the pulse shaper [3]. The wider the pulse, the higher the chance that consecutive pulses will interfere, producing pulse shift and, in the extreme, pulse pileup. These effects give what is called nonparalyzable dead time. The major sources of paralyzable and nonparalyzable dead time are listed in Table 5.1. Note that the largest source of dead time is the E1T vacuum tube employed in the counting process. The magnitude of a given effect can only be accurately quantified in the case of paralyzable dead time, which may be corrected for by using Equation 5.1. The effects of nonparalyzable dead time vary drastically with count rate, but nearly always gain significance as the count rate exceeds about 40,000 c/s for gas detectors and about 20,000 c/s for Si(Li) detectors.

Nonparalyzable dead time is generally not a problem when the scintillation detector is employed along with reasonably modern counting electronics. While the inclusion of dead-time correction circuits in modern systems has

DESIRED PROPERTIES OF AN X-RAY DETECTOR   125

Table 5.1. Sources of Dead Time

Source	Type	Magnitude
Transducer	Paralyzable	0.1–20 μs
Pulse shaper	Paralyzable	1–5 μs
Pulse pileup	Nonparalyzable	Large
Pulse shift	Nonparalyzable	Moderate
Coincidence	Nonparalyzable	Small
Counter tubes ($E1T$)	Paralyzable	30–50 μs
Shelving	Nonparalyzable	Moderate

almost completely eliminated dead-time problems with scintillation-counter-based counting chains, some care must be taken when experimental data recorded on modern (post-1980s) equipment are compared with older data. As an example, Figure 5.2 compares intensity data on Y_2O_3 recorded in the mid-1950s with similar data recorded on a post-1980 diffractometer. Since the dead-time loss is greatest at the highest count rate, the strongest lines in the pattern are reduced the most. Because it is common practice to normalize all intensities to the strongest line, the effect is to increase the relative intensities of the weaker lines in the pattern. One detector that still exhibits significant paralyzable and nonparalyzable dead time is the Si(Li) detector. At the time of preparation of this book (mid-1990s), the acceptable count rates measurable by a Si(Li)-based detector system are in the range 30,000–40,000 c/s. Great care must be taken in using these detector systems at higher count rates.

5.2.3. Energy Proportionality

A third important property of a detector is the *proportionality* of its signal to the energy of the X-ray. The size of an output pulse will be dependent on the current produced in the transducer, which is, in turn, related to the number of ionizing events leading to the formation of the pulse. Since the number of ionizing events is proportional to the energy of the incident X-ray photon, it follows that the size of the output voltage pulse is also directly proportional to the energy of the incident X-ray photon. This is illustrated in Figure 5.1c. The detector is said to be proportional when the size of the output pulse V is proportional to the energy E of the incident X-ray photon. The actual size of the output pulse will also depend upon the gain of any amplifiers that are used in the detector circuit. As will be seen in Section 5.4, the detector's energy proportionality and energy resolution permit electronic rejection of pulses caused by extraneous X-ray photons with energies other than the $K\alpha$ of interest.

5.2.4. Resolution

The fourth important property of a detector is its *resolution*. The resolution of a detector is a measure of its ability to resolve two X-ray photons of differing energy. As was previously stated, in a proportional detector an incident X-ray photon of energy E will produce an output pulse of voltage V. In practice, a number of photons of energy E will produce output pulses of average voltage V. Table 5.2 shows how the final number of electrons produced is derived for each of the commonly employed detectors. While the size of the output pulse from the detector (before any electronic amplification) is directly proportional to the final number of electrons, as shown in Figure 5.1d, the output pulses will have a spread δV, where δV is related to the resolution of the detector. As the value of the resolution decreases, the detector is better able to distinguish between X-ray energies. In reference to Figure 5.3, the percent resolution R of any detector is defined as

$$R(\%) = \frac{100\sqrt{W}}{V_{peak}}, \qquad (5.2)$$

where W is the full width at half-maximum (FWHM) of the pulse distribution produced by the detector (in volts), and V_{peak} is the voltage of the maximum in the pulse distribution.

The theoretical resolution of a detector is used to evaluate the performance of a particular detector. The resolution is given by [3]

$$R(\%) = \frac{100\,K\sqrt{I}}{\sqrt{E}}, \qquad (5.3)$$

where the E is the energy of the X-ray photon in keV, and I is the effective ionization potential in keV. For argon gas proportional counters the Fano factor K is 235 and defines the efficiency of the ionization process [4]. The term $(K\sqrt{I})$ is about 38. For scintillation counters the term is equal to 128. Thus, the

Table 5.2. Initial and Final Gain of Common Detectors

Detector Type	Useful Range (Å)	Energy/Ion Pair (eV)	Initial Electrons	Gain	Final Electrons
Gas Prop	1.5–50	26.4	305	6×10^4	2×10^7
Scintillation	0.2–2.0	350	23	10^6	2×10^7
Si(Li)	0.5–8.0	3.8	2116	1	2×10^3

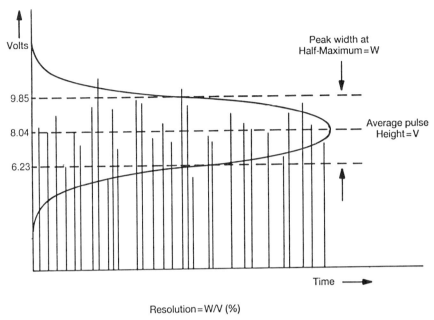

Figure 5.3. Detector resolution.

resolution for a scintillation detector is given by $128/\sqrt{E}$ and for a sealed xenon gas counter, $36/\sqrt{E}$.

5.3. TYPES OF DETECTOR

The three common types of detector employed in the conventional Bragg–Brentano diffractometer are the scintillation detector, the gas proportional detector, and the lithium-drifted silicon [Si(Li)] detector. Table 5.3 lists the more important properties of these three detectors for three different wavelengths, corresponding to the $K\alpha$ lines of chromium (2.291 Å), copper (1.542 Å), and molybdenum (0.710 Å). The $K\alpha$ lines of chromium and molybdenum represent the extremes of the wavelength range covered by conventional difffraction instruments, and, as has been previously mentioned, Cu $K\alpha$ radiation is by far the most popular radiation for powder measurements.

It will be seen from Table 5.3 that for Cr $K\alpha$, the longer of the three wavelengths, both the gas proportional and Si(Li) detectors have good resolution and quantum-counting efficiency. However, mainly because of problems with the Si(Li) detector at higher count rates, the gas proportional

Table 5.3. Properties of Common X-ray Detectors

Property	Scintillation			Xe Sealed Gas			Si(Li)		
	Cr	Cu	Mo	Cr	Cu	Mo	Cr	Cu	Mo
Quantum efficiency (%)	60	98	100	90	90	75	90	95	80
Linearity—loss at 40,000 c/s	Less than 1%			Up to 5%			Up to 50%		
Proportionality	Very stable			Pulse shift at high c/s			Pileup, etc., at moderate c/s		
Resolution (%)	55	45	31	17	14	10	3	2	1

detector (generally filled with a mixture of 90% xenon/10% bromine) is invariably the detector chosen for work with Cr $K\alpha$. At the short-wavelength end, the scintillation counter has by far the best linearity and quantum-counting efficiency of the three detectors and is usually the detector of choice. In the case of Cu $K\alpha$, the scintillation detector has the best overall counting efficiency characteristics and is usually the first choice, except in those cases where resolution is critical. Examples of this situation would include energy dispersive diffraction and where the diffractometer is being used without a monochromator.

5.3.1. The Gas Proportional Counter

When an X-ray photon interacts with an inert gas atom, the atom may be ionized in one of its outer orbitals, giving an electron and a positive ion, this combination being called an *ion pair*. For example, in the case of xenon (a typical counter gas), the energy required to remove an outer electron is about 20.8 eV. In the case of a gas counter detecting Cu $K\alpha$ radiation, each $K\alpha$ photon has a total energy of about 8.04 keV; thus each photon can produce $8040/20.8 = 387$ *primary* ion pairs. The actual number of electrons contributing to the measured signal after amplification will be proportional to, but much greater than, the 387 primary ion pairs. However, the resolution of the detector will be inversely proportional to the number of primary ion pairs. The pulses produced by this process are called the *photopeaks*. It can also happen that a second type of output pulse is produced, called an *escape peak*. The escape peak occurs when the energy E of the detected X-ray photons is high enough to remove an *inner* L or K orbital electron. The energy of the original photon is reduced by an amount equal to the ionization potential ϕ_L. Ionization of the detector gas then follows by exactly the same process as in the case of the photopeak, except that the energy available for ionization is now $E - \phi_L$ and the number of ion pairs will be $(E - \phi_L)/20.8$. In the case of Xe, ϕ_L is equal to 2.96 keV; thus the number of ion pairs contributing to the escape peak will be equal to $(8040 - 2960)/20.8 = 244$. The ratio of initial X-ray photons (which give rise to photopeaks) to the number that give rise to escape peaks approximates to the fluorescent yield value (see Section 1.4.3) of the appropriate level of the counter gas. In the case of xenon, the L fluorescent yield is about 0.10; thus about 10% of detected Cu $K\alpha$ photons contribute to the escape peak.

The basic configuration of the gas counter is shown in Figure 5.4. The counter itself consists of a hollow metal tube carrying a thin wire along its radial axis. The wire forms the anode, and a high potential of 1.5–2 kV is placed across it, with the casing of the counter being grounded. Electrons produced by the ionization process, described above, move toward the anode, while the positive ions go to the grounded casing. Under the influence of the positive field close to the anode, the electrons are accelerated, gaining kinetic energy. Part of this energy can be given up by ionizing other inert gas atoms that lie in the path of the electron as it moves toward the anode. Thus, each primary electron produced in the initial process may give rise to many

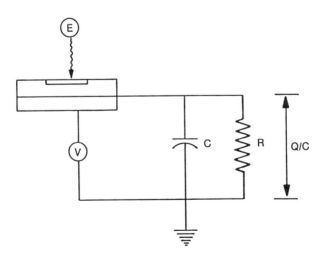

Figure 5.4. Schematic diagram of a proportional detector: Q is the charge of a single output pulse on the capacitor C.

secondary electrons, leading to an effect called *gas amplification*. The gas amplification may have a value between 10^3 and 10^5, depending upon the magnitude of the field (in turn, fixed by the voltage of the counter and the radii of the anode wire and the body of the counter). A burst of electrons reaching the anode causes a decrease in the voltage at the capacitor C; this decrease is also determined by the value of the resistance R. These voltage pulses are passed through suitable amplifiers to the scaling circuitry.

5.3.2. Position-Sensitive Detectors

Position-sensitive detectors (PSDs) [5] are finding increasing application in X-ray powder diffraction, mainly because of the advantage they offer in increased speed of data acquisition. In its simplest configuration, the PSD is essentially a gas proportional detector in which electron collection and pulse-generating electronics are attached to both ends of the anode wire. Figure 5.5 illustrates the principle of the PSD. The anode wire is made to be poorly conducting to slow down the passage of electrons. By measuring the rate at which a pulse develops (the *rise time*) at each end of the wire, it is possible to correlate the rise times with the position along the anode wire where the ionization originated. The use of PSDs in X-ray diffraction goes back to the mid-1970s, when they were applied mainly to specialized applications such as the measurement of residual stress [6]. Within a few years they were finding increasing application in commercial diffractometers [7, 8].

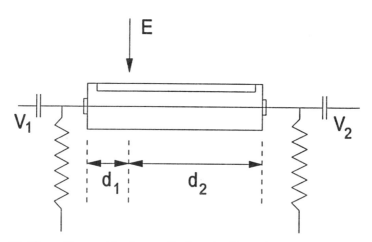

Figure 5.5. The position-sensitive detector: E = incoming X-ray photon; V_1 and V_2 = voltage at capacitors 1 and 2, respectively, d_1 and d_2 = distances from the entry point of the photon to sides 1 and 2 of the detector, respectively.

Since the PSD is able to record data from a range of angles at one time [9], it offers special advantages to those cases where speed of data acquisition is critical, for example, in the study of phase transformations and chemical reactions [10]. The development of the electronics required to permit a PSD to continuously scan two θ ranges beyond its aperture [11] have decreased powder pattern collection times to a few seconds. The effective angular resolution of the PSD is equivalent to a few hundredths of a degree 2θ; although this resolution is somewhat worse than can be obtained using a classical diffractometer, it is more than sufficient for most applications. It has also been shown [12,13] that a combination of the PSD and curved-crystal monochromator is particularly powerful for the quantitative analysis of very small (1–5 mg) amounts of material. The high quantum efficiency of PSDs has permitted the detection of 10 ppm concentrations of phase impurities [14].

5.3.3. The Scintillation Detector

Figure 5.6 shows a schematic diagram of a scintillation detector. In the scintillation detector, the conversion of the X-ray photon energies into voltage pulses is a double-stage process. In the first of these processes, the X-ray

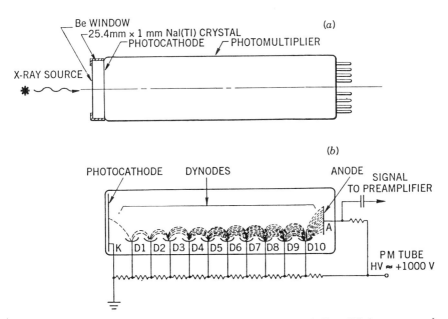

Figure 5.6. The scintillation counter. Reprinted from Jenkins et al. [3, p. 97], by courtesy of Marcel Dekker Inc.

photon is converted into flashes of blue light by means of a *phosphor*, or scintillator, which is a substance that has the ability to absorb radiation at a certain wavelength, then reemit radiation at a longer wavelength. For work in the x-ray region, it is common to use phosphors of sodium iodide doped with thallium. This phosphor emits light photons with wavelengths of around 4100 Å. In the second stage of the process, the blue light from the phosphor is converted to voltage pulses by means of a *photomultiplier* (PM) tube. Here the light photons fall onto an antimony/cesium photocathode, each producing a burst of electrons that are then focused onto a chain of typically 10 sequential photosurfaces called *dynodes*. Each of these dynodes has a successively higher potential, and the electrons produced at each dynode are accelerated to the next, such that at the following dynode more electrons can be produced by means of the kinetic energy gained by acceleration. This total effect gives an exponential increase or gain in the signal. After the last dynode, the electrons are collected by the anode and a voltage pulse is formed.

The number of electrons ejected from the photocathode is proportional to the number of blue light photons that struck it. The number of blue light photons, in turn, is proportional to the energy of the original X-ray that entered the crystal. Thus, the scintillation detector is a proportional detector. Unfortunately, due to a rather large number of losses, the electrical signal of an effective photon of light is only observed for about each 350 eV of incident X-ray energy. The overall resolution is on the order of 45% of the energy of the incident photon. Thus, the resolution is not such as to be able to resolve the difference between Cu $K\alpha$ (8.047 keV) and Cu $K\beta$ (8.9094 keV). This shortcoming is made up for to some extent by the high quantum efficiency and a low dead time of about 0.1 μs.

5.3.4. The Si(Li) Detector

Solid state detectors were developed toward the end of the 1960s, and initially their major application in analytical instrumentation was in the areas of X-ray fluorescence spectrometry and electron microscopy. By far the most common of these detectors is the Si(Li) detector diode, although Ge(Li) does offer some advantages for the measurement of short-wavelength radiation. In addition to their use in X-ray fluorescence spectrometry, there has been increasing interest in using these devices for stand-alone X-ray powder diffraction [15] and amorphous scattering studies.

As shown in Figure 5.7, the Si(Li) detector is a piece of *p*-type silicon that has been doped with lithium atoms to increase the electrical resistivity. The silicon single crystal is typically several millimeters thick and has an applied bias of 300–1000 V. A high concentration of lithium at the rear side of the crystal creates an *n*-type region, over which a gold contact layer is deposited.

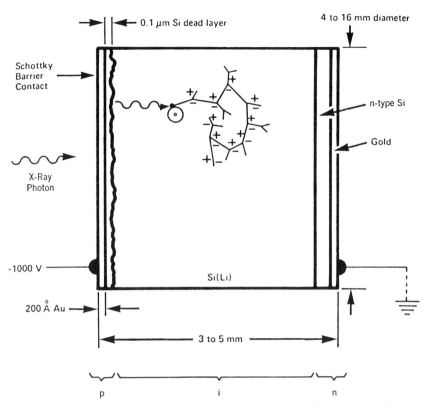

Figure 5.7. The Si(Li) detector. Reprinted from Jenkins et al. [3, p. 129], by courtesy of Marcel Dekker Inc.

A Schottky barrier contact is applied at the front side of the crystal, thus producing a *p-i-n*-type diode. An X-ray photon entering the detector produces a cloud of electron–hole pairs—the number of these pairs being proportional to the energy of the photon divided by the energy required to produce one pair (in the case of silicon, about 3.8 eV). The electrons are swept from the silicon by the potential difference across the diode, and a *charge-sensitive preamplifier* (i.e., a field effect transistor, FET) is then employed to generate a suitable signal to pass on via the main amplifier to the counting circuit. It is necessary to keep both the detector and the FET under cryogenic conditions to reduce noise and to minimize the motion of Li^+ ions. This is achieved by mounting the whole detector/preamplifier assembly on a liquid-nitrogen-cooled cold finger.

The function of the Si(Li) detector can be compared to a gas counter without gas amplification. In other words, whereas the actual number of primary electrons produced per event in a Si(Li) detector is significantly larger than in the gas counter (which means that the energy resolution will be better), the final number of electrons is much lower (which means that the signal will be weaker).

The advantages of the Si(Li) detector, feeding its output to a multichannel analyzer (MCA; see Section 5.4, below) include a high quantum-counting efficiency and a large collection angle. In addition, the ability to select energy (therefore, wavelength) ranges of interest offers some particular advantages in powder diffractometry [16]. A good example of this is the use of the Si(Li) detector to eliminate $K\beta$ radiation where a monochromator is not being employed (see Section 6.4). For many years, the need to run the Si(Li) detector under cryogenic conditions has somewhat limited its usefulness in diffraction applications. A major disadvantage of the Si(Li) detector for these applications remains its rather large dead time. Experiments have shown that dead-time effects are quite significant, even at relatively low count rates, causing the introduction of corresponding errors in the relative line intensities.

Energy Dispersive Diffraction. The availability of the Si(Li) energy-dispersive detector with the ability to discriminate between pulses differing by as little as 125 eV offers a means of observing a powder diffraction pattern without a diffractometer. There are three variables in the Bragg equation,

$$\lambda = 2d_{hkl} \sin \theta,$$

these being θ, d_{hkl}, and λ. Any one of these may be held constant in an experiment to display a plot of the other two. Thus, three types of experiments are possible:

1. Hold λ constant so that a scan of the diffraction angle shows the various d_{hkl}-values that diffract. This is a conventional powder diffraction experiment.

2. Hold d_{hkl} constant (by the use of a single crystal with fixed orientation), so that a scan of the diffraction angle shows the various wavelengths that are fluorescing from a sample. This is the conventional X-ray fluorescence experiment.

3. Hold θ constant and allow a polychromatic X-ray beam to strike the sample. In this case, each of the d_{hkl} values will diffract the λ that satisfies the Bragg equation for the fixed angle θ. An energy-resolving detector like the Si(Li) detector will permit the display of d_{hkl} as a function of λ. This is the technique known as *energy-dispersive powder diffraction*.

The original development of an energy-dispersive system to collect radiation from a crystalline specimen irradiated with the white radiation from an X-ray tube was carried out in the 1970s [17–19]. Since the individual sets of d-spacings will select certain narrow-wavelength portions of the primary spectrum to diffract at the angle of the detector, a correlation can be made between the measured energy of a given band and the d-spacing of the material. Although this technique has found some application in high-speed diffractometry, particularly using very bright synchrotron sources, the rather poor relative resolution of the energy-dispersive system has severely limited this application.

5.3.5. Other X-ray Detectors

Most Si(Li) detectors are designed to be sensitive to the 2–20 keV range. At the short-wavelength end of this region there is considerable loss of quantum-counting efficiency because of the rather low absorption cross section of silicon. Germanium has been substituted for silicon in applications involving the detection and measurement of shorter wavelengths, and the Ge(Li) detector is similar in its construction and operation to the Si(Li) detector. Early versions of the Ge(Li) detector had two major disadvantages. First, the high mobility of lithium in germanium caused the detector to be destroyed if it was accidentally warmed to room temperature. Second, the detector is characterized by rather dominant escape peaks that cause complications in the measured spectrum and strongly reduce the full energy-detection efficiency [20]. However, by the late 1970s the Ge(Li) detector was replaced by high-purity germanium (intrinsic germanium), which does not suffer from the shortcomings of Ge(Li), while retaining the advantage of high quantum-counting efficiency at shorter wavelengths.

A major disadvantage of the Si(Li) detector is the need to cool it with liquid nitrogen. Although some recent developments indicate that Peltier cooling may offer a long-term solution to this problem [21], the extensive use of Si(Li) detector systems has been inhibited by the mechanical inconvenience involved with having to move the entire Dewar assembly on the detector mount. For this reason, room-temperature solid state detectors have been sought. Cadmium telluride is one such detector and has been shown to operate successfully at room temperatures, albeit only for the shorter wavelength radiation. The major concern at the present time in the use of cadmium telluride detectors for X-ray diffraction applications is the high leakage current (and therefore higher background noise). This means that it is very difficult to obtain reasonable signal-to-noise ratios when one is using Cu $K\alpha$ radiation unless the detector area is very small. There have been attempts to solve this problem [22], but even with current limitations the cadmium telluride detector does have useful applications when used with Mo $K\alpha$ radiation or even with Cu $K\alpha$ where the beam size is limited, for example, in pole-figure devices.

One of the most promising room-temperature, solid state detection devices to be developed is the mercuric iodide detector. The room-temperature energy resolution of these detectors is comparable with that of cooled Si(Li). A major disadvantage has been the technical difficulty of growing commercial quantities of the mercuric iodide crystals, and for this reason there has been little or no application of these devices for X-ray diffraction experiments. Again, recent work has shown that this problem is solvable [23], and some success is now being enjoyed in the associated field of X-ray fluorescence spectrometry [24].

5.4. PULSE HEIGHT SELECTION

Since the size of a voltage pulse is proportional to the energy of an X-ray photon, when different wavelengths are incident upon a proportional detector, voltage pulses of different sizes will be generated. Figure 5.8 shows a hypothetical case involving three wavelengths: short-wavelength scatter λ_a at voltage V_a (16 V); the analytical wavelength λ_b at voltage V_b (8 V); and specimen fluorescence λ_c at voltage V_c (6 V). The background of the diffractogram is made up mainly of the wavelengths from short-wavelength scatter and specimen fluorescence. The lower portion of Figure 5.8 shows a representation of the three sizes of pulses over a period of 1 s. Since the wavelengths give voltage pulses of different sizes, it is possible to electronically discriminate these pulses. This technique is called *pulse height selection* (PHS), and the electronic device to carry it out is a *pulse height analyzer* (PHA). The principle by which pulse height selection works is shown in Figure 5.9. A pulse height selector contains three main components—two discriminators and an anticoincidence unit. A discriminator is a device that will give a symmetrical square-top output pulse for every input pulse above the voltage level at which the discriminator is set. As shown in Figure 5.9, the pulses from the detector are passed simultaneously into the first and second discriminators. In order to simplify the figure, just one of each size voltage pulse is shown. Discriminator number 1 is set at 7 V and discriminator 2 at 10 V. The smallest size input pulse (V_a) is of lower voltage than either discriminator and therefore, does not trigger an output from either discriminator. The middle-size pulse (V_b), with a size of 8 V, is between the set voltage levels of discriminators 1 and 2. Thus, only the first discriminator gives an output pulse, and this pulse is made positive. The largest size voltage pulse (V_c) is larger than both discriminator levels, and so each discriminator will generate an output pulse. However, pulses from the second discriminator are made negative. The outputs from both discriminators are fed to the anticoincidence unit, which adds them together. Since the V_c

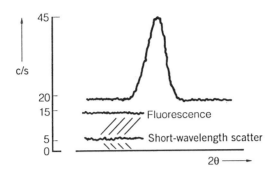

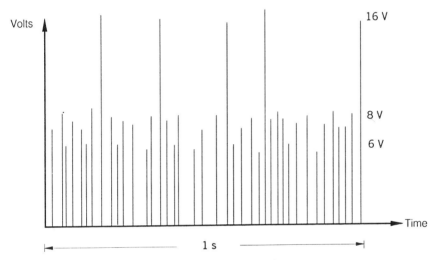

Figure 5.8. The pulse height selector.

pulse gave output from both discriminators they cancel out, leaving only the pulses from the middle-size pulse V_b. The limit defined by the two discriminators is called the *acceptance window* (usually simply called the "window"). In the example shown, pulses V_a and V_c both fall outside the window; hence only pulses V_b are passed on to the scalers.

Pulse height selection proves invaluable for the removal of such effects as sample fluorescence and background that may arise from shorter wavelengths from the X-ray tube continuum that pass the β-filter. It should be noted here that in the use of the Si(Li) and Ge(Li) detectors, the high energy resolution of these detectors allows a different electronic approach. In place of the PHA and

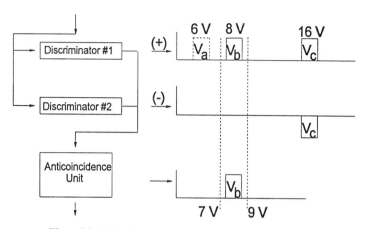

Figure 5.9. Anticoincidence system for pulse height selection.

the scaler timer, there is an MCA. This device acts like an audio equalizer, typically with 4000 separate filters. The 400–64,000 channels of an MCA each act as a separate scaler with its own pulse height selector. An MCA allows us to view photons of many different energies striking the detector. Each X-ray photon produces a voltage pulse that can be converted to a digital number via an analog–digital (A/D) circuit. This digit will be counted into one of the scalers in the MCA. For normal diffraction work we are only interested in photons of a single energy and therefore use the system previously described.

5.5. COUNTING CIRCUITS

The preamplified signal brought back from most X-ray detectors is proportional to the energy of the X-ray photon that caused it. The output from these detectors can be sent directly to a ratemeter or scaler/timer for display. The electronics associated with the detection of X-rays on most modern X-ray equipment are illustrated in Figure 5.10. The initial preamplified voltage pulse arriving at the signal-processing equipment is first sent to a linear amplifier to adjust its level to one appropriate for pulse height discrimination. After a pulse has been accepted by the PHA, it is passed on to two independent circuits: the first is a scaler/timer, which allows us to count the number of pulses (N) arriving in any time interval (t), and the second is a ratemeter (see Section 5.5.1, below). Because an independent measure of both N and t may be required, visual display systems generally incorporate separate scaler and timer units

COUNTING CIRCUITS

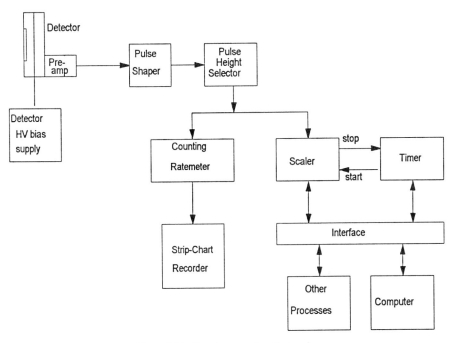

Figure 5.10. Signal-processing electronics.

that are coupled. Initiation of a count sequence starts the timer, and each pulse entering the scaler trips a series decade display (a scaling switch is often incorporated allowing the counting of every 10th, 100th, 1000th, etc. pulse). When the fixed-count method is employed, once a selected number of counts in the scaler is reached the scaler sends a stop pulse to the timer. Similarly, where the fixed-time method is used, the attainment of selected time on the timer causes a stop pulse to be sent to the scaler.

5.5.1. The Ratemeter

The *ratemeter* is a circuit that takes in the random arrival of pulses and puts out an average signal that can be displayed on a calibrated voltmeter or a strip-chart recorder. Since the detector circuitry gives information in digital form, some system of integration must be used if a continuous scan over an angular range is desired. For this purpose, a ratemeter circuit is used. A ratemeter is a simple tank circuit composed of a capacitor and resistor in parallel, as shown in Figure 5.11. In practice, it consists of pulse-amplifying and -shaping circuitry that produces pulses of fixed time and voltage. The stream of incoming pulses passes through a one-shot charge pulse generator,

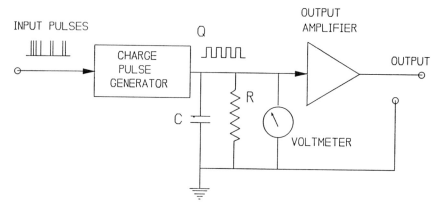

Figure 5.11. The ratemeter circuit.

triggering an equivalent number of output pulses, each containing a fixed charge Q to the capacitor C. Each pulse causes an increase in the voltage across the capacitor, which leaks off across the resistor R:

$$V_t = \frac{Q}{C}\exp\left(\frac{-t}{RC}\right), \qquad (5.4)$$

where V_t is the voltage pulse observed at the capacitor at time t.

For a mean counting rate of I c/s, the average voltage V across the resistor is equal to IQR. This voltage is read by a voltmeter as a calibrated counting rate and delivered to a strip-chart recorder through an output buffer amplifier. The function of the capacitor is to smooth the signal developed across the resistor. This smoothing effect is described by the time constant $\tau = RC$. The larger the time constant, the smoother will be the output voltage. By increasing the ratemeter time constant, measurement of the counting rate is averaged over a longer time period and the statistical noise is reduced. In practice, care must be taken in matching the time constant with the scanning speed.

5.6. COUNTING STATISTICS

The production of X-rays is a random process that follows the usual statistical laws. The random fluctuations expected in an X-ray counting system follow a Poisson distribution P:

$$P(N) = \frac{N_0}{N!}\exp(-N_0), \qquad (5.5)$$

where N_0 is the true value of the number of counts, and N is the number measured in any one experiment. When the value of the number of counts N is large, the standard deviation σ of N is simply $\sqrt{N}$. Since the Poisson distribution approximates the Gaussian distribution, the usual interpretation of confidence limits may be applied to this standard deviation. From the properties of the Gaussian distribution, there will be a 68.3% probability that any value of N lies between $N \pm \sigma$, a 95.4% probability that it lies between $N \pm 2\sigma$, and a 99.7% probability that it lies between $N \pm 3\sigma$. The number of counts collected is equal to the product of the counting rate R (c/s) and the count time t (s):

$$\sigma(N) = \sqrt{N} = \sqrt{(Rt)} \tag{5.6}$$

Thus, the random counting error is time dependent. In quantitative X-ray diffraction, allowance must be made for the fact that diffraction lines may be broadened by factors other than instrumental effects. Strain and small particle sizes are the most important of these noninstrumental factors. It is common practice, therefore, to integrate the total area under the peak to gain a measure of the peak intensity, rather than use peak heights. In this instance, the net error σ_{net} in counts is given by

$$\sigma_{net} = \frac{100\sqrt{(N_p + N_b)}}{N_p - N_b}, \tag{5.7}$$

where N_p is the total number of counts accumulated over the selected angular range, and N_b is the total number of counts under the background.

Figure 5.12 illustrates a diffraction peak as it would be step-scanned by a modern automated diffractometer. The count rate at each of the N_{pk} points at the peak is determined by dividing the measured intensity by the count time. Each point contains some contribution from the background. The average number of background counts recorded at each point is $(B_1 + B_2)/2$. The background count rates B_1 and B_2 are assumed to be equal to the first point (R_1) and the last $(R_{N_{pt}})$ measured. It is therefore important that the 2θ scan width be set wide enough not to exclude any of the peak intensity. At the same time it is important that the width be set narrow enough not to include any contributions from nearby peaks. The integrated intensity is the sum of each R_i value. The background-corrected integrated intensity is

$$I = \sum_{i=1}^{N_{pt}} R_i - N_{pt}\left(\frac{R_1 + R_{N_{pt}}}{2}\right), \tag{5.8}$$

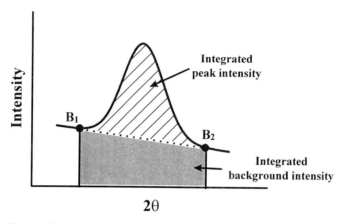

Figure 5.12. Determination of a background-corrected integrated intensity.

where N_{pt} is the number of points counted, and the R's are the count rates at each point in c/s. The variance of the background corrected intensity is

$$\sigma^2 = \sum_{i=1}^{N_{pt}} \sigma^2(R_i) + N_{pt}[(\sigma^2(R_1) + \sigma^2(R_{N_{pt}})] \tag{5.9}$$

In computations, it is convenient to carry only the σ^2 values because they propagate linearly and permit simple derivation of the errors associated with any other quantities to be determined from the intensity.

5.7. TWO-DIMENSIONAL DETECTORS

Historically, flat-plate (Laue) cameras have been used for crystal and fiber orientation measurements. Film techniques have the advantages of being low-cost, fairly easily adaptable to most X-ray sources and giving excellent resolution. However, while this two-dimensional technique has provided valuable data, it is rather time consuming and less convenient than electronic detection. The use of electronic two-dimensional detectors has provided a dramatic increase in the quality of data and the speed of data collection. One of the more common of the two-dimensional detectors is the *real-time image intensifier* [25]. In this system, X-rays scattered from a sample strike a phosphor (ZnS:Ni, GdOS$_2$:Tb, etc.) screen, resulting in the emission of visible light. This visible signal can either be observed directly or amplified via an electron lens–phosphor coupling system (image intensification), as shown in Figure 5.13. The resulting intensified image is then captured using a charge-

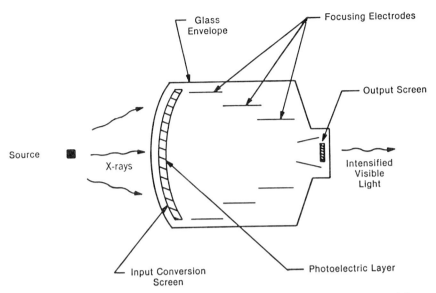

Figure 5.13. An image intensifier. From Jenkins [1, p. 22]. Copyright © 1988, John Wiley & Sons, Inc. Reprinted by permission of the publisher.

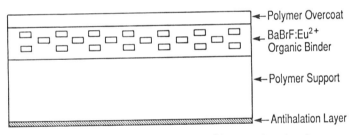

Figure 5.14. Cross section of a $BaBrF:Eu^{2+}$ storage imaging plate.

coupled device (CCD) video camera and processed using a computer. A video monitor can be used to show a real-time image or the image can be stored, for example, on videotape. For X-ray diffraction application, this system offers the advantages of moderate cost, real-time display of images, and the ability to computer enhance images. On the other hand, the system offers rather poor resolution compared with film methods and, at this stage of development, is restricted to rather small detection areas (5–8 cm).

A second device is the *two-dimensional position-sensitive detector*. This is similar to the PSD system already described in Section 5.3.2, except that

144 DETECTORS AND DETECTION ELECTRONICS

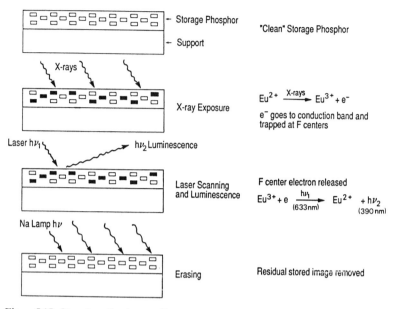

Figure 5.15. Steps in collecting, reading, and erasing a storage phosphor imaging plate.

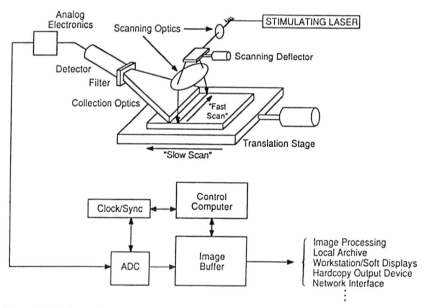

Figure 5.16. Schematic of the scanner used to read a stored image on a storage phosphor (ADC = analog–digital converter).

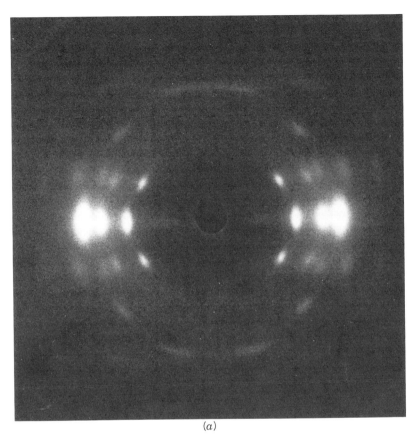

(a)

Figure 5.17. Two-dimensional images obtained using several different imaging devices: (a) X-ray, film; (b) real-time image intensifier; (c) two-dimensional position-sensitive detector; (d) storage phosphor imaging plate.

a two-dimensional array of detector wires is used to form an electronic grid. The two-dimensional detector offers the advantages of high-speed (minutes) digitized data collection, with good resolution. The major disadvantage, at this time, is the high cost (around $150,000).

A third system is the *storage phosphor imaging plate*. In this device, the X-rays strike BaBrF:Eu grains coated onto a polymer film base (Figure 5.14). The X-ray photons promote electrons in the phosphor to elevated energy levels to form a latent image. This latent image is then scanned by a red laser,

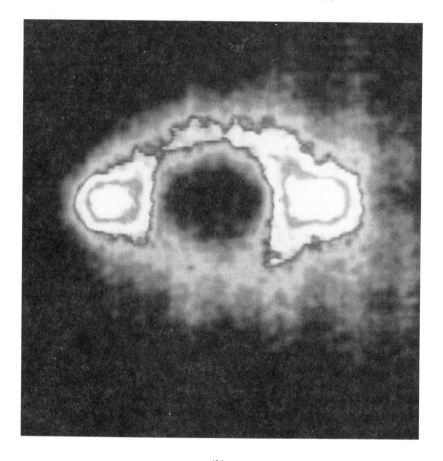

(b)

Figure 5.17. (Cont'd.)

causing the electrons to drop back to their ground state, emitting the stored energy as blue light. As in the case of the scintillation counter, the blue light can be collected by a photomultiplier to give a measurable signal (Figure 5.15). This signal is passed on to a computer, where the data are stored as a function of position and intensity. This system offers the advantages of good spatial resolution (50–100 µm), fast data collection (minutes), and intensity linear over about –5 orders of magnitude (as compared with 2 orders of magnitude for film methods). Unfortunately, at this time, suitable scanners cost more than

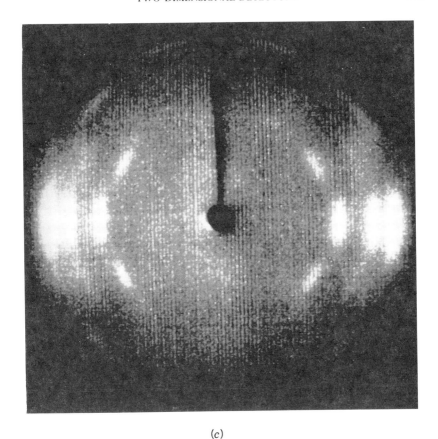

(c)

Figure 5.17. (Cont'd.)

$50,000. The scanner shown in Figure 5.16 [25] is an experimental device used to compare different methods of two-dimensional data storage from fiber orientation experiments. Figure 5.17 shows the results of such a comparison in which the sample was poly(ethylene terephthalate) (Mylar). The sample was stretched uniaxially to about five times its original length, and the sample-to-film distance was 5 cm. Currently, the image plate is being developed as a spherical 4π detector that can record the entire three-dimensional diffraction pattern simultaneously.

148   DETECTORS AND DETECTION ELECTRONICS

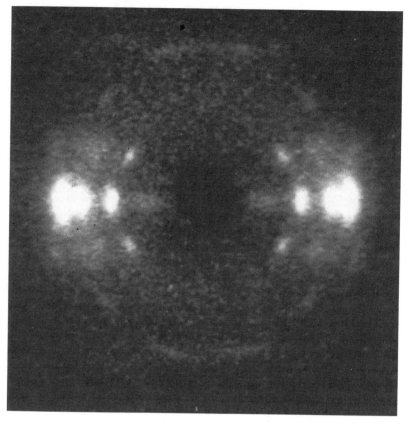

(d)

Figure 5.17 (Cont'd.)

REFERENCES

1. Jenkins, R. *X-Ray Fluorescence Spectrometry*, p. 59 et seq. Wiley (Interscience), New York, 1988.
2. McCarthy, G. J., private communication, 1986.
3. Jenkins, R., Gould, R. W., and Gedcke, D. *Quantitative X-Ray Spectrometry*, Chapter 4. Dekker, New York, 1981.
4. Jenkins, R., Gould, R. W., and Gedcke, D. *Quantitative X-Ray Spectrometry*, p. 144. Dekker, New York, 1981.
5. Gabriel, A., Dauvergne, F., and Rosenbaum, C. Linear, circular and two-

dimensional position sensitive detectors. *Nucl. Instrum. Methods* **152**, 191–194 (1978).

6. James, M. R., and Cohen, J. B. *The Application of a Position-Sensitive X-ray Detector to the Measurement of Residual Stresses*, Tech. Rep. No. 11, Northwestern University, Dept. Mater. Sci. and Mater. Res. Cent., Evanston, IL, 1975

7. Göbel, H. A new method for fast XRPD using a position-sensitive detector *Adv. X-Ray Anal.* **22**, 255–265 (1979).

8. Wölfel, E. R. A novel curved position-sensitive proportional counter for X-ray diffractometry. *J. Appl. Crystallogr.* **6**, 341–348 (1983).

9. Gabriel, A. Position-sensitive X-ray detector. *Rev. Sci. Instrum.* **48**, 1303–1305 (1977).

10. Snyder, R. L., Rodriguez, M. A., Chen, B. J., Göbel, H. E., Zorn, G., and Seebacher, F. B. Analysis of peritectic reactions and orientation by high temperature XRD and optical microscopy. *Adv. X-Ray Anal.* **35**, 623–632 (1992).

11. Göbel, H. E. The use and accuracy of continuously scanning position-sensitive detector in X-ray powder diffraction. *Adv. X-Ray Anal.* **24**, 123–138 (1982).

12. Foster, B. A., and Wölfel, E. R. Automated quantitative multiphase analysis using a focusing transmission diffractometer in conjunction with a curved position-sensitive detector. *Adv. X-Ray Anal.* **31**, 325–330 (1988).

13. Rassineux, F., Beaufort, D., Bouchet, A., Merceron, T., and Meunier, A. Use of a linear localization detector for X-ray diffraction of very small quantities of clay minerals. *Clays Clay Miner.* **36**, 187–189 (1988).

14. Snyder, R. L. Accuracy in angle and intensity measurements in X-ray powder diffraction. *Adv. X-Ray Anal.* **26**, 1–11 (1983).

15. Giessen, B. C., and Gordon, G. E. X-ray diffraction: New high speed technique based on X-ray spectrography. *Science* **159**, 973–975 (1968).

16. Bish, D. L., and Chipera, S. J. Comparison of a solid-state Si detector to a conventional scintillation detector–monochromator system in X-ray powder diffraction analysis. *Powder Diffr.* **4**, 137–143 (1989).

17. Drever, J. I., and Fitzgerald, R. W. Fluorescence elimination in X-ray diffractometry with solid state detectors. *Mater. Res. Bull.* **5**, 101–107 (1970).

18. Ferrel, R. E., Jr. Applicability of energy-dispersive X-ray powder diffractometry to determinative mineralogy. *Am. Mineral.* **56**, 1822–1831 (1971).

19. Laine, E., Lähteenmäki, I., and Hämäläinen, M. Si(Li) semiconductor in angle and energy dispersive X-ray diffractometry. *J. Phys. E.* **7**, 951–954 (1974).

20. Jenkins, R., Gould, R. W., and Gecke, D. *Quantitative X-Ray Spectrometry*, Sect. 4.3.1, p. 205 et seq. Dekker; New York, 1981.

21. Robie, S. B., and Scalzo, T. R. A comparison of detection systems for trace phase analysis. *Adv. X-Ray Anal.* **28**, 361–365 (1986).

22. Roth, M., and Burger, A. Improved spectrometer performance of cadmium selenide room temperature gamma-ray detector. *IEEE Trans. Nucl. Sci.* **NS-33**, 407–410 (1986).

23. Faile, S. P., Dabrowski, A. J., Huth, G. G., and Iwanczyk, J. S. Mercuric iodide (HgI_2) platelets for X-ray spectroscopy produced by polymer controlled growth. *J. Cryst. Growth* **50**, 752–756 (1980).
24. Kelliher, W. C., and Maddox, W. G. X-ray fluorescence analysis of alloy and stainless steels using a mercuric iodide detector. *Adv. X-Ray Anal.* **31**, 439–444 (1988).
25. Blanton, T., Eastman Kodak Co., private communication. Rochester, NY, 1993.

CHAPTER
6
PRODUCTION OF MONOCHROMATIC RADIATION

6.1. INTRODUCTION

With the growing need for high-quality X-ray powder diffraction data, one especially important parameter is the method by which the analytical wavelength is treated. Problems can occur in data treatment because of the polychromatic nature of the diffracted beam and the variability in the angular dispersion of the diffractometer. A combination of these two facts can lead to difficulty in manually assessing where the maximum of a peak occurs, especially in the range of angles from 30° to 60° 2θ. Of the many methods commonly employed to render the radiation monochromatic (or bichromatic), most will fit into two broad categories: instrumental methods and computer methods. In this chapter, the more important instrumental methods of monochromatization are presented [1]. Computer methods will be discussed later, in Section 11.3.3.

The more common instrumental methods of monochromatization include the following:

- Use of a β-filter
- Use of a proportional detector and pulse height selection
- Use of a Si(Li) solid-state detector
- Use of a diffracted-beam monochromator
- Use of a primary beam monochromator

The basic purpose of the monochromatization of the diffracted radiation is to obtain an experimental pattern from a single wavelength. Inspection of the Bragg law (Equation 3.1) reveals that each unique d-spacing will diffract different wavelengths at their own unique diffraction angles. Thus, if a pattern were measured using an X-ray beam containing two wavelengths, the observed pattern would, in fact, be two patterns (one for each wavelength) superimposed, one on top of the other, and clearly will be more difficult to interpret than a diffractogram from a single wavelength. While the problems associated with the interpretation of a diffractogram from two known wavelengths can be significant, much more of a problem occurs when addi-

Table 6.1. Characteristic Wavelength Values for Common Anode Materials[a]

	Wavelength (Å)		
Anode	$K\alpha_1$(100)	$K\alpha_2$(50)	$K\beta$(15)
Cu	1.54060	1.54439	1.39222
Cr	2.28970	2.29361	2.08487
Fe	1.93604	1.93998	1.75661
Co	1.78897	1.79285	1.62079
Mo	0.70930	0.71359	0.63229

[a]Relative intensities are shown in parentheses.

tional *unknown* wavelengths are diffracted. As was discussed in Section 1.4, the characteristic K-radiation emission from typical X-ray tube anode materials is much more complex than the simple α-doublet and β-doublet model generally employed in classical powder diffractometry. For most practical purposes, however, the copper K spectrum is generally considered to consist simply of two pairs of lines: the α_1, α_2 doublet, occurring from a $2p \rightarrow 1s$ transition, and the β_1, β_3 doublet, from a $3p \rightarrow 1s$ transition. Table 6.1 lists the values of the $K\alpha$ and $K\beta$ wavelengths for the more common X-ray tube target elements.

In most experimental work, the β-doublet intensity is typically reduced to less than a few percent of the α-doublet intensity by use of filtration, or it is removed by use of a diffracted-beam monochromator or a high-resolution energy-resolving detector. In each case, what remains is essentially bichromatic X-radiation. The exception to this occurs with the primary-beam monochromator, where only the α_1 is directed onto the specimen. In applying the usual bichromatic radiation to the measurement of interplanar spacings, the main problems are the angle-dependent dispersion of the α_1, α_2 doublet, and differences in peak asymmetry between the two. It follows from Bragg's law that an error in d-spacing is linearly related to an error in wavelength. For manual qualitative phase identification, an accuracy of about 5 parts per 1000 for $\Delta d/d$ is sufficient; and about twice this accuracy is required for computer search/matching. Provided that the true α_1 emission line is being used (i.e., either the diffracted beam is really monochromatic or the contribution from the α_2 is effectively removed from the diffraction profile), the wavelength need also be known only to about 2 parts per 1000. For accurate lattice parameter determination, it is common practice to use an internal standard and, in these cases, the wavelength value is to all intents and purposes calibrated out.

6.2. ANGULAR DISPERSION

Most powder diffraction work is carried out using the Cu $K\alpha_1$, $K\alpha_2$ doublet, and one of the greatest experimental inconveniences arising from this choice is the variable angular dispersion of the diffractometer. The angular dispersion of a diffractometer, $d\theta/d\lambda$, may be obtained by differentiating Bragg's law (Equation 3.1) to give

$$\frac{d\theta}{d\lambda} = \frac{1}{2d\cos\theta}, \tag{6.1}$$

$$\frac{d\theta}{d\lambda} = \frac{\tan\theta}{2d\sin\theta}, \tag{6.2}$$

$$\frac{d\theta}{d\lambda} = \frac{\tan\theta}{\lambda}. \tag{6.3}$$

As shown in these equations, the angular dispersion, $d\theta/d\lambda$, of a diffractometer increases with increasing 2θ; thus, the α doublet is unresolved at low 2θ values, whereas at high 2θ values it is completely separated. In the midangular range the $K\alpha_1$, $K\alpha_2$ lines are only partially resolved, leading to some distortion of the diffracted line profile, as well as giving difficulties with numerical techniques used to separate $K\alpha_2$ from $K\alpha_1$. The absolute value of the angular dispersion varies from about 100 eV at low 2θ values to about 2 eV at high 2θ values. The energy difference between the $K\alpha_1$ and $K\alpha_2$ lines of copper is about 20 eV; hence the separation of the doublet occurs somewhere in the midangular range of the diffractometer. Figure 6.1 shows angular dispersion as a function of 2θ for the $K\alpha_1$, $K\alpha_2$ lines. As a first approximation, the α-doublet is resolved when the angular dispersion is equal to the half-width of the diffraction line profile. Thus, with a line width of about 0.2°, the $K\alpha_1$, $K\alpha_2$ doublet starts to be resolved at about 50° 2θ. When manual peak-finding techniques are used, it is common practice to use the intensity-weighted geometric average of the $K\alpha_1$, $K\alpha_2$ wavelength, i.e.,

$$\frac{2 \times 1.54060 + 1.54439}{3} = 1.54186 \text{ Å},$$

as the experimental wavelength, at least until a 2θ value is reached where the $K\alpha_1$, $K\alpha_2$ doublet is sufficiently resolved to allow accurate measurement of the position of the $K\alpha_1$ line. One should note that the intensity tail of the $K\alpha_2$ under the $K\alpha_1$ will displace the observed maximum toward a higher angle.

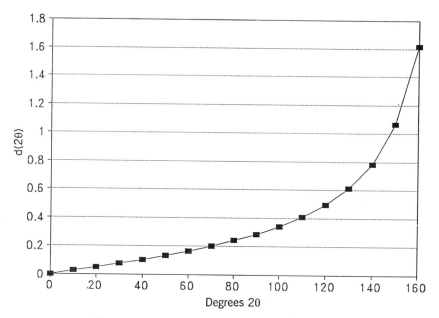

Figure 6.1. Angular dispersion of the copper $K\alpha$ doublet.

Where automated methods are used, $K\alpha_2$ stripping is generally employed over the whole range of the measured pattern.

The absolute energy difference between Cu $K\beta_1$ and Cu $K\beta_3$ is only about 2 eV. Thus, whereas the $K\alpha_1$, $K\alpha_2$ doublet is resolved at about 50° 2θ, the $K\beta_1$, $K\beta_3$ doublet always appears as a single reflection. This fact does offer some interesting experimental possibilities, since a diffractogram can also be measured with $K\beta$ radiation. Even though the Cu $K\beta$ lines are weaker than the corresponding Cu $K\alpha$ lines, the simplification in the measured pattern sometimes makes this intensity reduction worthwhile. It should also be noted, however, that the Cu $K\beta$ emission does contain a weak $K\beta_5$ line, which is a forbidden $3d \rightarrow 1s$ transition. This line is normally resolved from the $K\beta_1$, $K\beta_3$ doublet and typically has an intensity about 2% of the $K\beta_1$.

6.3. MAKEUP OF A DIFFRACTOGRAM

It is useful at this stage to review the makeup of a typical diffractogram, which has three major contributions: diffraction, scatter, and fluorescence. The diffractogram itself is made up of diffraction peaks superimposed upon background. As shown in Table 6.2, the desired peaks arise from diffraction of

MAKEUP OF A DIFFRACTOGRAM

Table 6.2. Contributions to a Diffractogram

Source	Gives—
Diffraction of required wavelength	Wanted peaks
Diffraction of other wavelength	Unwanted peaks
Coherent scatter from specimen	General background
Incoherent scatter from specimen	General background
Scatter from specimen support	Extra low-angle background
Fluorescence from specimen	General background

the assumed experimental wavelength, whereas the unwanted peaks arise from other (frequently unexpected) wavelengths. The background arises from coherent and incoherent scattering from the specimen, air scatter, and (especially at low values of 2θ) scatter from the sample support. A major function of the monochromatization system is to remove unwanted peaks and reduce background as much as possible.

6.3.1. Additional Lines in the Diffractogram

The main reason for the presence of additional weak lines and artifacts in a diffractogram is the impurity of the source. Probably in excess of 90% of all powder work today is done with Cu $K\alpha$ radiation. A copper tube emits a broad band of continuous (or white) radiation, along with intense characteristic α and β lines. However, the emitted spectrum may also contain contaminant lines from other elements from the X-ray tube target or tube window, as described in Section 4.6. Figure 6.2 shows a filtered and unfiltered spectrum from a cobalt target X-ray tube. It will be seen that, after filtration, the $K\beta$ line is significantly reduced (but not eliminated), but that the weak lines around the $K\alpha$ doublet remain. These are contamination lines from Fe $K\alpha$ (at about 134°), W $L\alpha$ (at about 27°), and W $L\beta$, and although in this example they are rather weak, their intensity will increase as the tube ages. Table 6.3 shows how the additional lines would manifest themselves in a diffractogram of α-quartz measured with, nominally, Cu $K\alpha$ radiation. The position of each of the first five reflections from quartz are shown (for Cu $K\alpha$), along with the angles at which the same five d-spacings would diffract Cu $K\beta$, W $L\alpha$, and W $L\beta$. The relative intensities of the additional lines will depend greatly on the type and efficiency of the monochromatization system employed, as well as on the age of the diffraction tube. For a reasonably new X-ray tube, a well-aligned monochromator should reduce all non-Cu $K\alpha$ lines to less than 1% of the strongest line in the pattern. Similarly, β-filtered radiation would typically give Cu $K\beta$ and W L lines at an intensity level of 1–2% of the equivalent Cu $K\alpha$ intensity.

156   PRODUCTION OF MONOCHROMATIC RADIATION

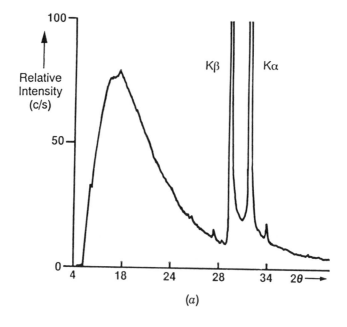

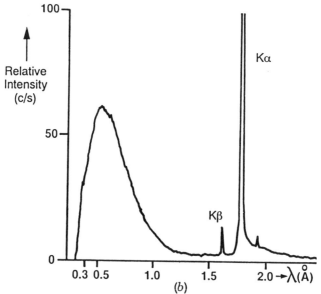

Figure 6.2. Unfiltered (a) and filtered (b) cobalt K radiation.

Table 6.3. Potentially Interfering Lines Observed in α-Quartz When Cu Kα Is Used

		2θ			
d-Value	(hkl)	Cu Kα	Cu Kβ	W Lα	W Lβ
4.257	(100)	20.850	18.823	19.967	17.231
3.342	(101)	26.652	20.044	25.515	22.116
2.457	(110)	36.542	32.917	34.959	30.245
2.282	(102)	39.456	35.522	37.737	32.627
2.237	(111)	40.284	36.261	38.526	33.302

6.3.2. Reduction of Background

Radiation striking the specimen or the specimen support may be diffracted, scattered, or produce secondary fluorescence radiation. X-radiation which eventually reaches the detector is further modified by whatever monochromatization process is employed. Although intensity is always of prime importance in powder diffraction measurements, the ease of interpretation of the resultant diffractogram is invariably dependent on the resolution of the pattern and the overall signal-to-noise ratio within that pattern. A pattern made up of broad and/or weak lines superimposed on a high, variable background is much more difficult to process than one in which the lines are sharp and well resolved and the background is low and flat. While the basic purpose of the monochromatization process is to produce (as much as possible) a single wavelength, the removal of white radiation and other characteristic lines from the source does much to reduce the effects of scatter and fluorescence discussed earlier.

When the radiation falling onto the specimen is energetic enough to excite characteristic lines from the elements making up the specimen, and where this fluorescence radiation falls within the acceptance range of the detection/monochromatization system, it too will contribute to the background. The intensity of fluorescence background is reasonably constant since the fluoresced photons are not diffracted and therefore are angle independent. Where a fixed divergence slit is employed, the irradiation length of the specimen decreases with increase of diffraction angle.

The low-angle region is a particularly difficult region in which to keep background to reasonable proportions. Figure 6.3 shows the low-angle region between 0° and 20°. It will be seen that the experimental curve is made up of diffraction peaks superimposed on background, which comes mainly from scatter from the specimen (plus perhaps from the specimen support). Much of the low-angle background from air scatter can be removed by use of a vacuum

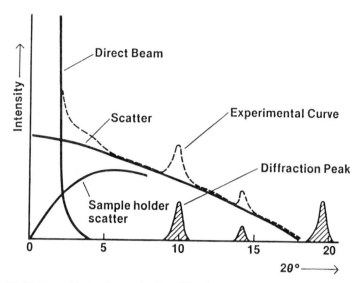

Figure 6.3. Makeup of the background at low diffraction angles. From R. Jenkins, Instrumentation. In *Modern Powder Diffraction* (D. L. Bish and J. E. Post, eds.), p. 30, Fig. 9. Mineralogical Society of America, Washington, DC, 1989. Reprinted by permission.

or helium path chamber around the specimen. However, such devices are not popular mainly because of the possibility of induced chemical changes, such as loss of water of hydration from the specimen under the reduced pressure. At very low (less than 5°) diffraction angles, the background is determined almost exclusively by the choice of divergence slit. As an example, with a $\frac{1}{2}°$ divergence slit, the direct beam would be observed at about 3° 2θ.

6.4. THE β-FILTER

The β-filter is a single-band bandpass device that is mainly used to improve the ratio of Cu $K\alpha$ to Cu $K\beta$. Section 1.6 discussed the way in which the mass attenuation coefficient varies with wavelength. It follows from that discussion that by passing a polychromatic beam of radiation through a carefully selected filter, preferential transmission of certain wavelengths can be achieved. Since the lowest absorption is always found immediately to the long-wavelength side of the absorption edge of the filter, by choosing a filter material with an absorption edge between the $K\alpha$ doublet and the $K\beta$ doublet of the X-ray tube target element, the α/β transmission ratio will be very high. For example, Figure 6.4 shows a superposition of the emission from a copper target X-ray

THE β-FILTER   159

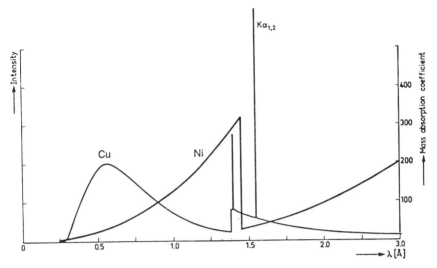

Figure 6.4. Use of the $K\beta$-filter. Reprinted from R. Jenkins and J. L. de Vries, *An Introduction to Powder Diffractometry*, p. 14, Fig. 12. Copyright © 1977, N. V. Philips, Eindhoven, The Netherlands.

tube and the mass attenuation curve for nickel. (Note that the sharp drop in the intensity of the continuum at 1.38 Å is due to self-absorption of the Cu target.) The nickel absorption edge (1.488 Å) lies between the Cu $K\alpha$ radiation (1.542 Å) and Cu $K\beta$ radiation (1.392 Å). Thus, the Cu $K\beta$ is strongly absorbed (mass attenuation coefficient = 286 cm²/g), and the Cu $K\alpha$ is weakly absorbed (mass attenuation coefficient = 49.2 cm²/g). Note, too, from Figure 6.4 that the mass attenuation for the short-wavelength continuum is very low. This results in much of the short-wavelength radiation passing the filter (see also Figure 6.2).

6.4.1. Thickness of the β-Filter

In the example just given, nickel has its K absorption edge at 1.488 Å. The mass attenuation coefficients of Ni for Cu $K\beta$ and Cu $K\alpha$ are 286 cm²/g and 49.2 cm²/g, respectively. The relative transmission efficiency of $K\alpha$ and $K\beta$ by a filter of given thickness can be readily calculated by use of Equation 1.10. From this equation it follows that the percentage transmission I is given by the expression

$$I = \exp\left[2 - \left(0.434 \frac{\mu}{\rho} \rho x\right)\right]. \tag{6.4}$$

160 PRODUCTION OF MONOCHROMATIC RADIATION

Table 6.4. Target Elements and Filters

Target	$K\alpha$ (Å)	β-Filter	Thickness (μm)	Density (g/cc)	$K\alpha$ (%)	$K\beta$ (%)
Cr	2.291	V	11	6.0	58	3
Fe	1.937	Mn	11	7.43	59	3
Co	1.791	Fe	12	7.87	57	3
Cu	1.542	Ni	15	8.90	52	2
Mo	0.710	Zr	81	6.5	44	1

Thus, in the case of a 15 μm thick nickel filter, having a density of 8.92 g/cm^3, the relative transmission I_t of Cu $K\alpha$ and Cu $K\beta$ is equal to

$$(I_t)K\alpha = \exp[2 - (0.434 \times 49.2 \times 8.92 \times 0.0015)] = 52\%$$

and

$$(I_t)K\beta = \exp[2 - (0.434 \times 286 \times 8.92 \times 0.0015)] = 2\%.$$

Table 6.4 lists the usual combinations of target element and β-filter and indicates the relative transmissions of $K\alpha$ and $K\beta$ radiation. It has been common practice in the industry for many years to employ β-filter thicknesses that give an integrated intensity ratio of $K\alpha/K\beta$ = between 50:1 and 25:1 at the detector [1, 2].

6.4.2. Use of Pulse Height Selection to Supplement the β-Filter

The β-filter is generally supplemented by some energy discrimination to remove the high-energy white radiation from the X-ray tube. The effectiveness of this white radiation removal depends upon the resolution of the detector. For Cu $K\alpha$, the resolution of the scintillation detector is about 3600 eV. Such a resolution would not allow the removal of radiation in the immediate vicinity of the Cu K lines but will permit the removal of the short-wavelength bremsstrahlung. In fact, quite a significant amount of short-wavelength radiation from the continuum will pass through the β-filter. For example, in the case of the 15 μm thick nickel filter already mentioned, whereas the transmission immediately to the short-wavelength side of the absorption edge was only 2%, at 1 Å the transmission is 22% and at 0.5 Å it is about 80%. Thus, the spectrum of the filtered radiation is typified by having a fair amount of short-wavelength radiation along with intense $K\alpha$ and weak $K\beta$. As the voltage on the X-ray tube is increased, the output of the continuum increases and spreads to shorter wavelengths. This in turn leads to an increase in the transmission of unwanted short-wavelength radiation. Hence, when the β-

filter is used on its own, it can never completely remove the continuous radiation unless the voltage employed on the X-ray tube is very low (i.e., less than about 30 kV). This is a particular problem when one is using Debye–Scherrer and Gandolfi cameras, where the transmitted white radiation gives rise to low-angle blackening of the film. It should also be noted that when gas proportional counters are employed, the counter itself gives significant gas discrimination against shorter wavelengths. Over recent years, the advent of much higher power X-ray sources generally means operation of the X-ray tube well in excess of 30 kV in order to give optimum intensity. For this reason, it is common practice to use a β-filter in combination with a proportional counter and a pulse height selector to give optimum separation of the continuum from the $K\alpha$.

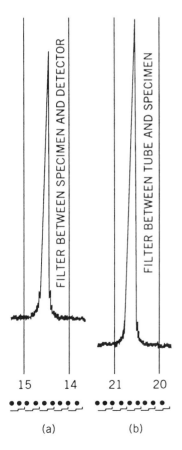

Figure 6.5. Position of the Ni β-filter when nickel-containing specimens are being measured. Reprinted from R. Jenkins and J. L. de Vries, *An Introduction to Powder Diffractometry*, p. 15, Fig. 14. Copyright © 1977, N. V. Philips, Eindhoven, The Netherlands.

6.4.3. Placement of the β-Filter

As far as removal of $K\beta$ and continuous radiation is concerned, it makes little difference whether the filter is placed before or after the specimen. However, the atoms in the specimen may be excited by the incident radiation, giving, in turn, fluorescence radiation from the specimen (specimen fluorescence), which can considerably increase the background level. Placing the filter between the detector and specimen will considerably reduce this background, and it is in general advisable to place the filter between specimen and detector. An exception to this general rule is found when a specimen is investigated that has a high percentage of the same element as the β-filter, for example, measurement of a nickel specimen with Cu $K\alpha$ radiation. Most fluorescent radiation produced from the nickel will pass the nickel β-filter. In this instance, it is better to place the filter between the X-ray tube and specimen to eliminate the Cu $K\beta$ radiation, which is very efficient in exciting nickel radiation. Figure 6.5 shows two diffraction scans over the nickel (200) line: (a) those made with the Ni β-filter between specimen and detector, and (b) those made with the β-filter between specimen and tube.

6.5. THE PROPORTIONAL DETECTOR AND PULSE HEIGHT SELECTION

Since the size of a voltage pulse is proportional to the energy of an X-ray photon, when different wavelengths are incident upon a proportional detector, voltage pulses of different sizes will be generated. In actual fact, pulses of a certain type, for example, the Cu $K\alpha$, do not all have exactly the same value since a statistical process is involved. As was discussed in Section 5.2.4 the distribution of pulses is defined by the resolution R of the detector given by

$$R = \frac{K}{\sqrt{E}}, \qquad (6.5)$$

where E is the energy of the incident X-ray photons in keV; K has a value of about 35 in the case of a xenon proportional counter and about 128 for a scintillation counter. The resolution of the detector is the peak width at half-height of the pulse amplitude distribution, expressed as a percentage of the average voltage level at which the pulses occur. While in principle one could use the property of a proportional counter to monochromatize X-radiation, in practice the resolution of the detector is insufficient to be really useful. As will be seen in Section 6.6, below, one exception to this is the Si(Li) detector. Table 6.5 quantifies the absolute energy resolutions of typical

Table 6.5. Resolution of Various Dispersion Devices at the Energy of Cu $K\alpha$

Device	Resolution (eV)	Resolution (%)
Detector alone:		
Scintillation counter	3638	45.3
Gas proportional counter	1086	13.5
Si(Li) detector	160	2.0
Monochromator:		
Ge(220) crystal	20	0.25
Graphite(002)	500	6.2

detectors and compares them with the resolution of typical monochromator crystals. It will be seen that the germanium crystal, which is typically used as a primary-beam monochromator, has an absolute resolution about 25 times better than pyrolytic graphite, which is typically used as a diffracted-beam monochromator. Only the Si(Li) detector has a resolution in the same absolute energy range as the monochromator crystals.

Even though it cannot monochromatize Cu $K\alpha$, the proportional detector and pulse height selector combination proves invaluable for the removal of such effects as harmonics from the monochromator crystal, sample fluorescence, and background that may arise from shorter wavelengths of the X-ray tube continuum which pass the β-filter. The principle of the pulse height selector was discussed in Section 5.4. Since the inherent resolution of the detector–pulse height selector combination is poor (15–50%, depending upon the detector employed and the energy of the measured radiation), it is never possible to use this combination on its own to completely separate the $K\alpha$ from the $K\beta$. However, a combination of β-filter plus pulse height selector gives radiation that, for most purposes, is sufficiently β free.

6.6. USE OF SOLID STATE DETECTORS

The resolution of a detector is inversely proportional to the number of initial ionizations produced within the detector. Solid state detectors such as the Si(Li) detector have a very low effective ionization energy per primary event and thus have a very good energy resolution. For example, the resolution of the Si(Li) detector is such that it can be used to separate Cu $K\alpha$ and Cu $K\beta$. An advantage, therefore, of using the Si(Li) detector is that one can easily select either of these wavelengths. This offers two major possibilities: first, one does not need a β-filter or crystal monochromator to select the $K\alpha$ wavelengths;

second, where advantageous, one can record the diffractogram with the $K\beta$ rather than the $K\alpha$ radiation.

One of the most successful recent applications of the Si(Li) detector has been as a replacement for the conventional scintillation detector in powder diffraction instrumentation. While the advantages of wavelength selectability and high quantum-counting efficiency are somewhat counteracted by large dead-time problems, it is clear that the Si(Li) detector has an important role to play in this area. Bish and Chipera [3] have summarized the advantages and disadvantages of such a replacement and have demonstrated the use of the wavelength selectability of the Si(Li) detector by comparing scans over the α-quartz quintuplet using the α_1, α_2 doublet and showing the same d-spacing region using the β_1, β_3 doublet. Use of β radiation can greatly simplify a diffraction pattern and, even though intensities are lower by about a factor of 6, the intensity loss can sometimes be justified. Because of the excellent energy resolution offered by the Si(Li) detector, backgrounds are typically lower than in conventional systems, leading to improved signal-to-noise ratios and improved detection limits.

6.7. USE OF MONOCHROMATORS

Each component wavelength of a polychromatic beam of radiation falling onto a single crystal will be diffracted at a discrete angle, in accordance with the Bragg law. This, in fact, forms the basis of the X-ray spectrometer used for elemental analysis (e.g., see Jenkins and de Vries [4]). The crystal monochromator used in X-ray powder diffractometry can be considered as a simple spectrometer. There are many physical forms of the crystal monochromator [5]. Figure 6.6 shows three common monochromator configurations that are used in modern powder diffractometers [6]. The upper configuration is the diffracted-beam monochromator in the parallel position. The center configuration is also a diffracted-beam monochromator but now in the antiparallel position. The lower configuration is the primary-beam monochromator. The diffracted-beam position is by far the most common, and the antiparallel position offers the major advantage of a lower torque on the detector arm [7]. In order that the diverging beam of diffracted radiation coming through the receiving slit should converge back through the monochromator slit, the crystal is bent. The bending may either be to the radius of the monochromator circle (the *Johann condition* [8]), or bent and ground to half of the radius of the monochromator circle (the *Johansson condition* [9]). In a parallel-beam arrangement the monochromator is flat. The monochromator is placed between the specimen and the detector at a θ such as to diffract the desired wavelength into the detector. Careful selection of crystal and a d_{hkl} based on its

(a) Diffracted-beam, parallel

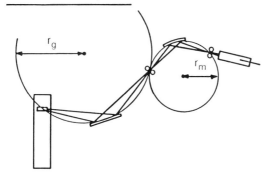

(b) Diffracted-beam, antiparallel

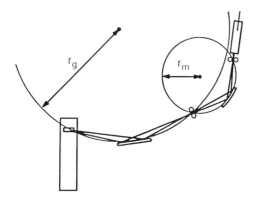

(c) Primary beam

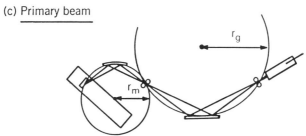

Figure 6.6. Monochromator configurations: in each case r_q is the radius of the goniometer circle and r_m is the radius of the monochromator circle.

orientation will eliminate all wavelengths except the $K\alpha_1$, but generally, for reasons of minimizing intensity loss, the $K\alpha_1$, $K\alpha_2$ doublet is accepted. Indeed, the obtaining of pure monochromatic radiation can only be achieved with considerable intensity loss. For example, the ability of the monochromator to separate the α_1, α_2 doublet can be defined in terms of the angular separation given by Equation 6.3. From that equation it is readily seen that the resolving power of a diffractometer is

$$\frac{\lambda}{\Delta\lambda} = \frac{\tan\theta}{\Delta\theta}. \tag{6.6}$$

where $\Delta\theta$ is the line breadth at half-maximum intensity.

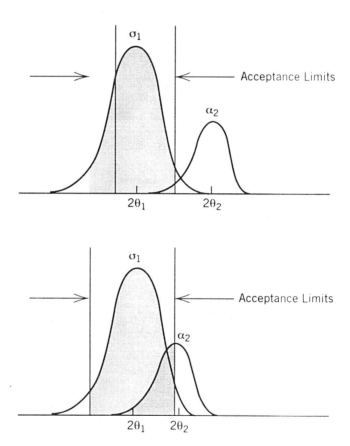

Figure 6.7. Effect of angular dispersion on the monochromator acceptance window. Reprinted from R. Jenkins, and J. L. de Vries, *An Introduction to Powder Diffractometry*, p. 17, Fig. 17. Copyright © 1977, N. V. Philips, Eindhoven, The Netherlands.

Table 6.6. Advantages and Disadvantages of Diffracted-Beam and Primary-Beam Monochromators

Monochromator	Advantages	Disadvantages
Diffracted-beam	Low intensity loss Removes fluorescence	Does not separate $K\alpha_1, K\alpha_2$ High torque on diffractometer
Primary-beam	Low background No torque problems	Sample fluorescence High intensity loss Alignment critical

The effect of the resolution of the crystal on the $K\alpha_1$, $K\alpha_2$ intensity ratio for a given slit system (i.e., given acceptance limits) is shown in Figure 6.7: the upper part shows a good value of $d\theta/d\lambda$ and almost complete separation of α_1, α_2 is possible; the lower part shows a poor value of $d\theta/d\lambda$ and the resulting contamination of the α_1 with α_2. In practice, good reflecting crystals (i.e., crystals giving high diffracted intensities) have a relatively large natural line breadth, giving poor resolution. Although this has little consequence in X-ray powder diffractometry (even when coupled with the poor inherent focusing of the Johann system), in cases where pure α_1 radiation may be required, alternative systems have to be employed. This, for example, might be the case in single-crystal diffractometry or in focusing cameras.

Even under the relatively poor resolution conditions employed with the diffracted-beam monochromator, overall intensity losses have in the past been quite high, and this probably accounts for the reluctance of early workers to employ them for routine use. However, the development in the late 1960s of high-efficiency crystals such as pyrolytic graphite [10] meant that net peak intensities obtained with and without monochromators are quite comparable, with the added advantage in the case of the monochromator of considerable background reduction (Table 6.6).

6.7.1. The Diffracted-Beam Monochromator

The diffracted-beam monochromator consists of a single crystal mounted behind the receiving slit, with a detector set at the correct angle to collect the wavelength of interest diffracted by the monochromator crystal (see Figure 6.8). The surface of the crystal, the goniometer receiving slit, and the detector slit all lie on the focusing circle of the monochromator. Soller slits may be placed somewhere near the position of the goniometer receiving slit. The monochromator is aligned by rocking the crystal about its center until only the required wavelength(s) can pass the detector slit. Figure 6.9 shows the

168   PRODUCTION OF MONOCHROMATIC RADIATION

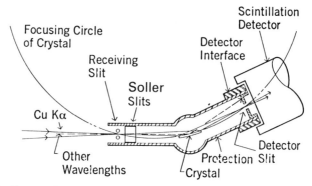

Figure 6.8. The crystal monochromator. Reprinted from Jenkins [7, p. 25], by permission.

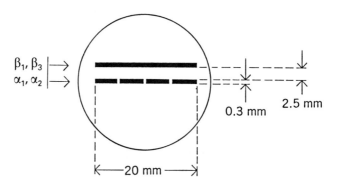

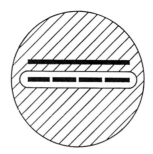

Figure 6.9. Intensity distribution at the detection window and passed by the detector slit. Reprinted from Jenkins [7, p. 28], by permission.

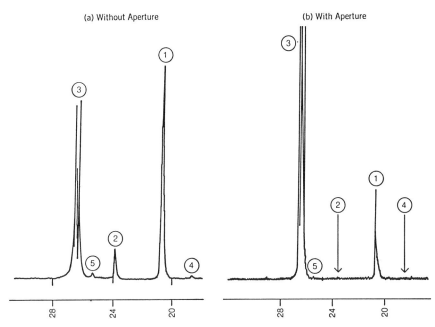

Figure 6.10. Effect of incorrect detector aperture setting on monochromatization. Reprinted from Jenkins [7, p. 28], by permission.

intensity distribution across the window of the detector in a typical diffractometer setup. The beam width is about 20 mm across, and the thickness of the beam about 0.3 mm. If the monochromator is correctly aligned, the α-doublet is exactly in the center of the window; as shown on the right-hand side of the figure, the detector slit is so positioned as to pass only the α radiation. With a 10 cm radius graphite monochromator crystal, the β radiation would lie about 2.5 mm to the low-angle side of the α-doublet. Thus, when used with a suitable slit or aperture, the monochromator has a sufficiently narrow bandpass to allow rejection of $K\beta$ radiation. Typical monochromator alignment problems occur due to incorrect superposition of slit and α radiation image. For example, if the slit is slightly rotated or the detector is slightly cocked, part of the β radiation image is also seen by the detector slit. The effect is to allow β radiation to be passed to the detector. As an example, Figure 6.10 shows a portion of the diffractogram of α-quartz measured with Cu $K\alpha$ radiation. Line 3 is the (110) reflection, and line 1 the (100) reflection. Without the detector aperture in place, the β radiation also reaches the detector. Line 2 in Figure 6.10 is the (110) $K\beta$ line, and line 4 is the (100) $K\beta$ line. Line 5 is the W $L\alpha$ reflection of the (110). As shown on the right-hand side of the figure, with the aperture correctly in place, all of the additional lines are removed.

6.7.2. The Primary-Beam Monochromator

While the primary-beam monochromator has been used for many years in camera applications such as the Guinier camera, until the late 1980s its use in powder diffractometers had been rather limited. The primary-beam monochromator has one major advantage over the diffracted-beam monochromator, this being its ability to almost completely isolate the $K\alpha_1$ line from the $K\alpha_2$. The use of pure monochromatic radiation yields diffractograms that are much *cleaner* than their bichromatic counterparts. In addition to its use in standard powder diffractometers, the primary-beam monochromator has found application with the position-sensitive diffractometer [11]. The alignment of the primary-beam monochromator is particularly critical. As an example, a small displacement in the position of the X-ray tube filament can cause a major misalignment and loss of intensity from the monochromator. Displacement of the filament can occur as the tube loading is varied. For this reason, it is important that the final monochromator adjustment be made at full X-ray tube loading.

6.8. COMPARISON OF MONOCHROMATIZATION METHODS

The β-filter is a single-band bandpass device that is mainly used to improve the ratio of Cu $K\alpha$ to Cu $K\beta$ to about 50:1. It is generally supplemented by pulse height energy discrimination to remove crystal harmonics and high-energy white radiation coming from the X-ray tube. The effectiveness of this white radiation removal depends upon the resolution of the detector. For Cu $K\alpha$, the resolution of the scintillation detector is about 3600 eV, and such a resolution would not allow the removal of radiation in the immediate vicinity of the Cu K lines. The Si(Li) detector is now being used as a replacement for the conventional scintillation detector in some powder diffraction instruments. While the advantages of wavelength selectability and high quantum-counting efficiency are somewhat counteracted by large dead-time problems, it is clear that the Si(Li) detector has an important role to play, especially for the detection/measurement of very weak peaks However, of the three monochromatization devices, the diffracted-beam monochromator is the most commonly employed in automated powder diffractometers, even though the primary-beam monochromator has by far the best resolving power.

Figure 6.11 shows the spectral distribution in the energy region around copper K radiation. The Cu $K\alpha$, $K\beta$ doublets are shown, along with other lines that could fall within the acceptance range of the monochromatization/detection device. Figure 6.11 also shows the relative bandpass regions for three of the monochromatization systems discussed. Because intensity is at

COMPARISON OF MONOCHROMATIZATION METHODS

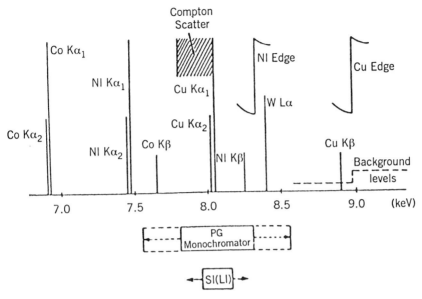

Figure 6.11. Relative bandpass regions of various monochromatization devices. From R. Jenkins, Instrumentation. In *Modern Powder Diffraction* (D. L. Bish and J. E. Post, eds.), p. 37, Fig. 14. Mineralogical Society of America, Washington, DC, 1989. Reprinted by permission.

a premium in most diffraction experiments, most diffracted-beam monochromators employ a pyrolytic graphite (PG) crystal, which has high reflectivity but poor dispersion relative to a "perfect" single crystal. The angular dispersion of the pyrolytic graphite crystal is less than 100 eV, but because its mosaic structure allows diffraction over a wide angular range, it has an effective bandpass of around 500–1000 eV, which, if correctly employed, is still sufficient to remove the $K\beta$ radiation. The actual bandpass value for a given monochromator will depend upon the widths and positions of limiting apertures that are placed somewhere on the focusing circle of the monochromator. It should also be appreciated that a slight mis-setting of the monochromator can displace its energy acceptance window to either lower or higher energy. From the foregoing, it is clear that even with monochromatizing devices correctly set up, it is possible that undesirable radiation can pass through the monochromator to the counting circuits. When the monochromatizing device is incorrectly set up, even more radiation may be passed.

Finally, it should be noted that computer searching for peaks is invariably more sensitive than manual techniques. The result may be that a $K\alpha_1$, $K\alpha_2$ doublet is recognized at a lower angle when computer methods are employed. This, in turn, can lead to problems during phase identification because all of the observed lines may not match the correct standard pattern selection. This problem can be made even worse when $K\alpha_2$ stripping or profile-fitting techniques are employed in the reduction of the experimental pattern and have not been employed in the measurement of the reference pattern(s).

REFERENCES

1. Parrish, W., and Roberts, B. W. Filter and crystal monochromator techniques. In *International Tables for X-Ray Crystallography*, Vol. III, pp. 73–88. Kynoch Press, Birmingham, England, 1962.

2. Parrish, W. *X-Ray Analysis Papers*, pp. 33–34. Centrex, Eindhoven, The Netherlands, 1965.

3. Bish, D. L., and Chipera, S. J. Comparison of solid state silicon detector to a conventional scintillation detector-monochromator system in X-ray powder diffraction analysis. *Powder Diffr.* **4**, 137–143 (1989).

4. Jenkins R., and de Vries, J. L. *Practical X-Ray Spectrometry*. Macmillan, London, 1967.

5. *International Tables for X-Ray Crystallography*, Vol. III, p. 73. Kynoch Press, Birmingham, England, 1962.

6. Klug, H. P., and Alexander, L. E. *X-Ray Diffraction Procedures*, 2nd ed., pp. 354–358. Wiley (Interscience), New York, 1974.

7. Jenkins, R. A new fixed-wavelength monochromator. *Norelco Rep.* **25**, 25–28 (1978).

8. Johan, H. H. Intense X-ray spectra obtained with concave crystals. *Z. Phys.* **69**, 185–206 (1931).

9. Johannson, T. New focusing X-ray spectrometer. *Z. Phys.* **82**, 507–528 (1933).

10. Gould, R. W., Bates, S. R., and Sparkes, C. J. Application of the graphite monochromator to light element X-ray spectroscopy. *Appl. Spectrosc.* **22**, 549–551 (1968).

11. Göbel, H. A new method for fast XRPD using a position-sensitive detector. *Adv. X-Ray Anal.* **22**, 255–265 (1979).

CHAPTER
7
INSTRUMENTS FOR THE MEASUREMENT OF POWDER PATTERNS

7.1. CAMERA METHODS

Historically, powder diffraction analysis began with the development of simple film cameras where the powder sample, in the form of a small diameter (< 1 mm) cylinder, was mounted at the center of a cylindrically shaped film (with the film diameter usually in the range of 50–150 mm). Powder cameras were first developed in the early 1920s, and—even though most powder diffraction work is carried out today with the powder diffractometer—there are several important film methods that are still popular because they offer valuable alternatives or complements to the diffractometer. Many factors can contribute to the decision to select a camera technique rather than the powder diffractometer for obtaining experimental data, including lack of adequate sample quantity, the desire for maximum resolution, and, possibly most important, the lower initial cost. Of the various camera types available, there are three commonly employed instruments: the Debye–Scherrer camera, the Gandolfi camera, and the Guinier focusing camera (see Figure 7.1). The Debye–Scherrer camera is the simplest of the powder cameras to use and probably in excess of 20,000 of these cameras have been supplied worldwide. This number should be compared with about 15,000 powder diffractometers and about 2000 Guinier cameras. It is useful to compare the different types of camera in terms of their operational differences, sample requirements, inherent resolution, and specimen limitations.

7.1.1. The Debye–Scherrer/Hull Method

The first X-ray diffraction camera was devised independently and almost simultaneously by Debye and Scherrer [1] in Switzerland in 1916 and Hull [2] in the United States. Figure 7.2 compares the optical arrangements of the *Debye–Scherrer camera* with the parafocusing diffractometer. The Debye–Scherrer camera uses a pinhole collimator (PC), which allows radiation from the source (S) to fall onto the rod-shaped specimen (SP). A secondary collimator (SC), or beam stop, is placed behind the sample to reduce secondary scatter of the beam by the body of the camera. The film (FM) is placed around

174 INSTRUMENTS FOR POWDER PATTERN MEASUREMENT

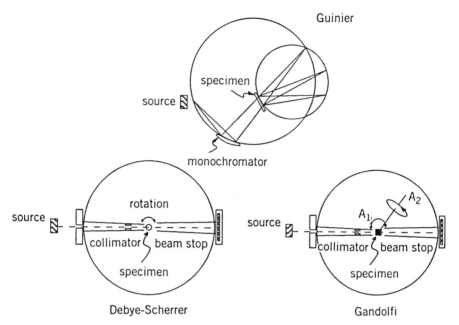

Figure 7.1. Cameras for X-ray powder diffraction.

the inside circumference of the camera body. A β-filter (FR) is typically placed between the source and the specimen to partially monochromatize the incident beam. Figure 7.3a shows an example of a Debye–Scherrer film. The powder diffractometer (see Figure 7.2) uses a line source (S) of radiation that diverges onto the flat surface of the specimen (SP) via a divergence slit (DS). A receiving slit (RS) is placed at the same distance from the center of the specimen as the source (F is the focusing circle). This distance then becomes the radius of the goniometer circle (G). A detector (D)—and often a monochromator—is placed behind the receiving slit to record the diffracted X-ray photons. In the Debye–Scherrer powder camera, the whole diffraction pattern is recorded simultaneously. In the powder diffractometer, the diffraction pattern is recorded sequentially by scanning the receiving slit through an angle 2θ while rotating the specimen through an angle θ. As indicated in Figure 7.2, the lines on the developed film from the camera correspond to the lines recorded on the powder diffractometer.

7.1.2. The Gandolfi Camera

Although the Debye–Scherrer camera is able to handle very small specimens, this fact itself can give rise to problems when the specimen is so small that

CAMERA METHODS

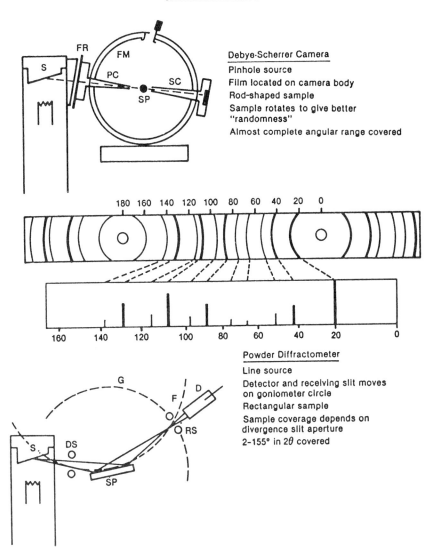

Figure 7.2. Measurement of powder patterns with the powder diffractometer and the Debye–Scherrer camera. From R. Jenkins, *X-Ray Fluorescence Spectrometry*, p. 39, Fig. 3.1. Copyright © 1988, John Wiley & Sons, Inc. Reprinted by permission of the publisher.

insufficient particles are present to give a random distribution. In some cases, one might even be asked to obtain a powder pattern from a small single crystal. The *Gandolfi camera* [3] (see Figure 7.1) offers a good alternative to the Debye–Scherrer camera in these instances, since this camera offers the additional feature of a second usable sample mount that is inclined at 45° to the

(a)

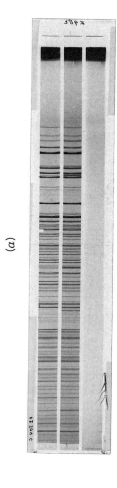

(b)

Figure 7.3. Examples of films taken with (a) the Debye–Scherrer camera and (b) the Guinier camera.

normal. sample rotation axis (A_1). During an exposure not only does the sample rotate about this 45°-inclined axis (A_2), but the entire sample mount rotates about the normal sample axis. This provides additional randomization when a powder pattern is obtained from a very small number of particles or from a single crystal, and adequate patterns can be obtained from single crystals or powdered samples down to 30 μm in size. Either horizontal or oblique sample mounts can be used with polycrystalline samples to give results identical to those obtained with the conventional Debye–Scherrer camera.

7.1.3. The Guinier Camera

The basic drawback to both Debye–Scherrer and Gandolfi cameras is their lack of resolution. Where moderate-to-high resolution is required it is general practice to use a focusing camera, of which the Guinier camera is the most common. Figure 7.3 shows examples of films taken with the Debye–Scherrer camera (a) and the Guinier camera (b). In addition to the difference in the sharpness of the lines, it will also be seen that the Guinier film shows *no* indication of $K\alpha_2$. The sources of error in the Guinier camera are not unlike those found with the Debye–Scherrer camera, of which camera radius, eccentricity, X-ray beam divergence, and film shrinkage are the most important factors. Additional problems arising from specimen displacement and transparency are best compensated by use of an internal standard. Where such techniques are employed, extremely accurate data can be obtained. The d-values obtained from diffractometers and Guinier cameras are usually of high quality, but in addition one can also generally assume that, even if an internal standard were not used, diffractometer and Guinier data are of higher quality than Debye–Scherrer data.

The Guinier geometry makes use of an incident-beam monochromator, which is simply a large single crystal cut and oriented to diffract the $K\alpha_1$ component of the incident radiation. In addition to diffracting a very narrow wavelength band, the monochromator curvature is designed to use the whole surface of the crystal to diffract simultaneously, thus yielding a large diffracted intensity. As a consequence of this curvature and the orientation of the crystal itself, the monochromator not only easily separates $K\alpha_1$ from $K\alpha_2$ but converts the divergent incident beam into an intense convergent diffracted beam focused onto a sharp line. As monochromators are on the order of 1 cm wide, they are designed to use the line focus of a fine-focus X-ray tube in order to gain the most intensity. In order to gain maximum advantage from this high inherent resolution, it is usual to employ single-emulsion film. Under optimum conditions, lines as close as 0.01 mm can be resolved. For a camera of 200 mm diameter, this corresponds to an ability to resolve lines from well-crystallized samples that are separated by as little as 0.005° 2θ.

7.2. THE POWDER DIFFRACTOMETER

The instrumentation required for X-ray powder diffractometry consists of three basic parts:

1. A source of radiation, consisting of an X-ray tube and a high-voltage generator
2. The detector and counting equipment
3. The diffractometer

Early designs of the powder diffractometers required relatively long data collection times and gave rather poor resolution. A number of ideas started to develop during the early 1940s as to how best to produce focusing and semi-focusing arrangements. Of the various geometric arrangements developed, two parafocusing geometries have become popular: the *Bragg–Brentano* system and the *Seemann–Bohlin* system. The majority of commercially available powder diffractometers employ the Bragg–Brentano arrangement since this is generally a reasonable compromise between mechanical simplicity and performance. Focusing geometries were first employed in powder cameras, and the original idea was proposed independently by both Seemann [4] and Bohlin [5], even though the focusing arrangement proposed was not truly focusing because of the finite width of the source and specimen. Thus, the word *parafocusing* was coined to describe the geometric arrangement, and Seemann–Bohlin parafocusing cameras have been available for many years. While the first powder diffractometers were nonfocusing [6], parafocusing optics were soon employed [7] to improve performance, in general, and resolution, in particular. There are five basic movements in any powder diffractometer geometry:

1. Angular motion of the X-ray tube
2. Angular motion of the specimen
3. Angular motion of the receiving slit
4. Linear motion of the tube to specimen dimension
5. Linear motion of the receiving slit to specimen dimension

Figure 7.4 shows the basic movements of these goniometer parameters to form the common arrangements. In the Bragg–Brentano arrangement, (e.g., see Parrish et al. [8]), either the tube is fixed and the specimen and receiving slit (RS) vary in the ratio of θ_1 to $2\theta_2$ ($\theta:2\theta$) or the specimen is fixed and the tube and receiving slit each vary as θ ($\theta:\theta$). In each case the tube–specimen distance (r_1) and the specimen receiving slit distance (r_2) are fixed and equal to each

THE POWDER DIFFRACTOMETER

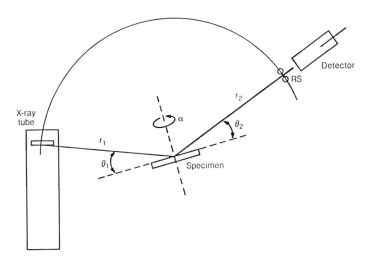

Type	Tube	Specimen	Receiving Slit	r_1	r_2
Bragg-Brentano $\theta:2\theta$	Fixed	Varies as θ*	Varies as 2θ	Fixed	$=r_1$
Bragg-Brentano $\theta:\theta$	Varies as θ	Fixed *	Varies as θ	Fixed	$=r_1$
Seeman-Bohlin	Fixed	Fixed *	Varies as 2θ	Fixed	Variable
Texture Sensitive (Ladell)	Fixed	Varies as θ precesses about α	Varies as 2θ	Fixed	Variable

*Generally fixed, but can rotate about α or rock about goniometer axis.

Figure 7.4. Common mechanical movements in powder diffractometers.

other. While the $\theta:2\theta$ system is by far the most commonly used (due mainly to mechanical simplicity), the $\theta:\theta$ is especially useful in cases where there is an advantage in holding the position of the specimen fixed, e.g., in cases where special attachments are used to vary specimen temperature, environment, etc. In the Seemann–Bohlin [9] arrangement, θ_1 is fixed and θ_2 varies as the distance (r_1) varies. The fourth arrangement noted in Figure 7.4 is a rather special arrangement proposed by J. Ladell (see Greenberg [10]) called the *texture-sensitive geometry*. This geometry is similar to the Seemann–Bohlin system except that the specimen is precessed about α as it moves through θ. This system is somewhat like the *Schultz pole figure geometry* [11] and offers the advantage that it collects diffracted photons from essentially all crystallites—independent of their orientation. Even though most commercial diffractometers still employ the Bragg–Brentano geometry as it was developed in the early 1940s, many other specialized designs have evolved from the original concept. One of the more important examples of such new designs is the *parallel beam diffractometer* [12, 13].

7.3. THE SEEMANN–BOHLIN DIFFRACTOMETER

Figure 7.5 shows the basic arrangement of the Seemann–Bohlin parafocusing geometry as typically used both in cameras and in diffractometers. In the camera system (Figure 7.5a), radiation from a point source S diverges onto the curved surface of the specimen SP. The diffracted radiation from each set of (hkl) planes converges onto the focusing circle at T_1, T_2, T_3, etc. The distances from specimen to film are D_1, D_2, and D_3, respectively. A flat specimen is typically employed in the diffractometer system (Figure 7.5b) because of the difficulty of maintaining a curved sample such that it is co-concentric with the focusing circle whose radius changes with 2θ. Other than that, the optical arrangement is essentially the same as in the camera. There are a number of different configurations used in the Seemann–Bohlin diffractometer geometry, including the incident- or diffracted-beam monochromator, and with the specimen stationary, in reflection, or in transmission. However, it is common practice to fix the angle of the specimen, giving a constant irradiation area. Use of a small θ value is especially useful for the measurement of thin-film specimens [14].

The advantages offered by the Seemann–Bohlin geometry include higher absolute intensities without loss of resolution, and the ability to work with a fixed specimen. The major disadvantage in the use of a powder diffractometer with the Seemann–Bohlin geometry is the difficulty in mounting the specimen such that it is co-concentric with the focusing circle. Also, some Seemann–Bohlin diffractometers are limited so as not to be able to access the low 2θ region. Additional mechanical problems may arise because of the large and variable distance from specimen to detector, especially where a diffracted-beam monochromator is employed. Use of an incident-beam monochromator makes it possible to move the specimen closer to the source slit, but this configuration is not effective in removing specimen fluorescence, and backgrounds may be high. The correction factors employed in Seemann–Bohlin diffractometers are rather more complex than in the simpler Bragg–Brentano arrangement [15].

7.4. THE BRAGG–BRENTANO DIFFRACTOMETER

Although there are many manufacturers of powder diffractometers, the majority of commercial systems employ the Bragg–Brentano parafocusing geometry. A given instrument may provide a horizontal or vertical $\theta:2\theta$ con-

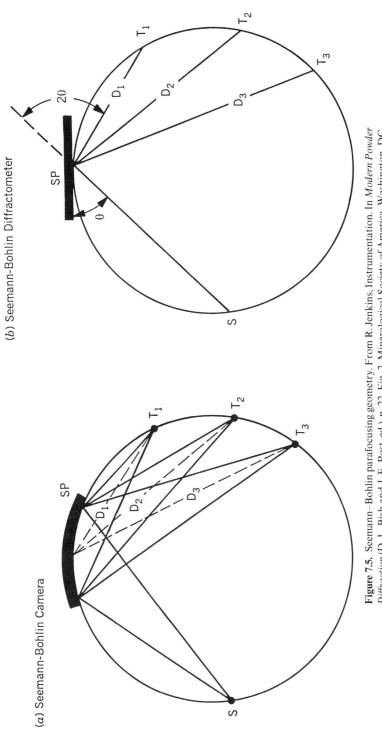

(a) Seemann–Bohlin Camera

(b) Seemann–Bohlin Diffractometer

Figure 7.5. Seemann–Bohlin parafocusing geometry. From R. Jenkins, Instrumentation. In *Modern Powder Diffraction* (D. L. Bish and J. E. Post, ed.), p. 22, Fig. 2. Mineralogical Society of America, Washington, DC, 1989. Reprinted by permission.

(a) Vertical θ:θ

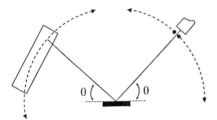

(b) Vertical θ:2θ

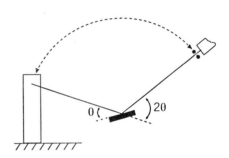

(c) Horizontal θ:2θ

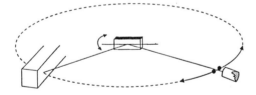

Figure 7.6. Various configurations of the Bragg–Brentano parafocusing diffractometer. From R. Jenkins, Instrumentation. In *Modern Powder Diffraction* (D. L. Bish and J. E. Post, eds.), p. 22, Fig. 3. Mineralogical Society of America, Washington, DC, 1989. Reprinted by permission.

figuration or a vertical $\theta:\theta$ configuration (Figure 7.6). The vertical $\theta:\theta$ and $\theta:2\theta$ systems are generally most advantageous for handling powder samples at room temperatures, but the horizontal system offers special advantages where a heavy sample or bulky accessories are placed at, or beyond, the receiving slit position. A two-dimensional view of the geometric arrangement is shown in Figure 7.7. A divergent beam of radiation coming from the line of focus (F) of

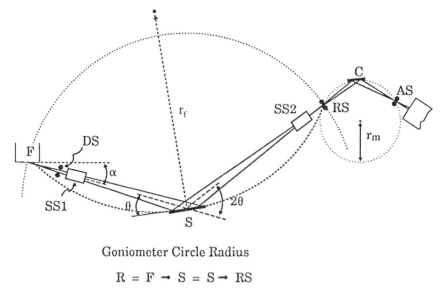

Goniometer Circle Radius

R = F → S = S → RS

Figure 7.7. Geometric arrangement of the Bragg–Brentano diffractometer.

the X-ray tube passes first through a divergence slit (DS), then through a parallel plate set collimator (Soller slits) (SS1), before striking the specimen (S) at an angle θ. The diffracted rays leave the specimen at an angle 2θ to the incident beam (and θ to the specimen surface), pass through a second parallel plate collimator (SS2), through the receiving slit (RS), to the detector. A diffracted-beam monochromator, consisting of a crystal (C) and a detector slit (AS), may be placed between the receiving slit and the detector. In order to establish the parafocusing condition, the axes of the line focus of the X-ray tube and of the receiving slit are at equal distances from the axis of the goniometer. The monochromator is generally a poorly focusing device in which the receiving slit, the crystal, and the detector slit all lie on the circumference of the monochromator circle of radius (r_m). The X-rays are collected by a suitable radiation detector, usually a scintillation counter or a sealed gas proportional counter.

The receiving slit assembly and the detector are coupled and move around a goniometer circle of radius R, centered about the specimen, in order to scan a range of 2θ (Bragg) angles. The source-to-specimen distance and the specimen-to-receiving-slit distance are both equal to R. For $\theta:2\theta$ scans, the goniometer rotates the specimen about the same axis as that of the detector, but at half the rotational speed, in a $\theta:2\theta$ motion. The surface of the specimen thus remains tangential to the focusing circle r_f. In addition to being a device that accurately sets the angles θ and 2θ, the goniometer also acts as a support for all of the various slits and other components that make up the diffractometer. The purpose of the parallel-plate collimators is to limit the axial divergence (i.e., divergence across the specimen along the diffractometer axis, to be discussed in Section 7.5.1) of the beam and hence partially control the shape of the diffracted line profile. It follows that the center of the specimen surface must be on the axis of the goniometer, and this axis must also be parallel to the axis of the line focus, divergence slit, and receiving slit.

Two circles (actually cylinders) are generated by the Bragg–Brentano arrangement, the goniometer circle and the focusing circle. A cross section of these circles is shown in Figure 7.8. The goniometer circle has a fixed radius R. The specimen lies at the center of this circle, while the source and receiving slits lie on its circumference. Radiation from the tube is directed by the divergence slit onto the surface of the specimen, at an average angle of θ. Radiation is diffracted by the specimen at an angle 2θ to the incident beam, and the receiving slit is placed at this same angle (the "Bragg angle") to collect the diffracted X-ray photons. The source F, the specimen S, and the receiving slit RS all lie on the circumference of the focusing circle of radius r_f. It will be apparent from Figure 7.8 that since the distances from source to specimen and from specimen to receiving slit are fixed, the radius r_f must vary with the diffraction angle. There is a simple relationship between the goniometer and focusing circles:

$$r_f = \frac{R}{2 \sin \theta}. \tag{7.1}$$

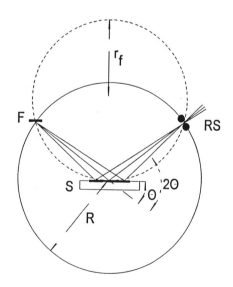

Figure 7.8. The focusing and goniometer circles.

Some of the mechanical requirements of the parafocusing geometry are fulfilled in the design and the construction of the goniometer, whereas the others are met during the alignment procedure, for example, the takeoff angle α of the beam, which is the angle between the anode surface and the center of the primary beam. Depending on the target of the X-ray tube and the voltage on the tube, the intensity increases and resolution decreases with increasing takeoff angle. For normal operation, the takeoff angle is typically set at 3–6°. In the alignment of the diffractometer great care must be taken in the setting of the mechanical zero of the goniometer and the $\theta:2\theta$ positions, since errors in these adjustments each may introduce errors in the observed 2θ values. Even with a properly aligned diffractometer, all measurements will be subject to certain errors (the more important of these will be discussed below), including the axial divergence error, the flat specimen error, the transparency error, and the specimen displacement error [8,16]. Other inherent errors may also be present but are generally insignificant for most routine work. These include refraction and effects of focal line and receiving slit width.

Most mechanically driven $\theta:2\theta$ goniometers employ gears with parallel axes (see Figure 7.9) [17]: these may be either (a) *spur* gears, (b) *helical* gears, or (c) *herringbone* gears. A spur gear is a disk of metal or hard plastic that rotates about the axes of the cylinder. Parallel teeth are cut around the circumference of the cylinder, each tooth in a common plane with the axis. This type of gear performs well at moderate speeds with medium pressure exerted upon the teeth, and it generally provides the simplest and most economical solution to the coupled $\theta:2\theta$ motion. A helical gear resembles a spur gear, except that the teeth are spiraled around the body rather than formed parallel to the axis of the gear body. Spiraling of the teeth provides a somewhat smoother operation allowing higher speeds to be obtained, with better life and the ability to carry greater loads. However, the diagonal teeth result in an axial component of force (end thrust) when power is transmitted through the gear set. The herringbone gear resembles two helical gears placed side by side, having opposite spiraling directions so that the teeth come together to form a chevron pattern. Spiraling of the teeth in both directions neutralizes end thrust and obviates the need for thrust bearings. A disadvantage of the herringbone gear system is that the gears must be aligned very carefully so that each half tooth takes an equal share of the load. Herringbone gears are significantly more expensive to make, and to our knowledge at least one major equipment manufacturer has changed from the originally specified herringbone gears to spur gears.

The *worm* gear (Figure 7.9d) is a special type of helical gear in which the teeth are curved to the edge to partly envelop the worm. Due to the wedgelike action of the worm thread on the gear teeth, worm gears give the quietest operation and provide the widest range of speed reductions. More recent

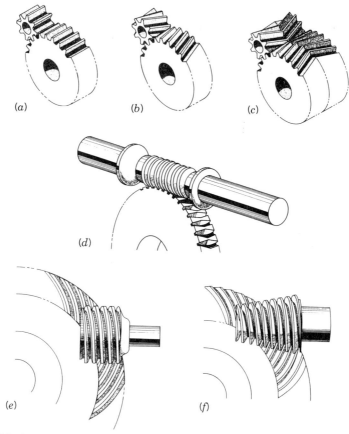

Figure 7.9. Gear systems employed in commercial diffractometers: (a) spur; (b) helical; (c) herringbone; (d) worm; (e) helicon; (f) spiroid.

innovations in the gear industry are the *helicon* (Figure 7.9e) and *spiroid* (Figure 7.9f) gears. These gears are extensions of the helical gear, but ones in which the worm meshes with the face of the gear, rather than its edge. The advantage of this type of design is that there is gear contact over the entire length of the pinion, yielding better overall accuracy. As was pointed out earlier, the objective of a goniometer is to accurately position the 2θ axis in order to measure a d-spacing via an angle. In order to improve the positioning capability of a goniometer, either the gearing and bearing precision can be increased or some type of readout may be used to measure the actual angular position. In principle, one could avoid the use of high-precision gears by employing accurate absolute angular encoders directly on the output θ and 2θ

shafts. In practice, high-accuracy encoders have been too costly to make them a viable alternative, and to date no commercially available diffractometer employs such encoders. However, even an encoder with an accuracy of only 18 arc seconds (0.005° 2θ) would offer potential improvements in angular positioning capability. The development of precision stepper motors in recent years has permitted the attachment of independent stepper motors to the θ and 2θ motions and elimination of the gearing.

7.5. SYSTEMATIC ABERRATIONS

Three inherent systematic errors, particularly noteworthy with the parafocusing geometry [18] are axial-divergence error, flat-specimen error, and specimen-transparency error. Briefly, axial divergence occurs because the X-ray beam diverges out of the plane of the focusing circle. Axial divergence is controlled by limiting divergence in the incident and diffracted beams by using collimators SS1 and SS2 (as shown in a three-dimensional view in Figure 7.10). Axial divergence introduces asymmetric broadening in the diffracted profile, particularly at low diffraction angles. The flat-specimen error occurs because the surface of the specimen is flat rather than conforming to the curvature of the focusing circle. The flat-specimen error also causes asymmetric broadening of the diffracted line profile. The transparency error occurs because the incident X-ray photons penetrate a significant depth into the specimen, rather than being diffracted purely from the specimen surface. Accordingly, the effective diffracting surface lies somewhat below the physical surface of the specimen and hence is below the focusing circle.

7.5.1. The Axial-Divergence Error

The axial-divergence error is due to divergence of the X-ray beam along the axis of the diffractometer in the plane of the specimen. This divergence occurs

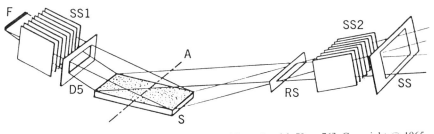

Figure 7.10. Bragg–Brentano geometry. Reprinted from Parrish [9, p. 76]. Copyright © 1965, Centrex, Eindhoven, The Netherlands.

because an extended X-ray source is used, typically a 12 mm long line focus from a sealed X-ray tube. Referring again to Figure 7.10, we can see that radiation from the line source F diverges through the divergence slit DS to the specimen S. The divergence slit does not, however, limit the divergence of the beam in the plane of the specimen, and so this axial divergence is controlled by placing a Soller slit, SS1, consisting of thin molybdenum foils of fixed length and fixed spacing, between the source and the divergence slit. A second collimator, SS2, is generally placed either between the specimen and the receiving slit RS or between the receiving slit and the detector (SS is the scatter slit). In addition to producing some asymmetric broadening of the diffraction profile in the low 2θ direction, axial divergence introduces a decreasing negative error in 2θ up to 90°, then an increasingly positive error beyond 90°. The form of the axial-divergence error for a single set of Soller slits [19] is as follows:

$$\Delta 2\theta = -h^2(K_1 \cot 2\theta + K_2 \operatorname{cosec} 2\theta)/3R^2, \qquad (7.2)$$

in which h is the axial width of the specimen, and R the radius of the goniometer circle; K_1 and K_2 are both constants determined by the specific collimator characteristics. As an example [20], for two 2 cm long, 0.5 mm spacing collimators and a 173 mm radius goniometer circle, Equation 7.2 reduces to

$$\Delta 2\theta = -0.01125 \cot 2\theta + 0.00188 \operatorname{cosec} 2\theta. \qquad (7.3)$$

As shown in Figure 7.11, the divergence allowed by a Soller collimator will depend upon its spacing s and its length L. Each slice of the incident beam emerging between adjacent plates of the Soller slit (SS1 in Figure 7.10) acts like the beam in a Debye–Scherrer camera, producing its own cone of diffraction. Figure 7.12 shows one of these Debye cones, while Figure 7.13 shows the adjacent cones produced by the extended source at (a) low 2θ angle and (b) high 2θ angle. The increasing radius (f_n) of the Debye ring with diffraction angle causes a variation in the intensity across the receiving slit, leading to the angular error shown in Equations 7.2 and 7.3. It may be seen in Figure 7.13 that there is a shift in the centroid of the intensity distribution with diffraction angle. In addition to the shifted centroid, the effect of axial divergence causes the low-angle side of a diffraction profile to rise more slowly than the high-angle side, resulting in a significant profile asymmetry that is more pronounced at low diffraction angle. This is the principal cause of profile asymmetry in powder diffractometry.

Most Soller collimators typically subtend an angle of $\pm 2.5°$ and will remove much of the axial divergence. A diffracted-beam monochromator will

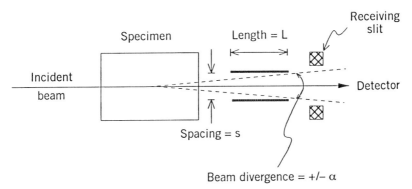

Figure 7.11. Axial divergence.

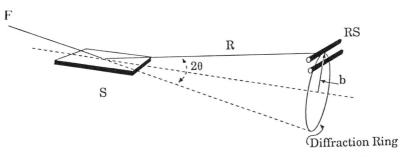

Figure 7.12. Intersection of the receiving slit RS with the diffraction cone. *Key*: F = line source; S = specimen; R = radius; b = radius of the diffraction ring.

also reduce the axial divergence. For this reason, some instrument configurations employing a diffracted-beam monochromator do not use a second Soller collimator. The effect of the axial divergence on the profile shape can only be minimized by decreasing the axial divergence allowed by the collimator—either by decreasing the blade spacing in the collimator block or increasing the collimator length. Either process will also lead to a lowering of the absolute count rate from a given specimen. The use of very narrow collimators also significantly reduces the intensity of radiation [21], so different manufacturers approach the problem in different ways. Most manufacturers place fixed Soller collimators between the tube and specimen and either in front of or behind the receiving slit [22]. The collimator divergence is then chosen to give an optimum intensity and line shape. Some manufacturers allow the user the choice of easily removing the second collimator (between specimen and receiving slit) if intensity is at a premium. Removing the

190   INSTRUMENTS FOR POWDER PATTERN MEASUREMENT

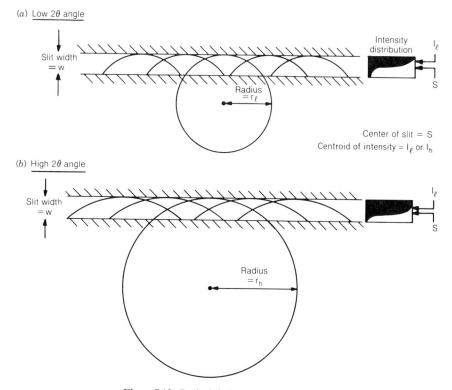

Figure 7.13. Peak shift due to axial divergence.

diffracted-beam Soller slits typically doubles the X-ray intensity, increases the background, and introduces additional asymmetry into the diffracted profile. A third alternative employed by at least one manufacturer is to use rather long collimators to reduce the axial divergence to around $\pm 1°$. This gives a very sharp, more symmetrical diffraction line but with considerable loss of intensity. This last approach is likely to become more popular as the intensity from X-ray tubes increases.

Although a number of effects combine to produce asymmetric diffraction profiles, with the effect increasing at low angles, axial divergence is by far the most important contributor. Figure 7.14 shows the low-angle diffraction pattern of silver behenate, which has been proposed as a low-angle calibration standard. The pronounced asymmetric distortion on the low-angle side of each profile requires that the intensity centroid be used to locate the diffraction center rather than the peak maximum.

SYSTEMATIC ABERRATIONS 191

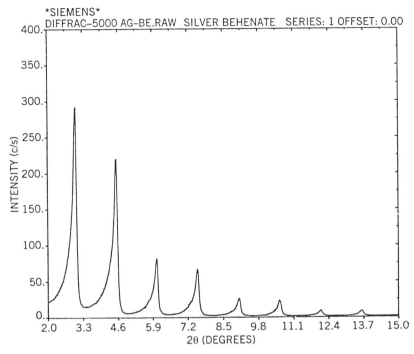

Figure 7.14. The typical asymmetric diffraction profiles due primarily to the axial divergence distortion in a sample of silver behenate.

7.5.2. The Flat-Specimen Error

Figure 7.15 illustrates the flat-specimen error and shows that the specimen surface S forms a tangent to the focusing circle whose radius is r_f, rather than being co-concentric with the goniometer focusing circle. The figure also shows that the extreme edges of the irradiated specimen surface lie on a different focusing circle of radius r'_f, which effectively lowers the expected angle of the diffracted beam, leading in turn to a negative systematic error in the observed 2θ maximum. The overall effect is that the peak intensity occurs at a point d_1, which is lower in angle than the position of the receiving slit RS (F is the line source). Because of the distortion of the average focusing circle, the flat-specimen error also causes asymmetric broadening of the diffracted line profile toward low 2θ angles. Thus, the magnitude of the flat-specimen error is dependent both on the aperture of the divergence slit (since this fixes the irradiation length of the specimen) and the diffraction angle. The form of the

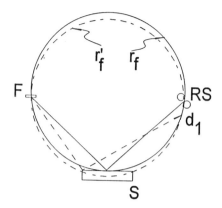

Figure 7.15. The flat specimen error.

flat-specimen error is as follows [23]:

$$\Delta 2\theta = -\frac{\alpha^2 \cot \theta}{343.8}, \qquad (7.4)$$

where α is the angular aperture of the divergence slit in degrees, and θ is expressed in radians. The factor 343.8 derives from a number of constants for a given diffractometer geometry [9]. These constants include the radius of the goniometer circle, collimator configuration, etc. Since the radius of the focusing circle decreases with increase in Bragg angle, the flat-specimen error increases with 2θ. Where fixed divergence slits are employed, the practical compromise is to use a slit which is wide enough to give reasonable intensities, while still keeping the flat specimen aberrations within acceptable limits. For most qualitative work, where scans are started at around $8°2\theta$, this generally means using a 0.5° divergence slit. Table 7.1 lists the angular aperture in degrees for

Table 7.1. **Maximum Irradiation Lengths (mm) with Various Divergence Slit Apertures and X-ray Wavelengths** [a]

Div Slit	$2\theta_{min}$	Mo $K\alpha$	Cu $K\alpha$	Cr $K\alpha$
0.25°	4.6°	8.86	19.24	28.58
0.50°	9.2°	4.45	5.61	14.35
1.00°	18.4°	2.22	4.83	7.18
2.00°	37.2°	1.11	2.42	3.59
4.00°	78.0°	0.56	1.22	1.81

[a] Specimen length is 20 mm.

full X-ray beam coverage of the specimen length, assuming a goniometer circle radius of about 17 cm. Since there are critical mechanical difficulties in actually making a slit with an aperture of much less than 0.25°, this is the smallest aperture fixed divergence slit commonly employed.

7.5.3. Error Due to Specimen Transparency

Figure 7.16 illustrates what happens when the effective diffracting surface lies below the focusing circle. In practice, there are two common effects that may cause such a circumstance to come about. The first of these is specimen transparency. The specimen-transparency error occurs because the incident X-ray photons penetrate to many atomic layers below the surface of the analyzed specimen. Accordingly, the average diffracting surface lies somewhat below the physical surface of the specimen. This error increases with decreasing absorption of the X-ray beam by the specimen, i.e., with decreasing linear attenuation coefficient. The form of the transparency error is

$$\Delta 2\theta = \frac{\sin 2\theta}{2\mu R}, \qquad (7.5)$$

in which μ is the linear attenuation coefficient of the specimen for the diffracted wavelength, and 2θ is expressed in radians. For organic materials and other

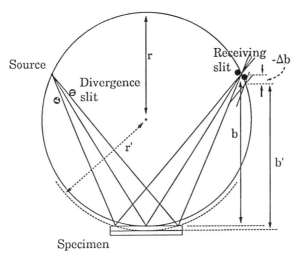

Figure 7.16. Specimen transparency and specimen displacement. *Key:* r = radius of focusing circle; r' = effective radius; b = true height of receiving slit above specimen surface; b' = effective height.

194 INSTRUMENTS FOR POWDER PATTERN MEASUREMENT

low-absorbing specimens where the linear attenuation coefficients are very small, the transparency error can lead to angular errors of as much as a tenth of a degree and extreme angular asymmetry of the profiles. For this reason, low-absorbing specimens are often deliberately prepared in the form of a thin film, by using a zero-background holder (Section 9.6.4), to reduce the penetration effect [24].

7.5.4. Error Due to Specimen Displacement

Referring back to Figure 7.16, we note that the second circumstance which may cause the effective specimen surface to lie below (or above) the focusing circle is the effect of specimen displacement. The specimen-displacement error arises due to practical difficulties in accurately placing the sample on the focusing circle of the goniometer. The form of the specimen-displacement error is as follows:

$$\Delta 2\theta = -\frac{2s\cos\theta}{R} \text{ (in radians),} \tag{7.6}$$

or

$$\Delta 2\theta = -\frac{114.59s\cos\theta}{R} \text{ (in degrees),} \tag{7.7}$$

where s is the displacement of the specimen from the focusing circle, and again θ is expressed in radians. Although the sample holder itself can be accurately referenced against a machined surface on the goniometer, it is difficult to pack the powdered sample into the holder such that its surface is level with the surface of the holder. The specimen-displacement error is typically the largest of the errors found in the recording of experimental data. The major effects of specimen displacement are to cause asymmetric broadening of the profile toward low 2θ values and to give an absolute shift in peak 2θ position, equal to about $0.01°$ 2θ for each 15 μm displacement.

Based on the foregoing discussion, one would intuitively expect some direct relationship between the specimen-displacement and transparency errors, and this is indeed the case. The mass absorption law described in Section 1.6 in terms of the linear attenuation coefficient μ is

$$I = I_0 \exp(-\mu x). \tag{7.8}$$

For 99% absorption (i.e., $I = 1$, and $I_0 = 100$), it follows that $\mu = 0.5x$. Since the penetration depth p is equal to $x\sin\theta$, we have

$$p = \frac{\sin\theta}{2\mu}. \tag{7.9}$$

Since the displacement s is equivalent to the penetration depth p, combination of Equations 7.7 and 7.9 gives

$$\Delta 2\theta = \frac{2\cos\theta \sin\theta}{R} \frac{}{2\mu}, \qquad (7.10)$$

or

$$\Delta 2\theta = \frac{\sin 2\theta}{2\mu R}, \qquad (7.11)$$

which is the form of the transparency error for a *thick* specimen given in Equation 7.5.

7.6. SELECTION OF GONIOMETER SLITS

7.6.1. Effect of Receiving Slit Width

Figure 7.17 illustrates the broadening of the line due to axial divergence and the role of the receiving slit. In this simplified schematic view, four points $a, b, c,$ and d on the specimen are point sources for diffracted photons. The cone of diffraction from each point source is allowed to diverge by an amount dependent upon the length and spacing of the Soller slits. Each point source is the center of the Debye ring, so at the receiving slit the four beams become four arcs. The receiving slit width is chosen to embrace all, or at least most, of these arcs. Figure 7.18 shows the effect of choosing different-size receiving slits on the peak shape and peak intensity, for a partially resolved doublet from α_1, α_2 radiation. If the beam width at the receiving slit were 0.1 mm, choice of a 0.1 mm receiving slit should allow virtually all of the diffracted photons to completely pass the slit. Use of a narrower slit, say of 0.05 mm, would have little effect on the resolution but would lower the peak intensity by about a factor of 2. Conversely, using a slit that is too wide would only marginally affect the intensity but would deteriorate the peak shape. In each instance, the reason is that the only *extra* photons allowed by the too-wide receiving slit are the *nonfocused* (cross-fired) photons.

It should be clear from the foregoing discussion that there is only one *optimum*-size receiving slit for any particular combination of X-ray tube focus, takeoff angle, and goniometer circle radius. In general, as the intensity increases, the resolution of the lines decreases, and vice versa (as illustrated in Figure 7.19). Here a series of measurements of peak intensity and width have been taken [9], using different receiving slit widths. Intensity data are shown both in terms of relative counts (I_{rel}) and normalized counts (I_{norm}). When the beam width is close to the size of the receiving slit (in this instance 0.15°),

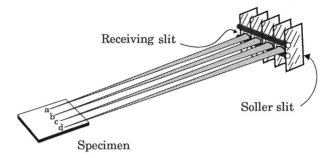

Figure 7.17. Influence of axial divergence on beam width.

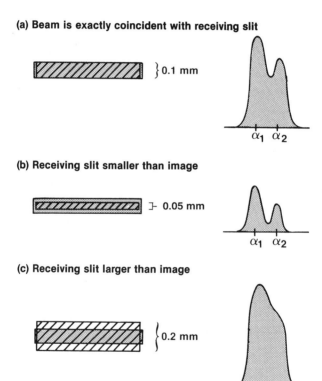

Figure 7.18. Selection of receiving slit width. Reprinted from R. Jenkins and J. L. de Vries, *An Introduction to Powder Diffractometry*, p. 22, Fig. 31. Copyright © 1977, N. V. Philips, Eindhoven, The Netherlands.

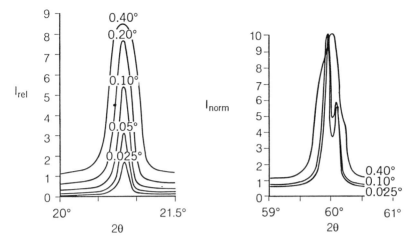

Figure 7.19. Effect of receiving slit on the profile shape. Reprinted from Parrish [9, p. 115]. Copyright © 1965, Centrex, Eindhoven, The Netherlands.

optimum intensity and resolution are obtained. When the receiving slit is smaller than the beam width, the intensity is reduced in rough proportion to the increase in the slit size, with marginal improvement in the resolution of the α_1, α_2 doublet. Thus, going from a 0.1° receiving slit to a 0.05° receiving slit reduces the intensity by about a factor of 2. When the receiving slit is larger than the beam width, only a marginal increase in intensity is seen and the resolution of the α_1, α_2 is very poor. The decrease in resolution arises from the inclusion of extra diffracted photons coming from angles slightly smaller or larger than the correct diffraction angles. One problem in using a fixed receiving slit to cover an angular range is that the radii of the Debye rings increase with increase in the Bragg angle, as shown in Figure 7.20. The overall effect is to give a varying illumination across the width of the slit.

7.6.2. Effect of the Divergence Slit

The function of the divergence slit is to limit the vertical divergence of the X-ray beam so that as much of the sample surface is irradiated as possible but, at the same time, irradiation of the sample support is avoided. Where a fixed divergence slit is employed, there is an associated fixed divergence of the X-ray beam. As the angle θ of the specimen is varied, the range of interception of the diverging beam by the specimen and its support also changes. Figure 7.21 shows the three conditions that generally predominate at low 2θ values [25]. Here is shown a specimen of length L_S mounted in a holder of length L_H. The

198   INSTRUMENTS FOR POWDER PATTERN MEASUREMENT

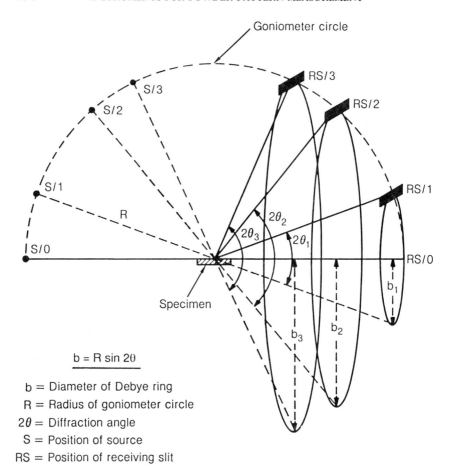

$b = R \sin 2\theta$

b = Diameter of Debye ring
R = Radius of goniometer circle
2θ = Diffraction angle
S = Position of source
RS = Position of receiving slit

Figure 7.20. Effect of diffraction angle on axial divergence.

beam from the source F passes through the divergence slit. The center line of the beam makes an angle θ with the sample surface and a length L_I of the specimen is irradiated. Three different values for the angle θ are illustrated. At a moderate value of θ, say 40°, only the specimen is irradiated. Increasing the value of L_I to a higher angle with the same divergence slit aperture means that less of the specimen surface is illuminated. Decreasing θ to a lower angle (i.e., 20° in Figure 7.21) allows the condition where both specimen and support are in the beam. This leads to scatter of the beam, with subsequent increase in the background intensity. As more of the sample support is irradiated, the degree of scatter and the intensity of the background increases. By the time θ de-

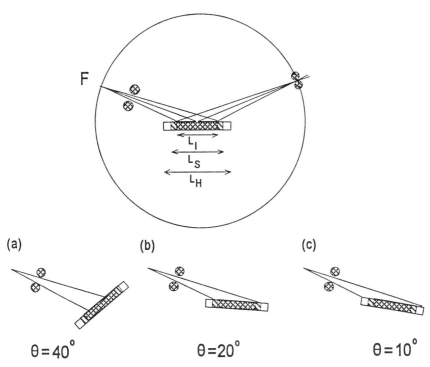

Figure 7.21. Effect of the angle of the beam on the irradiation surface. From R. Jenkins, Instrumentation. In *Modern Powder Diffraction* (D. L. Bish and J. E. Post, eds.), p. 30, Fig. 8. Mineralogical Society of America, Washington, DC, 1989. Reprinted by permission.

creases to 10°, both specimen and support are bathed in the X-ray beam. Decreasing the value of θ even further brings about a fourth case, not illustrated in Figure 7.21, where the length of the sample holder is not enough to intercept the whole beam. Under these conditions, as the angle decreases, less holder intercepts the beam and the amount of scatter decreases. One effect of a badly chosen set of conditions (too wide a divergence slit or too small a sample holder length), is that a small (scatter) peak may appear in the diffractogram at around 4–5° 2θ, which can easily be confused with a broad peak or perhaps a basal plane reflection for a clay at about 20 Å.

Because of the situation just described, the choice of the divergence slit is quite critical, especially at low 2θ values. The majority of commercially available powder diffractometers employ fixed divergence slits. Typically a number of slits are supplied with the instrument, and the operator must choose which one to use. Figure 7.22 shows plots of length of specimen irradiated as a function of diffraction angle for a typical X-ray powder

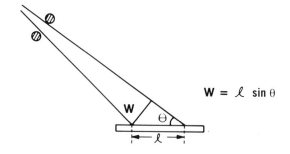

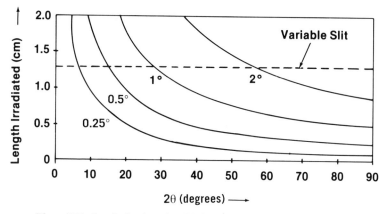

Figure 7.22. Irradiation lengths with fixed and variable divergence slits.

diffractometer. The upper portion of the figure shows a diverging beam incident upon the specimen surface. The irradiation length ℓ can be calculated from the maximum beam width W and the incident angle θ. The irradiation length of the specimen varies approximately with the sine of the angle θ. A 20 mm long specimen will be completely covered by the beam at 18° 2θ when a 1° divergence slit is employed. The angle of complete coverage would be about 9° for a 0.5° divergence slit and so on. In the case of the 1° divergence slit, not only is the specimen completely irradiated below 18° 2θ, but scatter increases from the specimen holder. This is not to say that one could not use a 1° divergence slit below 18°2θ, but simply that care must be exercised at these low angles because backgrounds will be high and *ghost peaks* may occur, as described above. Using the 1° divergence slit at angles higher than 18°2θ will cause less and less of the specimen area to be irradiated as 2θ is increased.

In an attempt to avoid high-angle intensity loss and to enable the operator to work at very low angles without having to put up with high background

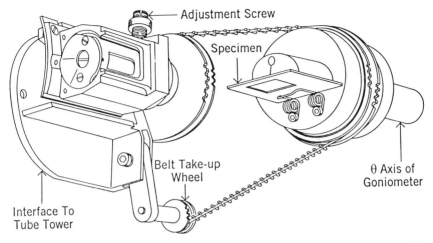

Figure 7.23. The original theta compensating divergence slit. Reprinted from Jenkins and Paolini [26, p. 14], by permission.

values, some diffractometers may be equipped with a variable divergence (θ-compensating) slit [26]. The variable divergence slit is typically a pair of molybdenum rods that are mounted on a drum driven by the θ-axis of the goniometer (see Figure 7.23). As the diffraction angle changes, so too does the angle of view subtended by the slit. This gives a constant specimen irradiation length of about 13 mm (see also Figure 7.22). The original variable divergence slit was belt driven by the θ-axis of the goniometer. More recent versions are independently driven by a stepper motor and thus may be used as either fixed or variable slits. The illumination length L from a variable divergence slit is given by

$$L = \frac{2R \sin\theta \tan(\delta/2)}{\sin^2\theta - \cos^2\theta \tan^2[\theta(\delta/2)]}, \quad (7.12)$$

where δ is the divergence angle, and R is the goniometer radius. When δ is small ($< 4°$), the second term in the denominator is insignificant and Equation 7.12 reduces to

$$L = \frac{2R \tan(\delta/2)}{\sin\theta}. \quad (7.13)$$

As an example, for a 17 cm radius goniometer, $L = 1.89$ for $2\theta = 18°$ and $\delta = 1°$. A well-aligned variable slit allows work at angles very close to the direct

X-ray beam—perhaps as low as 0.5° 2θ. A disadvantage of the variable slit system is that the intensities it gives do not agree with those obtained with a fixed divergence slit system (where the irradiated volume is constant) and need to be corrected by multiplying by the sine of the diffraction angle [27]. It follows from Equation 7.11 that the ratio of intensities from fixed (I_{fds}) and variable (I_{vds}) slits is

$$\frac{I_{\text{fds}}}{I_{\text{vds}}} = \frac{L_{\text{fds}}}{L_{\text{vds}}} = \frac{2R\tan(\delta/2)}{L_{\text{vds}}\sin\theta}.$$ (7.14)

REFERENCES

1. Debye, P., and Scherrer, P. Interference of X-rays—employing amorphous substances. *Phys. Z.* **17**, 277–283 (1916).
2. Hull, A. W. A new method of X-ray crystal analysis. *Phys. Rev.* **10**, 661–696 (1917).
3. Gandolfi, G. Discussion upon methods to obtain X-ray powder patterns from a single crystal. *Mineral. Petrog. Acta* **13**, 67–74 (1967).
4. Seemann, H. Focusing X-ray spectroscope. *Ann. Phys. (Leipzig)* [4] **59**, 455–464 (1919).
5. Bohlin, H. New method for X-ray crystallography of pulverized substances. *Ann. Phys. (Leipzig)* [4] **61**, 421–439 (1920).
6. Le Galley, D. P. Type of Geiger–Muller counter suitable for the measurement of diffracted Mo $K\alpha$ X-rays. *Rev. Sci. Instrum.* **6**, 279–283 (1935).
7. Parrish, W., and Gordon, S. G. Precise angular control of quartz-cutting by X-rays. *Am. Mineral.* **30**, 326–346 (1945).
8. Parrish, W., Hamacher, E. A., and Lowitzsch, K. The Norelco X-ray diffractometer. *Philips Tech. Rev.* **16**, 123–133 (1954).
9. Parrish, W. Advances in X-ray diffractometry of clay minerals. *X-Ray Analysis Papers* (W. Parrish, ed.), pp. 105–129. Centrex, Eindhoven, The Netherlands, 1965.
10. Greenberg, B. The Ladell diffractometer: Geometry and applications. *Adv. X-Ray Anal.* **36**, 631–640 (1993).
11. Schultz, L. G. A direct method of determining preferred orientation of a flat reflection sample using a Geiger counter X-ray spectrometer. *J. Appl. Phys.* **20**, 1030–1033 (1949).
12. Goehner, R. P., and Eatough, M. O. A study of grazing incidence configurations and their effect on X-ray diffraction data. *Powder Diffr.* **7**, 2–5 (1992).
13. Schuster, M., and Göbel, H. E. Parallel-beam coupling into channelcut monochromators using curved graded multilayers. *J. Phys. D* **28**, A270–A275 (1995).
14. Huang, T. C., and Parrish, W. Characterization of thin films by X-ray fluorescence and diffraction analysis. *Adv. X-Ray Anal.* **22**, 43–63 (1979).

15. Parrish, W., and Mack, M. Seemann–Bohlin X-ray diffractometry. I. Instrumentation. *Acta Crystallogr.* **23**, 687–692 (1967).
16. Wilson, A. J. C. *Mathematical Theory of X-Ray Powder Diffraction.* Philips Technical Library, Eindhoven, The Netherlands, 1963.
17. Jenkins, R., and Schreiner, W. N. Considerations in the design of goniometers for use in X-ray powder diffractometers. *Powder Diffr.* **1**, 305–319 (1986).
18. Parrish, W., and Wilson, A. J. C. Precision measurement of lattice parameters of polycrystalline specimens. In *International Tables for X-ray Crystallography* (J. S. Kasper and K. Lonsdole, eds.), Vol. II, pp. 216–234. Kynoch Press, Birmingham, England, 1959.
19. Klug, H. P., and Alexander, L. E. Diffractometric powder technique. In *X-Ray Diffraction Procedures*, 2nd ed., pp. 302–303. Wiley, New York, 1974.
20. Parrish, W., and Lowitzsch, K. Geometry, alignment and angular calibration of X-ray diffractometers. *Am. Mineral.* **44**, 765–787 (1965).
21. Stoecker, W. C., and Starbuck, J. W. Effect of Soller slits on X-ray intensity in a modern diffractometer. *Rev. Sci. Instrum.* **36**, 1593–1598 (1965).
22. Klug, H. P., and Alexander, L. E. Profiles and positions of diffraction maxima. In *X-ray Diffraction Procedures*, 2nd ed., Chapter 5. Wiley, New York, 1974.
23. Wilson, A. J. C. Geiger-counter X-ray spectrometer: Influence of size and absorption coefficient of specimen on position and shape of powder diffraction maxima. *J. Sci. Instrum.* **27**, 321–325 (1950).
24. Misture, S. T., Chatfield, L., and Snyder, R. L. Accurate powder diffraction patterns using zero background holders. *Powder Diffr.* **9**, 172–179 (1994).
25. Jenkins, R., and Squires, B. Problems in the measurement of large d-spacings with the parafocusing diffractometer. *Norelco Rep.* **29**, 20–25 (1982).
26. Jenkins, R., and Paolini, F. R. An automatic divergence slit for the powder diffractometer. *Norelco Rep.* **21**, 9–16 (1974).
27. Bowden, M. E., and Ryan, M. J. Comparison of intensities from a fixed and variable divergence X-ray diffractions experiments. *Powder Diffr.* **6**, 78–81 (1991).

CHAPTER
8
ALIGNMENT AND MAINTENANCE OF POWDER DIFFRACTOMETERS

8.1. PRINCIPLES OF ALIGNMENT

As described in the previous chapter, there are many sources of experimental error that may be encountered in the measurement of a typical X-ray powder pattern. Table 8.1 summarizes the more common types of systematic error, some of which are due to the basic instrument geometry, some due to misalignment, some due to specimen preparation and presentation, and some due to data processing. In order to obtain adequate experimental accuracy, it is important that the source and magnitude of these various errors be understood. The first step in the setting up and calibration of any powder diffractometer is the basic alignment of the diffractometer. In this chapter the general principles of alignment, and the errors which may accrue during an alignment procedure, will be described. Many newer diffractometers offer some measure of automated or semiautomated alignment. Since these procedures are typically instrument specific, the reader is referred to descriptions in the manufacturers' manuals.

The alignment of the diffractometer is a complex procedure since the specimen, detector, X-ray tube anode, and slits each have to be correctly set with respect to each other, in all three dimensions. As indicated in earlier chapters, there are a number of different types of powder diffractometer, and each of these requires a specific and detailed alignment procedure, often involving unique alignment tools. However, by far the most common of the diffractometer geometries is the Bragg–Brentano parafocusing system. Because this is so, it is possible to outline a *global* alignment scheme and use this to point out potential pitfalls and problem areas in the alignment of a given goniometer, and also to indicate which steps are more critical than others. A vertical $\theta{:}2\theta$ goniometer, in which the basic alignment tools are a single-knife-edge zero gauge and a double-knife-edge 2:1 gauge, will be used as the example.

As shown in Table 8.2, there are essentially seven basic steps in alignment of a typical powder diffractometer. While all seven steps are important, the three most important and most time consuming are those listed as steps 2, 3, and 4, i.e., the setting of the takeoff angle of the X-ray tube, the setting of the

Table 8.1. Main Sources for Error in X-ray Powder Diffractometry

1. Specimen displacement
2. Instrument misalignment
3. Mis-setting of the 2:1 adjustment
4. Error in the zero 2θ position
5. Specimen transparency
6. Flat specimen error
7. Axial X-ray beam divergence
8. Peak distortion due to $K\alpha_1$ and $K\alpha_2$ wavelengths
9. Poor selection of time constant in ratemeter recording
10. Poor selection of step width in step scanning
11. Incorrect application of data smoothing

Table 8.2. Basic Steps in Aligning a Diffractometer

1. The rough xyz alignment
2. Setting of the takeoff angle
3. Setting of the mechanical zero
4. Setting of the 2:1 adjustment
5. Alignment of the divergence slit assembly
6. Tuning of the monochromator
7. Final adjustments and instrument checks

mechanical zero, and the setting of the 2:1 rotation axes of the goniometer. The effect of varying the takeoff angle is to change the intensity and line width (and perhaps line shape) of the diffracted profiles, and generally this has little effect on the accuracy of the experimental 2θ values. However, mis-setting of either the mechanical zero or the 2:1 goniometer setting can introduce significant systematic errors into the observed 2θ values. In addition, mis-setting of the 2:1 can affect both line shapes and intensities. Other mechanical errors [1,2] such as the horizontal accuracy of slit placement, eccentricity of gearing, and temperature sensitivity of components [3] are usually beyond the direct control of the user and will not be discussed here.

Sections 8.1.1 through 8.1.6 below outline the first six basic alignment steps, and Section 8.2 discusses routine alignment checks.

8.1.1. The Rough xyz Alignment

Figure 8.1 illustrates the three (generally) separate assemblies that make up a typical vertical $\theta{:}2\theta$ powder diffractometer:

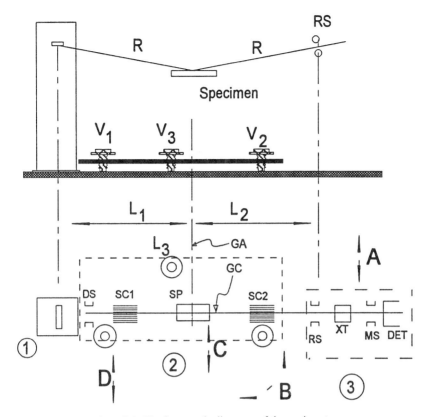

Figure 8.1. The first rough alignment of the goniometer.

1. The X-ray tube mounted inside a tube tower, fixed in turn to a tabletop, which usually acts as the top of the high-voltage generator
2. The goniometer, which provides the basic 2:1 movement of specimen and receiving slit, and also acts as the support for the divergence slit assembly (DS), the specimen holder (SP), and two Soller collimators (SC1 and SC2)
3. The detector/monochromator assembly, consisting of the receiving slit (RS), the monochromator crystal (XT), the monochromator slit (MS), and the detector (DET)

Note that assembly 3 is typically attached to the 2θ drive of the goniometer by a *detector arm*.

In the first rough setting of the *xyz* directions of the goniometer, it is most helpful to ensure that the tabletop is level before any further setup is attempted.

208   ALIGNMENT AND MAINTENANCE OF DIFFRACTOMETERS

This is usually achieved by use of leveling screws on the legs of the table and using a bubble gauge to establish a level working surface. Once this is done, the goniometer leveling screws (V_1, V_2, and V_3) can be used to ensure that the base of the goniometer assembly is also level (see Figure 8.1). The next step is to ensure that the distance from source to specimen (L_1), and from specimen to receiving slit (L_2) are equal and set to the goniometer circle radius (R)—typically in the range 17–20 cm. Note that this is not a terribly critical adjustment; an accuracy of ± 1 mm is generally good enough for the distances L_1 and L_2. It is important, however, to make sure that any radiation protection provided by the manufacturer fits at the selected value of the goniometer circle radius.

The final step in this first rough alignment, and generally the most difficult and time consuming, is to ensure that the center of the tube focus, the center line of the slits and the specimen, and the center line of the monochromator/detector assembly all lie on a single straight line. By far the easiest way of achieving this is (as indicated in Figure 8.1) to first (A) align the center line of the detector/monochromator assembly to the divergence and Soller slits; then second (B,C) to align the center of the specimen to this same center line; and third (D) to bring the goniometer and the detector/monochromator assemblies to the center of the X-ray tube. As indicated in the figure, these steps are carried out by the movements A, B, C, and D and are generally executed in that order.

8.1.2. Setting the Takeoff Angle

Figure 8.2 illustrates the basis of the procedure for setting the takeoff angle. As shown in the figure, the takeoff angle α is defined as the angle between the surface of the X-ray tube target and a line drawn through the center of the target and the center of the goniometer axis. When the X-ray tube is mounted

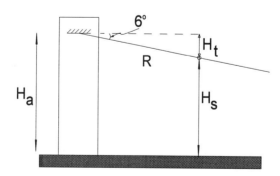

Figure 8.2. Setting the takeoff angle of the X-ray rube.

in the tube tower, the height of the surface of the X-ray tube anode above the tabletop is defined as H_a. The height of the central axis of the goniometer above the tabletop is given as H_s. The distance from the center of the tube target to the center of the goniometer (i.e., the goniometer circle radius) is equal to R. The center of the goniometer lies below a horizontal line from the surface of the X-ray tube anode by an amount H_t. As can be seen in Figure 8.2, since H_a and R are fixed, α must be set by the selection of H_s, i.e.,

$$\alpha = \sin^{-1}(H_t/R) = \sin^{-1}\left(\frac{H_a - H_s}{R}\right). \tag{8.1}$$

The choice of takeoff angle is determined by a number of factors including the desired instrumental resolution, the desired intensity, the required beam width at the receiving slit, and the focal spot dimensions of the X-ray tube. As α is increased, the number of photons available will also increase; however, the number of cross-fired (i.e., out-of-focus) photons will increase as well, causing a decrease in resolution. Figure 8.3 shows the effect of the takeoff angle on intensity at various X-ray tube operating potentials. It is clear that the resolution loss in going from 6° to 10° is not justified by the relatively small increase in intensity, whereas the increase in intensity in raising α from 3° to 6° is substantial. In practice, the takeoff angle is generally between 2° and 8°, with 6° being the preferred setting for general-purpose use. Angles below about 2° will cause a high dependence of X-ray intensity on anode surface roughness,

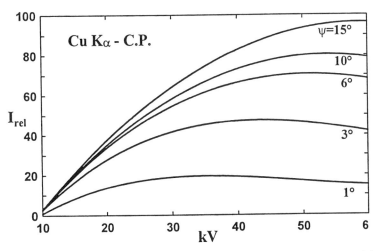

Figure 8.3. The effect of takeoff angle on intensity at various tube operating potentials.

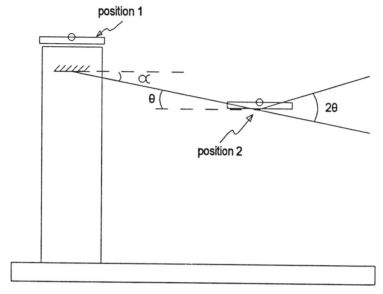

Figure 8.4. Checking the takeoff angle of the X-ray tube.

and angles higher than 8° are difficult to reach because of mechanical limitations and the size of the window in both the X-ray tube and tube tower. For a fine-focus tube, α is typically set at about 6°, and for a normal focus tube, at about 4°. With the typically employed Soller collimators, these settings will give a beam width at the receiving slit of about 0.2 mm.

The current value of the takeoff angle can very easily be checked for vertical tube diffractometers by using a bubble gauge. As illustrated in Figure 8.4, the bubble gauge is first placed on the top of the tube tower (position 1) to ensure that the tube tower is approximately parallel to the ground. By use of a flat surface in the sample position, the bubble gauge is now placed at the sample position (position 2). The $\theta:2\theta$ movement of the goniometer is rocked up and down until the bubble gauge indicates that the sample position is now parallel to the top surface of the X-ray tube tower. In this position, the angle θ is exactly equal to the takeoff angle; thus, half the reading of the goniometer 2θ gives the takeoff angle α.

8.1.3. Setting the Mechanical Zero

The mechanical zero of a goniometer is the angle at which a single line bisects the center of the receiving slit, the goniometer rotation axis, and the center of the projected source from the X-ray tube. The correct setting is usually

PRINCIPLES OF ALIGNMENT

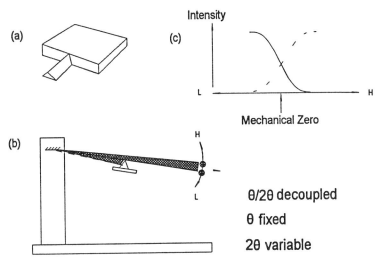

Figure 8.5. Setting the mechanical zero of the diffractometer.

established by use of a single knife-edge or slit placed on the goniometer rotation axis. An error of $x°(\pm)$ in mechanical zero will produce an equivalent error of $x°(\pm)$ in all observed 2θ values. Figure 8.5 shows the principle of the method of setting the mechanical zero using a single knife-edge. For this adjustment, the θ:2θ movement are decoupled and the θ position is kept fixed. A single knife-edge (Figure 8.5a) is placed at the sample position of the goniometer, ensuring that the centerline of the slit is exactly at the center of the goniometer axis. As shown in Figure 8.5b, with the knife-edge pointing upward, the slit will now cut off the lower half of the beam coming from the X-ray tube. Moving 2θ from a low (L) position to a high (H) position will cause the receiving slit to sweep through the X-ray beam, producing an angle/intensity curve like that shown by the dashed line in Figure 8.5c. If the knife-edge is now inverted 180° so that it points down, a similar movement of the goniometer from L to H will produce an angle/intensity curve like that shown by the solid line in Figure 8.5c. The correct mechanical zero is the point at which the two intensity curves intersect.

Note that for an ideal alignment the curves should cross at about 50% of the intensity maximum of the curves. An intensity value higher or lower than about 50% is probably due to one or more of the following causes:

1. The zero alignment is not correct.
2. The assumed center of the goniometer sample holder shaft is incorrect.

3. The surface of the knife-edge gauge is not flat.
4. The knife-edge gauge is not correctly placed at the goniometer axis.

8.1.4. Setting the 2:1

The basic function of the goniometer is to rotate the receiving slit at twice the angular speed of the specimen. Should this 2:1 relationship be in error, a cyclic error will be introduced into the observed 2θ value. The 2:1 alignment is generally made with a double-knife-edge slit at the specimen position and is typically the last major step made during the alignment process. Errors in the 2:1 setting are very common in powder diffractometers, and they introduce a significant and completely unnecessary variable error in the observed 2θ maxima from experimental patterns. Figure 8.6 shows the use of the double-knife-edge gauge for the setting of the 2:1.

In the setting of the 2:1, the θ:2θ movement of the goniometer is coupled and a screw or micrometer adjustment of the θ is made. A double-knife-edge gauge is used for this alignment, as shown in Figure 8.6a. The double-knife-edge gauge is placed in the sample position of the goniometer, as shown in Figure 8.6b. The gauge cuts off the lower portion of the radiation coming from the X-ray tube. Rocking the gauge in either a clockwise (Cl) or counter-clockwise (CCl) direction will cause the beam intensity through the receiving slit to be

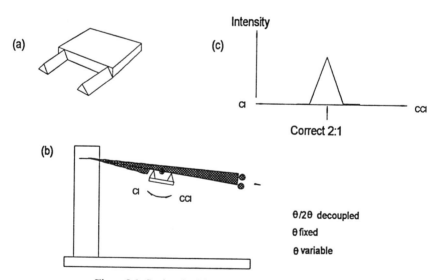

Figure 8.6. Setting the 2:1 movement of the goniometer.

decreased (see Figure 8.6c). The correct 2:1 setting occurs when the recorded intensity value is a maximum.

8.1.5. Alignment of the Divergence Slit

In most powder diffractometers the divergence slit assembly is adjustable to allow the correct centering of the divergent beam onto the specimen surface. This adjustment is illustrated in Figure 8.7. The radiation from the X-ray tube passes through the divergence slit to the specimen. Rocking the divergence slit assembly through the arc a–b causes the zero line through the center of the slit to move back and forth over the surface of the specimen. When a fluorescent screen is placed in the specimen position, the image will be seen to move toward the X-ray tube as the divergence slit assembly is lowered (direction a) and away from the X-ray tube as the divergence slit is raised (direction b). The assembly is rocked up and down until the fluorescence image is exactly centered about the centerline of the specimen. It may happen that the fluorescence image on the screen is distorted to a trapezoidal rather than a rectangular image. This is an indication that the surface of the X-ray tube anode and the sample surface are not parallel, generally because the

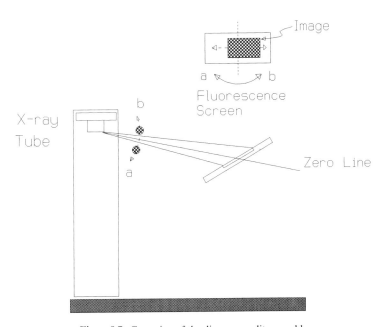

Figure 8.7. Centering of the divergence slit assembly.

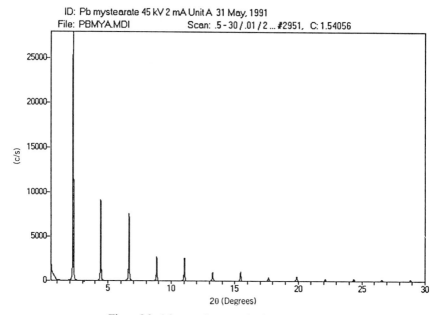

Figure 8.8. A low-angle scan using lead mystearate.

goniometer has not been correctly leveled at the outset. This is a serious problem that should be rectified, even though it probably means restarting the alignment from scratch.

If a variable (theta-compensating) divergence slit has been used, it is useful to check the low-angle alignment by running a scan with a specimen of a large d-spacing single crystal or pseudocrystal. A good standard for this purpose is a lead stearate monolayer, similar to those used in X-ray fluorescence spectrometry [4]. Figure 8.8 shows such a scan obtained using lead mystearate, $2d = 80.5$ Å. Thus, for Cu $K\alpha$ radiation the angles of the various orders (n) can be calculated as

$$2\theta(n) = 2\sin^{-1}\left(\frac{n\lambda}{2d}\right) = 2\sin^{-1}(0.01916n). \quad (8.2)$$

The values for the first 10 orders are given in Table 8.3.

8.1.6. Tuning of the Monochromator

There are many different varieties of diffracted beam monochromator in use today, but most of them employ a pyrolytic graphite crystal bent to a 4 inch

Table 8.3. The First 10 Reflections from Lead Mystearate

n	2θ	n	2θ
1	2.20	6	13.20
2	4.40	7	15.41
3	6.60	8	17.63
4	8.79	9	19.85
5	10.99	10	22.09

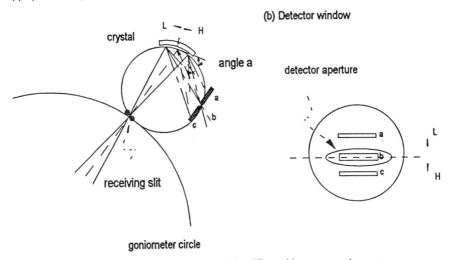

Figure 8.9. Geometric arrangement of the diffracted beam monochromator.

radius, and a detector aperture or slit. These are configured as shown in Figure 8.9a. The (002) plane of graphite is typically employed, and this has a 2d-spacing of 6.715 Å. Each unique wavelength falling onto the crystal will be diffracted at a unique angle 2θ in accordance with Bragg's law. For example, Table 8.4 lists a few typical wavelengths and the values of 2θ at which they will diffract.

The detector, with the aperture forming an entrance slit, is placed at the required angle to allow wanted radiation (indicated by b in Figure 8.9a) to pass, at the same time preventing unwanted radiation (indicated by a and c in Figure 8.9a) from passing to the detector. Adjusting the angle of either the crystal mount or the detector (frequently achieved by swiveling the detector

Table 8.4. Angles at Which Pyrolytic Graphite Will Diffract Various Wavelengths

Radiation	λ (Å)	Angle (°2θ)
Cu $K\beta$	1.392	11.96
Cu $K\alpha$	1.542	13.28
Fe $K\alpha$	1.936	16.76

about its mount) effectively moves the aperture from low (L) to high (H), as illustrated in Figure 8.9b. Note that there is a relatively small (1.34°) angular difference between Cu $K\alpha$ and Cu $K\beta$. As was discussed in Section 6.7, with a 4 inch radius monochromator circle, this translates to about 2–3 mm across the surface of the detector aperture, making the setting of the crystal and/or detector aperture rather critical.

A common error is to allow the detector aperture to be twisted away from its correct position of exactly bisecting the horizontal axis of the detector window. Such a mis-setting somewhat reduces the amount of α radiation entering the detector but, even more important, also allows β radiation to enter the detector. In order to finally check the correct setting of the monochromator, one should perform a short scan over a range of angles, as was previously illustrated in Figure 6.10, in order to ensure that no β radiation is entering the detector.

8.2. ROUTINE ALIGNMENT CHECKS

The only way to ensure that the alignment of the powder diffractometer is within acceptable limits (and that it remains within acceptable limits) is to perform an initial alignment check, followed by regular, less-detailed alignment checks. The first of these should be performed whenever a major realignment has been attempted, and its purpose is to establish the magnitude of potential alignment errors in the diffractometer. The second, which should be carried out on a regular (weekly) basis, is to ensure that the integrity of the alignment has been maintained. Each of these checks can be performed with a suitable instrument reference standard, and α-quartz is a popular choice for this purpose. The reference standard is made by cementing a solid piece of polycrystalline α-quartz into a sample holder and surface grinding to a flatness of better than ± 10 μm. Such a procedure will ensure that the sample displacement error is minimal.

The last step in the alignment of a diffractometer is to establish an external standard curve for the instrument. Mallory and Snyder [5] first proposed the development of the external standard curve, both as an alignment check and as an automated method to correct for all instrumental aberrations. An external standard with well-known lattice parameters is chosen so that Equation 3.10 may be used, along with Bragg's law (Equation 3.1), to compute the theoretical locations for the observable diffraction lines. The external standard's diffraction pattern is determined over the full 2θ range, and the peak locations are determined in exactly the same manner as will be used in routine analysis. The difference between the observed and calculated 2θ values is then plotted vs. 2θ. As an example, Figure 8.10 shows a typical calibration curve for an α-quartz specimen measured over the range of 20–110° 2θ. The horizontal line at $\Delta 2\theta = 0.0$ represents the "theoretical curve" that would be obtained in the absence of all errors and distortions. The "practical curve" is obtained by fitting a least squares polynomial through the observed points. This *calibration curve* includes all peak shifts due to such aberrations as axial divergence and flat-specimen error but does not incorporate the specimen displacement that will vary from one specimen to another [6]. While the calibration curve can be predicted from the geometric forms of the aberrations discussed in Section 7.5 [7], it is best to determine this curve experimentally after each instrumental alignment. When aligning the diffractometer, it is useful to allow a range in the deviation of the fit of the data points to the calibration curve, thus defining an acceptable range of spread. The dashed lines on Figure 8.10

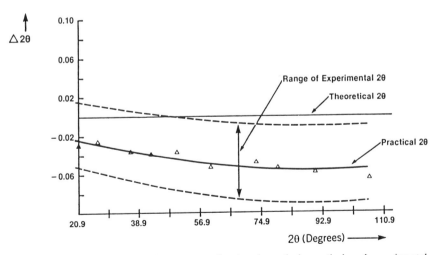

Figure 8.10. A calibration curve for α-quartz, showing theoretical, practical, and experimental curves. From R. Jenkins, Instrumentation. In *Modern Powder Diffraction* (D. L. Bish and J. E. Post, eds.), p. 26, Fig. 5. Mineralogical Society of America, Washington, DC, 1989. Reprinted by permission.

show limits of ±0.03°, but the value chosen in a particular laboratory will, of course, depend upon the accuracy being sought. In the example shown, the "practical curve" does not exactly bisect the error limit lines because these curves are calculated from theory whereas the practical curve is a least squares fit to actual experimental data. Once a good calibration curve has been obtained, it is important to repeat alignment checks on a regular basis to make sure that the proper alignment is maintained. The same α-quartz instrument reference standard can also be employed for this purpose, by measuring a few selected line positions and intensities.

A deterioration in alignment will probably manifest itself in one or more of the following five factors:

1. An overall loss in intensity
2. An angle dependent loss in intensity
3. A systematic shift in all 2θ values
4. An angle-dependent shift in 2θ values
5. Loss in resolution (i.e., distortion in peak shape)

Table 8.5. Typical Calibration Data

Test No. 1: Position and intensity of (101) reflection

Peak angle(101)	= 26.64° 2θ
Intensity(101)	= 39,600 c/s

Test No. 2: Scan over the quartz quintuplet

Resolution (H_1/H_2)	= 2.0
Peak angle $K\alpha_1(203)$	= 68.12° 2θ
Peak intensity $K\alpha_1(203)$	= 4,420 c/s

Test No. 3: Count rates obtained with different receiving slits

Slit width (mm)	Intensity (c/s)
0.05	9,100
0.10	20,050
0.20	39,500
0.30	45,310

A series of simple measurements on the instrument reference standard will allow the extent of each of these five factors to be quantified. These measurements are listed in Table 8.5, along with data from a well-aligned powder diffractometer. By measuring intensities and 2θ values at low [the $K\alpha_1(101)$ line] and moderate [the $K\alpha_1(203)$ line] angles, it can easily be judged whether any intensity loss, or 2θ shift, has occurred for all lines or varies with the 2θ value. An overall residual error for the intensities being produced by a diffractometer is [8]

$$R_I = \sum \frac{I_{obs} - I_{calc}}{I_{obs}}, \qquad (8.3)$$

where I_{obs} is the observed intensity and I_{calc} is the intensity calculated from Equation 3.28, the sum being taken over all observed lines in the pattern.

The instrumental resolution is checked by scanning over a series of partially resolved lines and observing any change in peak separation. Figure 8.11b shows a scan over the so-called *quartz quintuplet*, which is referred to in Table 8.5. The quintuplet is actually six lines comprising three pairs of $K\alpha_1$, $K\alpha_2$ doublets [for the (212), (203), and (301) d-spacings], but the $K\alpha_1(301)$ and the $K\alpha_2(203)$ are unresolved. A slow scan is made over the quintuplet, and then three measurements are taken from the scan: the intensity and peak position of the $K\alpha_1(203)$, and the resolution value H_1/H_2. Here H_1 is defined as the average height of the three troughs between the line pairs [$K\alpha_1(212)$, $K\alpha_2(212)$; $K\alpha_2(212)$, $K\alpha_1(203)$; and $K\alpha_1(301)$, $K\alpha_2(301)$], and H_2 is defined as the intensity of the $K\alpha_1(203)$ line. In all cases, the intensities are corrected for background. Figure 8.11a shows a scan over the quartz quintuplet for a diffractometer exhibiting moderate resolution. In this example, the height of the $K\alpha_2(212)$ is 2320 c/s over an average background of 130 c/s. The average height of the three troughs is 1220 c/s, giving a resolution of $(2320 - 130)/(1220 - 130) = 2.01$.

A third and final test should be carried out to ensure that the optimum receiving slit is being used. In the data shown in Table 8.5, the beam width at the divergence slit is about 0.2 mm. Using a receiving slit smaller than 0.2 mm reduces the X-ray intensity through the slit in proportion to the slit width. A slit that is wider than the beam width gives a little more intensity but much poorer resolution. Thus, in the given example, the optimum receiving slit width is 0.2 mm. In addition to giving a quantitative measure of instrument performance, alignment checks can also be used as indicators of change in instrument performance. Section 8.4 will deal with such use of these data.

The curve shown in Figure 8.10 is a measure of the quality of the 2θ alignment. Figure 8.12 shows a curve that measures the quality of the intensity alignment (or *angular sensitivity*). The curve shown is a plot of the intensity

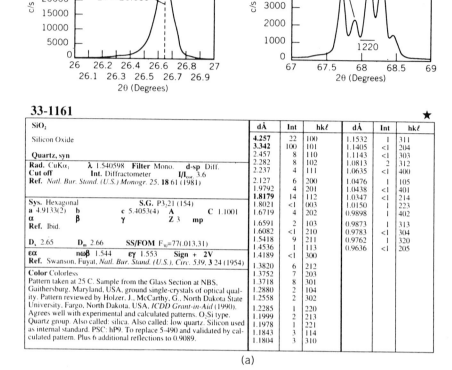

Figure 8.11. Use of the quartz quintuplet as a measurement of instrument performmance.

values for 14-selected lines from an α-alumina reference standard (National Institute of Standards and Technology, SRM 1976 *Instrument Sensitivity Standard*).[1] The intensity values are plotted as ratios of the measured relative line intensities (strongest line = 100) to the certified relative values. Also shown on the curve are the relative standard deviations of the fit and the average ratio value. Round-robin tests [9] have shown that specific models of powder diffractometers may exhibit a 5–15% intensity variation across the 2θ range due to differences in the optical arrangements of specific instruments. Beyond this, however, much larger variations may be observed. As examples, Figure

[1] Available from NIST, Gaithersburg, Maryland.

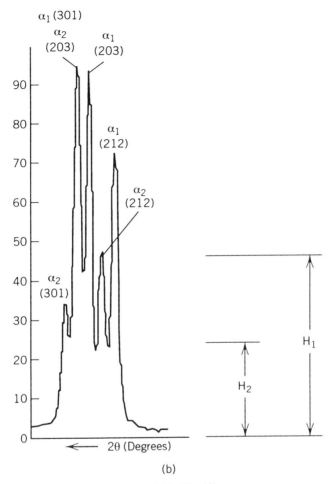

Figure 8.11. (*Cont'd.*)

8.13 shows sensitivity curves for two diffractometers that are not performing correctly. Figure 8.13a shows a sensitivity curve that drops off sharply with diffraction angle. Closer inspection of this particular diffractometer revealed that the variable divergence slit that was used on the diffractometer had stuck in the open position. Figure 8.13b shows a curve with a large random spread in the data. Examination of this system showed that the drive mechanism on the variable divergence slit was sticking.

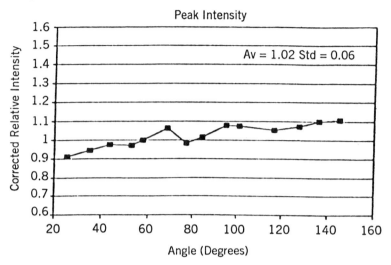

Figure 8.12. A diffractometer sensitivity curve.

8.3. EVALUATION OF THE QUALITY OF ALIGNMENT

As has been shown in the previous section, there are several useful techniques available for the evaluation of the alignment quality. In practice, two difficulties generally present themselves to the typical experimenter: the first problem is to establish just how close to the *average performance* is the instrument under evaluation; the second is to ascertain what is the *average* performance of a diffractometer: The first problem can be solved by running well-established instrument standards as indicated above. A useful means of solving the second problem is to use the *round-robin* approach to gauge the average performance of a given equipment type, i.e., by comparing data sets or many such instruments. A number of these round-robin tests, have been run in recent years [9–14].

As far as the accuracy of angular data are concerned, it has been shown on a number of occasions that, where internal standards are not used, the average 2θ precision in a given laboratory is typically ± 0.01 (1σ) 2θ and between laboratories the precision is about five times worse, or around $\pm 0.05°$ 2θ. On the other hand, where internal standards are used, *both* inter- and intralaboratory precision is about $\pm 0.01°$ 2θ.

In the setting up of instrumental parameters for a given series of experiments, there are many variables under the control of the operator, such as source conditions, receiving slit width, scan speed, and step increment. It is

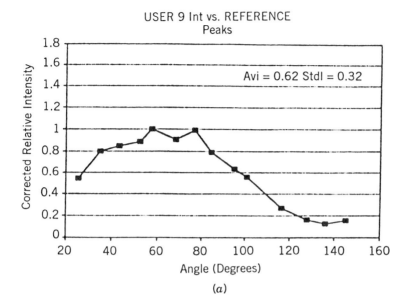

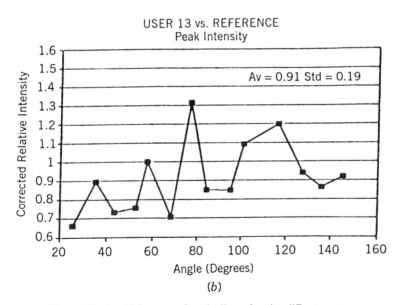

Figure 8.13. Sensitivity curves from badly performing diffractometers.

desirable to have a quantitative or semiquantitative means of evaluating these parameters. One helpful technique is to use a figure of merit (FOM) to provide guidelines in the establishment of optimum conditions [15]. Such figures of merit based on counting statistical limitations have long been used in, for example, X-ray fluorescence spectrometry [16]. However, in X-ray fluorescence, the line broadening parameters are determined purely by the fixed collimators and line shapes are easily correlated with diffraction angle. In X-ray diffractometry, the situation is much more complicated and conventional figures of merit are not directly applicable.

Jenkins and Schreiner addressed this problem in their review of the data from an intensity round-robin test [17] and suggested an instrument parameter FOM for X-ray powder diffraction. This FOM has the form

$$\text{FOM} = M\left(\frac{W}{M+4B}\right)^{1/2}, \tag{8.4}$$

where M is the maximum peak count rate above background; W is the full width at half-maximum (FWHM, in hundreths of a degree); and B is the average background count rate (as illustrated in Figure 8.14). The derivation of this FOM is based on the well-known counting statistical equation for the calculation of the net counting error $\sigma(n)$:

$$\sigma(n) = \frac{100[(N_p + N_b)^{1/2}]}{N_p - N_b}, \tag{8.5}$$

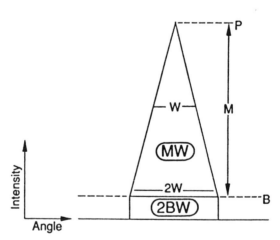

Figure 8.14. Derivation of the Jenkins–Schreiner FOM.

where N_p is the integrated intensity of the peak and background, and N_b is the background intensity. From this expression it follows that the minimum value of $\sigma(n)$, the value of

$$\frac{N_p - N_b}{(N_p + N_b)^{1/2}},$$

should be as large as possible. This ratio is therefore taken as the basis of the FOM. Figure 8.14 illustrates the case of a simple peak in which the fwhm is given by W and the width at the base as $2W$. All intensities are given in counts/second. The area under the peak is equal to MW and that under the background is $2BW$. Thus, for the peak alone it follows that

$$N_p = MW + 2BW = W(M + 2B). \tag{8.6}$$

Similarly, for the background $N_b = 2BW$, and thus, $(N_b - N_b) = MW$. Substituting these data in Equation 8.6 gives

$$\text{FOM} = \frac{MW}{[W(M + 4B)]^{1/2}}. \tag{8.7}$$

Application of the FOM to a wide range of data from different diffractometer configurations and types has shown that the larger the value of the FOM, the statistically better are the net count data. As an example, Table 8.6 shows data from 15 diffractometers that vary in configuration and represent diffractometer systems from four different manufacturers, with X-ray sources that include normal- and broad-focus sealed tubes, a rotating anode tube, and a synchrotron. Data are given for peak, background, and peak width at half-maximum. An FOM has been calculated for each unit using Equation 8.7. As is fairly common with a comparison of this type, the range of numbers is considerable: the peak counts vary from 61,378 to 5559—a factor of about 11; the background varies from 4827 to 64—a factor of about 75. Peak width at half-maximum values vary from 0.09° 2θ to 0.36° 2θ. While the variations in peak and background are very large, note that the range in the figures of merit is less than 4, with the range of 125.1 to 29.0. This is an important point, especially in quantitative analysis, since the net statistical counting error is related to the FOM, and not the absolute value of peak and/or background counts. For this reason, the use of the peak/background ratio as a measure of optimum measurement conditions is little more than a poor approximation to the ideal situation.

226 ALIGNMENT AND MAINTENANCE OF DIFFRACTOMETERS

Table 8.6. Figure-of-Merit Data from Various Diffractometers

Unit Number	Peak Counts	Background Counts	Peak Width	Figure of Merit
1	61,378	1,077	0.32	125.1
2	47,247	829	0.34	122.2
3	28,960	4,827	0.36	80.0
4	26,840	395	0.32	113.5
5	25,852	507	0.34	90.3
6	21,352	251	0.25	71.3
7	18,300	201	0.25	66.2
8	14,918	178	0.23	57.5
9	13,684	127	0.18	48.4
10	13,062	124	0.09	34.5
11	12,816	279	0.22	51.0
12	11,356	167	0.26	52.9
13	9,934	242	0.35	54.9
14	6,081	83	0.22	35.8
15	5,559	64	0.16	29.0

8.4. TROUBLESHOOTING

An ongoing problem for the analyst is to ensure that the diffractometer is always working under optimum conditions. When problems are identified, it is useful to ascertain where the problem is occurring and, if possible, to correct the error. It has been shown that a given set of 2θ values may vary from the *correct* values for a combination of three major reasons:

1. Inherent aberrations due to the specific geometry in use
2. Specimen displacement and/or transparency errors
3. Misalignment problems

Once a given instrument reference standard is chosen, for example, α-quartz, it is a simple matter to calculate the magnitudes of the errors listed in items 1 and 2, above, for any selected angle(s). In our laboratories a *spreadsheet* program is used for this purpose, and the data shown in Table 8.7 were obtained by the use of this program. The specific parameters for a given equipment geometry can be specified, as follows: the radius (R) of the goniometer circle; the divergence slit aperture (Div Slit); the axial divergence constant terms $[K(1)$

Table 8.7. Magnitude of Systematic Errors[a]

No.	Raw Data		Corrections				Corrected Data	
	2θ	d-Value	Transp.	Axial Diverge	Flat Spec	Displ	2θ	d-Value
1	5.000	17.660	−0.0014	−0.1502	−0.0666	0.0199	4.868	18.137
2	10.000	8.838	−0.0028	−0.0746	−0.0332	0.0199	9.942	8.889
3	15.000	5.901	−0.0042	−0.0492	−0.0221	0.0199	14.966	5.915
4	20.000	4.436	−0.0054	−0.0364	−0.0165	0.0199	19.978	4.441
5	25.000	3.559	−0.0065	−0.0286	−0.0131	0.0199	24.985	3.561
6	30.000	2.976	−0.0074	−0.0232	−0.0109	0.0198	29.989	2.977
7	35.000	2.562	−0.0082	−0.0193	−0.0092	0.0198	34.992	2.562
8	40.000	2.252	−0.0087	−0.0163	−0.0080	0.0198	39.995	2.252
9	45.000	2.013	−0.0090	−0.0139	−0.0070	0.0198	44.997	2.013
10	50.000	1.8227	−0.0090	−0.0119	−0.0062	0.0198	49.999	1.8227
11	55.000	1.6682	−0.0088	−0.0102	−0.0056	0.0198	55.001	1.6682
12	60.000	1.5406	−0.0083	−0.0087	−0.0050	0.0198	60.003	1.5405
13	65.000	1.4336	−0.0075	−0.0073	−0.0046	0.0197	65.005	1.4336
14	70.000	1.3430	−0.0065	−0.0061	−0.0042	0.0197	70.007	1.3429
15	75.000	1.2654	−0.0052	−0.0050	−0.0038	0.0197	75.010	1.2652
16	80.000	1.1984	−0.0037	−0.0039	−0.0035	0.0197	80.012	1.1982
17	85.000	1.1402	−0.0020	−0.0029	−0.0032	0.0197	85.015	1.1400
18	90.000	1.0894	−0.0000	−0.0019	−0.0029	0.0196	90.018	1.0892
19	95.000	1.0448	0.0021	−0.0009	−0.0027	0.0196	95.021	1.0446
20	100.000	1.0056	0.0044	0.0001	−0.0024	0.0196	100.024	1.0054
21	105.000	0.9709	0.0068	0.0011	−0.0022	0.0195	105.027	0.9708
22	110.000	0.9404	0.0093	0.0021	−0.0020	0.0195	110.031	0.9402
23	115.000	0.9133	0.0118	0.0032	−0.0019	0.0195	115.034	0.9132
24	120.000	0.8895	0.0143	0.0043	−0.0017	0.0194	120.038	0.8893
25	125.000	0.8684	0.0169	0.0056	−0.0015	0.0194	125.042	0.8683
26	130.000	0.8499	0.0193	0.0070	−0.0014	0.0194	130.046	0.8498
27	135.000	0.8338	0.0216	0.0086	−0.0012	0.0193	135.050	0.8336
28	140.000	0.8197	0.0238	0.0105	−0.0011	0.0193	140.054	0.8196
29	145.000	0.8077	0.0259	0.0128	−0.0009	0.0192	145.058	0.8076
30	150.000	0.7975	0.0277	0.0157	−0.0008	0.0192	150.063	0.7974

[a] Parameters: W'length = 1.540598; Lin Mu = 50.00; R = 17.30; Rho = 2.00; Displ = 0.0030; Div Slit = 1.00; $K(1) = 0.01125$; $K(2) = 0.00188$.

and $K(2)$]; the wavelength of the radiation used (W'length); and the linear absorption coefficient (Lin Mu) and density (Rho) of the instrument reference standard. A value can also be added for the specimen displacement (Displ), which allows the magnitude of this error to be calculated for different values of specimen displacement. The spreadsheet then calculates the individual contributions to the total error for each 2θ value, plus the total error, in terms of both 2θ and d-spacing. Raw data can be entered either as 2θ or as d-spacing. If an

ALIGNMENT AND MAINTENANCE OF DIFFRACTOMETERS

Table 8.8. Misalignment Indicators Using Calibration Data[a]

No.	T_1	T_2	R_1	R_2	RES	Probable Cause
1	Same	High	Same	Same	Same	2:1 error
2	High/low	High/low	Same	Same	Same	Zero error or sample displacement
3	Same	Same	Same	Low	Same	Divergence slit
4	Same	Same	Low	Low	Same	Monochromator or PHS[b]
5	Same	Same	Low	Low	Less	X-ray tube aging
6	Same	Same	Low	Low	Low	Wrong receiving slit

[a] Here T_1 is the 2θ value of the low-angle line and R_1 is the corresponding count rate; T_2 and R_2 are similar values for the high-angle reflection. RES refers to the resolution value obtained from the quartz quintuplet.
[b] PHS = pulse height selector.

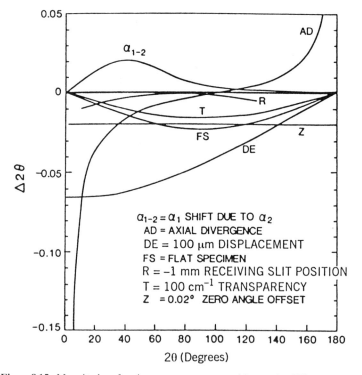

Figure 8.15. Magnitudes of various errors encountered in powder diffractometers.

instrument alignment check is now run, as described previously in Section 8.2, study of the experimental data in light of data as indicated in Table 8.7 will reveal whether problems exist. Table 8.8 gives some rough indications of typical errors that are encountered in the laboratory. The columns of data for T_1, T_2, R_1, R_2, and RES, are based on the tests previously described in Section 8.2.

As a further aid in the evaluation of experimental data, it is useful to examine a graphic representation of the shift in peak position as a result of these and other errors for α-quartz [18]. Such a representation is shown in Figure 8.15. Curves are given for axial divergence, flat specimen (for a constant 13 mm sample illumination length), transparency, and a $\pm 0.02°$ mis-setting of the 2θ dial. Also shown is the effect of a minus 1 mm error in the specimen–receiving slit distance, a 100 μm displacement error, and the shift in the α_1 peak position due to the presence of the α_2 line. These values are of the magnitude typically found in most analytical laboratories. Note that the displacement error has by far the greatest influence.

REFERENCES

1. Jenkins, R., and Schreiner, W. N. Considerations in the design of goniometers for use in X-ray powder diffractometers. *Powder Diffr.* **1**, 305–319 (1986).
2. Parrish, W., and Wilson, A. J. C. Some new minor geometric aberrations in powder diffractometry. *Acta Crystallogr.* **7**, 622 (1954).
3. Brown, G., Wood, I. G., and Nicholls, L. Thermal and mechanical instabilities in the alignment of Bragg–Brentano parafocusing powder diffractometers. *Powder Diffr.* **2**, 7–21 (1987).
4. Jenkins, R., Hom, T., Villamizar, C., and Schreiner, W. N. Calibration of the powder diffractometer at low values of 2θ. *Adv. X-Ray Anal.* **25**, 289–294 (1982).
5. Mallory, C. L., and Snyder, R. L. The control and processing of data from an automated X-ray powder diffractometer. *Adv. X-Ray Anal.* **22**, 121–132 (1979).
6. Snyder, R. L., Hubbard, C. R., and Panagiotopoulos, N. C. A second generation automated powder diffractometer control system. *Adv. X-Ray Anal.* **25**, 245–260 (1982).
7. Jenkins, R. Effects of diffractometer alignment and aberrations on peak positions and intensities. *Adv. X-Ray Anal.* **26**, 25–33 (1983).
8. Snyder, R. L. Accuracy in angle and intensity measurements in X-ray powder diffraction. *Adv. X-Ray Anal.* **26**, 1–11 (1983).
9. Jenkins, R., and Schreiner, W. N. Intensity round-robin report. *JCPDS–ICDD Methods and Practices Manual*, Sect. 13:2. Newtown Square, PA, 1989.
10. Jenkins, R. A round-robin to evaluate computer search/match methods for qualitative powder diffractometry. *Adv. X-Ray Anal.* **20**, 125–137 (1976).

11. Jenkins, R., and Hubbard, C. R. Preliminary report on the design of the second round-robin to evaluate search/match methods for qualitative powder diffractometry. *Adv. X-Ray Anal.* **22**, 133–142 (1978).
12. Edmonds, J. W., Bfowa, A., Fischer, C., Foris, C. M., Goehner, R., Hubbard, C. R., Jenkins, R., Schreiner, W. N., and Virser, J. Round-robin on cell parameter refinement. In *JCPDS–ICDD Methods and Practices Manual*, Sect. 13:1. International Centre for Diffraction Data, Newtown Square, PA, 1986.
13. Fawcett, T. G. Establishing a peak profile calibration standard for powder diffraction analysis. *Powder Diffr.* **3**, 209–219 (1988).
14. Calvert, L. D., Sirianni, A. F., Gainsford, G. J., and Hubbard, C. R. A comparison of methods for reducing preferred orientation. *Adv. X-Ray Anal.* **26**, 105–110 (1982).
15. Jenkins, R., and Gilfrich, J. V. Figures-of-merit, their philosophy, design and use. *X-Ray Spectrom.* **21**, 263–269 (1992).
16. Jenkins, R., and de Vries, J. L. *Practical X-Ray Spectrometry*, 2nd ed., p. 103. Springer-Verlag; New York, 1970.
17. Jenkins, R., and Schreiner, W. N. Intensity round robin report. *Powder Diffr.* **4**, 74–100 (1989).
18. Schreiner, W. N., Jenkins, R., Surdukowski, C., and Villamizar, C. Systematic and random powder diffractometer errors relevant to phase identification. *Norelco Rep.* **29**, 42–52 (1982).

CHAPTER
9
SPECIMEN PREPARATION

9.1. GENERAL CONSIDERATIONS

The aim of any diffraction experiment is to obtain the best possible data within the appropriate constraints of the relevant circumstances, so that the data can be correctly interpreted and analyzed. One of the major problems in achieving this goal is the preparation of the specimen. Various methods of specimen preparation have been devised [1,2], and the success of a given diffraction experiment will invariably depend on the correct choice of preparation method for the sample being analyzed and for the instrument conditions being used for the analysis. Because of the many problems associated with the form of the analyzed material, it is useful to differentiate between the sample and the specimen. The *sample* is generally considered to be the material submitted for analysis. It is also generally assumed that the sample has been correctly taken from its source (i.e., *sampled*) by the submitter so as to give a representative measure of the problem at hand. It is then up to the diffractionist to take an aliquot of the sample and prepare a *specimen* for analysis. This specimen preparation may involve a number of steps such as drying, grinding, sieving, dilution, and mounting. It is tacitly assumed that the analysis of the specimen represents the analysis of the sample. While this assumption is generally true, it may not always be so, and the analyst must be on guard to ensure that all reasonable precautions are taken to avoid contamination or phase changes during the process of specimen preparation.

A diffraction pattern contains a good deal of information, of which three parameters are of special interest:

- The position of the diffraction maxima
- The peak intensities
- The intensity distribution as a function of diffraction angle

These three pieces of information can, in principle, be used to identify and quantify the contents of the sample, as well as to calculate the material's crystallite size and distribution, crystallinity, stress, and strain. The ideal specimen preparation for a given experiment depends largely on the

232   SPECIMEN PREPARATION

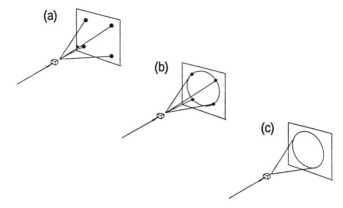

Figure 9.1. Single-crystal (a), oriented-powder and (b), and random-powder (c) diagrams.

information desired. A sample that is used only for the identification of its constituents may be quite different from one used to measure strain, which in turn may be different from one used in quantitative analysis.

The biggest overall problem in specimen preparation is due to preferred orientation. Figure 9.1 illustrates, in a simplified manner, the diffraction patterns that would be obtained from (a) a single crystal, (b) an oriented powder, and (c) a random powder. In each instance, the specimen is placed between a monochromatic beam of radiation and a piece of photographic film. In the case of the single crystal, because only certain crystal planes are in the position to diffract radiation onto the film, the pattern will appear as a series of spots on the film, the position of the spots being dependent upon the structure and orientation of the crystal. Changing the position of the crystal will bring other planes into diffracting position, and the pattern of spots will change. In the case of a random powder, whatever the orientation of the specimen, there is always a sufficient number of crystallites to diffract from the appropriate d-spacing(s). In the case of a nonrandom powder, however, there are more crystallites in certain orientations and fewer in others. Thus, the pattern obtained is somewhere between the single-crystal and random-powder patterns. As described in Chapter 3, particle and crystallite size, sample size, sample position, crystallite orientation, and absorption will affect the quality and appearance of any diffraction pattern [3]. As a rule of thumb, the best specimen preparation methods are those that allow the analyst to obtain the desired information with the *least* amount of sample treatment. This is especially important in the analysis of unknown materials, pharmaceuticals, and organics where the materials being analyzed may be sensitive to grinding, humidity, or the atmospheric environment. Extensive preparation may change the specimen so that the analyst is no longer analyzing the original problem. A good example is the case of hydrates and polymorphs in formulation. The presence of hydrates and/or polymorphs may change the melting, solubility, and blending characteristics of the formulation. In such a case, the analyst must ensure that the specimen preparation method does not dehydrate the sample or change the polymorphic composition. This may eliminate specimen preparation methods involving grinding, spray-drying, and vacuum-drying and may necessitate that the sample be sealed in a controlled environment.

9.2. COMPOSITIONAL VARIATIONS BETWEEN SAMPLE AND SPECIMEN

Care must be taken in any specimen preparation method to ensure that the physical state and composition of the prepared specimen be as close as possible to that of the original sample. In practice, this ideal situation may be difficult to attain, and some of the possible causes for compositional variations between specimen and sample are given in Table 9.1. As an example, there are a number of problems that may occur during a grinding process. These include the introduction of amorphism or strain; decomposition of the specimen due to local heating, loss of water and/or CO_2; solid state addition or replacement reactions; and contamination. During the actual analysis, there may be compositional variations induced by irradiation. Weak ligands (such as those

Table 9.1. Possible Causes for Compositional Variations Between the As-received Sample and the Prepared Specimen

Source of Problem	Manifestation of Problem
Induced by grinding	Amorphism Strain Decomposition Solid state reaction Contamination
Induced by irradiation	Polymerization Decomposition Amorphism
Special problems	Environmental conditions: loss of water in vacuum High-temperature decomposition

present in organometallic compounds) may rupture. Polymeric materials such as nylon may cross-link, resulting in an increased molecular weight. However, it should be noted here that the production of *color centers* in certain materials on X-irradiation (e.g., KCl turns violet) are compositional variations at the atomic level and do not affect *bulk analysis* in any way. Finally, there may be special problems due to local environmental conditions.

9.3. ABSORPTION PROBLEMS

Although Bragg–Brentano geometry exposes a constant volume of sample to the beam, causing the absorption to be the same at all 2θ angles (as described in Section 1.6), absorption considerations affect the beam penetration depth and must still be considered. In nonparafocusing geometries, the effects of absorption can be extreme. If the composition of the sample is known, the analyst can either approximate or exactly calculate the absorbing ability of the sample to be analyzed using the relationship given in Equation 1.10. As an example, the mass attenuation coefficient for Cu $K\alpha$ radiation by calcium carbonate, where W is the weight fraction, is given by

$$\left(\frac{\mu}{\rho}\right)(CaCO_3) = \left(\frac{\mu}{\rho}\right)W_{Ca} + \left(\frac{\mu}{\rho}\right)W_C + \left(\frac{\mu}{\rho}\right)W_O$$
$$= (162 \times 40/100) + (4.6 \times 12/100) + (11.5 \times 48/100)$$
$$= 70.9 \, cm^2/g.$$

In the example, the numbers 162, 4.6, and 11.5, are the respective mass attenuation coefficients for Ca, C, and O. Appendix B gives a tabulation of the mass attenuation coefficients of the known elements for several types of radiation wavelengths.

Table 9.2 shows Cu $K\alpha$ X-ray penetration depths, at three different diffraction angles, for a typical high-absorbing material (MoO_3), for a typical medium-absorbing material ($CaCO_3$), and for a typical low-absorbing material [aspirin (acetylsalicyclic acid)]. It can be seen that the penetration depths vary widely—from a few micrometers to nearly 200 µm. Absorption effects should also be considered if a binder or a substrate is used in the diffraction experiment. Insufficient sample intensity can be an indication of small crystallite size or high sample absorption. The analyst may want to analyze a thinner sample to see if absorption is a problem, or a thicker sample that may place more crystallites in the X-ray beam. Sample absorption can be reduced by changing the sample density in the X-ray beam by adding diluents. In transmission experiments, the thickness and elemental composition of the

Table 9.2. Penetration Depths (in μm) of Cu $K\alpha$ Radiation for Various Materials

Material	μ/ρ	ρ	20°	40°	60°
MoO_3	92.7	4.71	1.8	2.7	3.6
$CaCO_3$	39.9	2.71	7.4	11	15
Aspirin	7.0	1.40	82	121	161

substrate have to be carefully considered so that the substrate does not absorb or block the incoming X-ray beam. Some workers use very thin (75 μm) Al foil, and others use plastic films such as Mylar, Kapton, or other hydrocarbons as sample substrates. As mentioned previously, the analyst has to be careful that the diffraction pattern of the substrate does not grossly interfere with the pattern of the sample.

9.4. PROBLEMS IN OBTAINING A RANDOM SPECIMEN

9.4.1. Particle Inhomogeneity

Because the X-ray penetration into a specimen is frequently small (see Table 9.2), special problems can occur where the individual particles are large relative to the penetration depth of the X-ray beam. Since the attenuation of an X-ray beam follows an exponential law, irregular absorption can occur from a heterogeneous particle. For example, if one were analyzing specimens of chalcopyrite ($CuFeS_2$) from an open-cast copper mine where the ore had been exposed to the air, particle heterogeneity could occur due to oxidation of the sulfide at the surface of the individual particles. Figure 9.2 illustrates the case of three particles, each within the analyzed layer of the specimen. At 40° 2θ the X-ray path length is about 30 μm. The first particle (A) is made up of pure $CuFeS_2$; the second particle (B) is made of $CuFeS_2$ but surrounded by a layer of $CuFe_2O_4$; the third particle (C) is made up of pure $CuFe_2O_4$. For the sake of discussion, let us assume an average particle size of 20 μm. Let us also assume that the surface layer of $CuFe_2O_4$ on the heterogeneous particle is 5 μm thick and the center of the particles has a diameter of 10 μm of $CuFeS_2$. Since these two compounds have different mass attenuation coefficients ($CuFeS_2 = 143.2$ and $CuFe_2O_4 = 116.1$ for Cu $K\alpha$ radiation), the percentage of the intensity transmitted by each particle can be easily calculated by use of Equation 1.10. These values are 32, 35, and 40%, respectively. Clearly, the measured X-ray intensity will be dependent not only on the difference in the mass attenuation

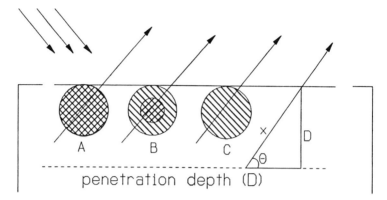

Figure 9.2. Effect of particle heterogeneity on X-ray intensity.

coefficients of the two phases (this effect is called the *absorption effect*) but also on the heterogeneity of the individual particles (this effect is called the *particle inhomogeneity effect*).

9.4.2. Crystal Habit and Preferred Orientation

Preferred orientation, introduced in Section 3.9.1, is the most serious cause of powder diffraction intensity distortion. Figure 9.3 shows a series of Debye rings from a specimen exhibiting preferred orientation. Because of the orientation, the intensity distribution around the Debye rings is not constant, but

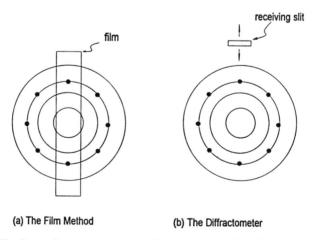

(a) The Film Method   (b) The Diffractometer

Figure 9.3. The Bragg–Brentano geometry allows only a portion of the Debye ring to be measured.

rather manifests itself as a distribution of spots on the rings. The position of the spots is determined by the structure and orientation of the individual crystallites. In the Debye–Scherrer film method (Figure 9.3a), a film is placed in the path of the diffracted beam and, where the film intersects the Debye rings, blackening of the film will result. Where there is an uneven, or *spotty*, intensity distribution, this will be seen on the developed film. In the case of the powder diffractometer (Figure 9.3b), a receiving slit moves across the rings, every time the slit crosses a Debye ring, an intensity signal will be measured. However, unlike the film method, any intensity variations in the ring will not be directly observable because, as illustrated in Figure 9.4, everything falling within the

(a) Randomly oriented specimen

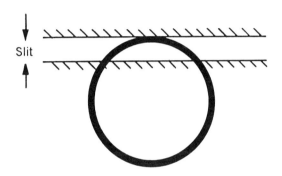

(b) Specimen exhibiting preferred orientation

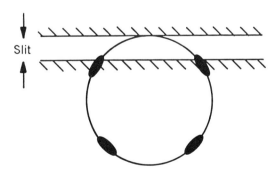

Figure 9.4. Random and preferred orientation.

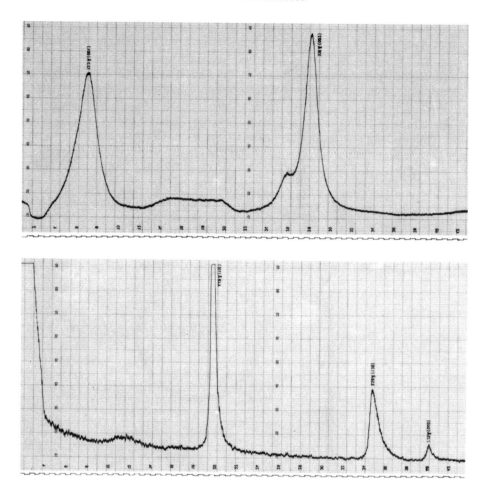

Figure 9.5. Reflection (top) and transmission (bottom) patterns for oriented montmorillonite clay. Reprinted from R. Jenkins and J. L. de Vries, *An Introduction to Powder Diffractometry*, p. 27, Fig. 38. Copyright © 1977, N. V. Philips, Eindhoven, The Netherlands.

slit aperture is integrated to give the measured signal. Figure 9.4a shows the intensity as it would be observed in a randomly oriented specimen, and Figure 9.4b shows the intensity from an oriented specimen. It will be apparent from the foregoing discussion that if the intensity distribution around the Debye ring is variable, the receiving slit may pass too little or too much intensity depending or what part of the ring the slit intersects. Such is the case in a specimen exhibiting preferred orientation.

Figure 9.5 shows two diffractograms from a specimen showing a high degree of preferred orientation [4]. The specimen in question is a montmorillonite clay that, because of its plate-like habit, shows very strong basal plane (001) orientation. The upper part of Figure 9.5 is the diffraction pattern measured in the normal reflection mode, and the lower part has been obtained in transmission. Study of the reflection diffractogram shows that both of the strong lines in the pattern are (00*l*) reflections—in fact the (001) and (002). These are the only strong lines observed because only the (00*l*) planes are ever in the right position to satisfy the Bragg relationship. By measuring the same material in transmission mode, only (*hk*0) reflections are allowed. The *random* diffraction pattern would be some composite of the observed reflection and transmission patterns.

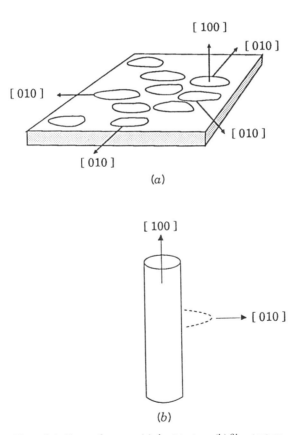

Figure 9.6. Types of texture: (a) sheet texture; (b) fiber texture.

Preferred orientation does not have to be (00*l*)-type orientation because not all crystal habits are plate-like. Different types of materials show different habits, or *texture*. Figure 9.6 shows two examples. *Sheet* texture is akin to the plate-like texture shown by some minerals and, for example, may be shown by metals that have been subject to rolling. This texture can be measured by using a *texture goniometer* to reveal information about the degree or effectiveness of the rolling process. *Fiber* or *rod* texture may be exhibited by fibrous materials such as asbestos minerals and certain organic and natural fibers. Again, a *fiber camera* can be used to evaluate the degree of orientation and stress of a given fiber. In standard powder diffractometry, highly textured materials invariably present special problems in specimen preparation due to the physical problem in attempting to force textured materials to take up a random orientation. In order to fully describe the texture in a specimen, a full-pole figure would need to be determined.

9.4.3. Particle Statistics

In order to obtain accurate intensity measurements, it is necessary to prepare a random specimen. The question now arises as to whether one can quantify a *random specimen*. The standard deviation is dependent on the number of crystallites (N_s) contributing to the intensity of the measured line and is given by $\sqrt{N_s}$, assuming a Poisson distribution. A number of factors must be considered in computing N_s [5]. The relative standard deviation is given by $100/\sqrt{N_s}$. Here N_s is simply the ratio of the irradiated sample volume to the crystallite volume, with due allowance for the crystallite packing efficiency, the vertical divergence of the incident beam, and the intersection of the diffraction cone and the receiving slit. The irradiation volume V_i of a specimen is the product of the irradiation width, the irradiation length, and the penetration depth d of the X-ray beam. For a Bragg–Brentano diffractometer the volume irradiated is independent of diffraction angle, so a convenient 2θ of $18°$ will be used here. The irradiation length L_i of a specimen is fixed for a given diffractometer geometry and is typically between 8 and 15 mm. The irradiation length for a fixed divergence slit is related to the divergence slit opening δd, in degrees. Thus, the following relationship holds:

$$\frac{(L_i)_1}{(L_i)_2} = \frac{\sin\theta_2}{\sin\theta_1}\frac{\delta d_1}{\delta d_2}. \tag{9.1}$$

As an example, for a goniometer radius of 170 mm, the irradiation length of a specimen is 2.0 cm for a divergence slit aperture of $1°$, at a diffraction angle of $18°\ 2\theta$. Thus, the irradiation length in centimeters can be calculated using the

Table 9.3. Diffraction Angles and Irradiation Lengths

Div slit →	0.5°	1.0°	2.0°
2θ	Irradiation	Length (cm)	
10	1.8	max	max
20	0.9	1.8	max
30	0.6	1.2	max
40	0.4	0.91	1.8
50	0.3	0.74	1.4
60	0.3	0.62	1.2
70	0.27	0.54	1.0
80	0.25	0.49	0.98
90	0.22	0.44	0.88

following relationship:

$$L_i = \frac{2.0 \delta d \sin 9}{\sin \theta}. \tag{9.2}$$

Table 9.3 lists typical irradiation length values for 0.5, 1°, and 2° divergence slits.

The basic absorption equation discussed in Section 1.6 related the intensity of an X-ray beam before (I_0) and after (I) passing through an attenuator of thickness x, density ρ, and mass attenuation coefficient μ/ρ:

$$I = I_0 \exp\left[-\left(\frac{\mu}{\rho}\right)\rho x\right], \tag{9.3}$$

Rewriting Equation 9.3 in the dimensions of $\log_{10}$, we have

$$2.303 \log\left(\frac{I_0}{I}\right) = +\frac{\mu}{\rho}\mu x. \tag{9.4}$$

For 99% absorption, i.e., $I_0 = 100$ and $I = 1$, it follows that

$$x = \frac{4.6}{(\mu/\rho)\rho}. \tag{9.5}$$

As was shown in Figure 9.2, the penetration depth D is related to the mass attenuation coefficient, the specimen density, and the diffraction angle. Thus,

$$D = \frac{4.6 \sin \theta}{(\mu/\rho)\rho}. \tag{9.6}$$

For the purpose of the following calculation, 1.0 cm will be used for the irradiation length. Thus, the irradiation volume V_i in cubic centimeters is given by

$$V_i = 1.0 \frac{[2.0 \delta d \sin 9]}{\sin \theta} \frac{4.6 \sin \theta}{(\mu/\rho)\rho}, \tag{9.7}$$

or in cubic micrometers by

$$V_i = 1.439 \times 10^{12} \frac{\delta d}{(\mu/\rho)\rho}. \tag{9.8}$$

If it is assumed that the particles are cubic, with a side of a μm, then the volume of one particle will be a^3 μm^3. The total number of particles being irradiated is given by $N_t = V_i/a^3$. Of this number only a small fraction N_s will contribute to the measured intensity. Thus,

$$N_s = N_t F_F A_D A_S, \tag{9.9}$$

where F_F is the filling factor (or packing fraction), and A_D and A_S are correction factors for vertical divergence and receiving slit apertures. Particles with reflecting planes parallel to the specimen surface will diffract, as will those within a slight variation $\pm \Delta$, where Δ is the angle at which the X-ray tube focus is seen by the specimen, i.e.,

$$A_D = \frac{2 \sin 6°}{170} = \frac{1}{850} \quad \text{(radians)}$$

for a goniometer circle radius of 170 mm and a takeoff angle of 6°. Since both plane and counterplane can diffract, the effective multiplicity is 2 and the fraction of total particles that can diffract is given by

$$A_D = \frac{1}{850} \frac{2 \times 2}{\pi} \cos \theta = 0.001498 \cos \theta. \tag{9.10}$$

This diffracted intensity is spread out over a cone with an opening angle of 2θ.

PROBLEMS OBTAINING A RANDOM SPECIMEN

The cross section of this cone at the receiving slit is

$$A_S = \frac{W}{2\pi 170 \sin 2\theta}, \quad (9.11)$$

where W is the width of the receiving slit in millimeters. Thus,

$$A_S = \frac{W}{1.069 \sin 2\theta}. \quad (9.12)$$

Substituting these data in Equation 9.9 gives

$$N_s = 9.96 \times 10^5 \frac{1}{\sin \theta} \frac{F_F \delta dW}{(\mu/\rho)\rho a^3}. \quad (9.13)$$

The data are calculated for a specimen of α-quartz. These data may be compared with data from the classic work of Alexander and Klug [6] shown in Table 9.4. It will be seen that the relative standard deviation of the measured intensity due to particle statistics is less than a few percent when the particle size is less than 5 µm; however, the statistical error increases rapidly as the particle size exceeds about 10 µm [7].

Table 9.4. Intensity Measurement on Different-size Fractions of < 325 Mesh Quartz Powder

Specimen No.	15–50 µm	5–50 µm	5–15 µm	< 5 µm
1	7,612	8,688	10,841	11,055
2	8,373	9,040	11,336	11,040
3	8,255	10,232	11,046	11,386
4	9,333	9,333	11,597	11,212
5	4,823	8,530	11,541	11,460
6	11,123	8,617	11,336	11,260
7	11,051	11,598	11,686	11,241
8	5,773	7,818	11,288	11,428
9	8,527	8,021	11,126	11,406
10	10,255	10,190	10,878	11,444
Mean area	8,513	9,227	11,268	11,293
Mean deviation	1,545	929	236	132
Mean % deviation	18.2	10.1	2.1	1.2

Source: Data from Alexander and Klug [6].

9.5. PARTICLE SEPARATION AND SIZE REDUCTION METHODS

The easiest way to reduce particle size effects is to grind the sample in a mortar and pestle. If sizing is required, it may be accomplished by sieving; however, sieving below about 20 μm requires special sieves and must be done as a wet slurry. The JCPDS Associateship[1] at the former National Bureau of Standards, [now the National Institute of Standards and Technology (NIST)] developed a procedure for the production of reference standards. The procedure is based on the use of two different specimens, the optimum particle size to obtain accurate d-spacing being different from the optimum size needed for the most accurate intensities. Thus, one specimen is used for obtaining d-values, and the other specimen, having a smaller particle size (10 μm), is used for intensity measurements.

The particle size needed will often determine the extent of grinding. The analyst must be careful that the grinding procedure does not decompose the sample. If the material is soft (e.g., cellulose), grinding can reduce the crystallinity of the sample. Care must be taken when several mixed materials are being ground since the harder materials in the mixture may grind softer materials and the grinding process may destroy the softer material's crystallinity without reducing the size of the hard materials. In most cases, when working with mixtures and unknowns, the analyst may have a trade-off between optimized particle size and varying sample crystallinity.

If the sample is to be ground, care must be taken that the grinding apparatus does not contaminate the sample. Mortars and pestles made from boron carbides usually reduce the probability of contamination. Soft materials such as organic compounds, malleable metals, and fibrous or plate-like compounds may be difficult to prepare with a mortar and pestle. In order to analyze a phase of interest or to identify phases in low concentrations, it may be necessary to physically or chemically treat a sample to concentrate the material(s) of interest. Separation methods include purification by crystallization, sieving, habit, magnetic separation, and microscopic separation. In general, it is a good practice to wet-grind less than 0.1 g of the sample using a solvent that will not dissolve the sample. Wet-grinding is essential to reduce the particle size below about 20 μm, while grinding the entire sample ensures that phase separation will not occur in preparing the specimen.

9.6. SPECIMEN PREPARATION PROCEDURES

X-ray powder diffraction is a very versatile technique that has been applied to a myriad of different problems. It is not surprising, therefore, that many different methods of specimen preparation have been developed. The following paragraphs will summarize the more common of these methods. For further information, the reader is referred to several excellent review articles that appraise and compare some of the more common specimen preparation procedures [1, 2, 8–10]. In qualitative and quantitative analyses of unknown materials, the analyst must reduce preferred orientation so that the diffraction pattern can be reduced and matched to standards for identification. If several sharp intense peaks are observed that do not match the ICDD Powder Diffraction File (PDF) reference pattern intensities of a material, then the sample may be oriented and the analyst should prepare the sample differently. Orientation can be detected by running the sample in one position (without a sample spinner) and then moving the sample through 90° in its own plane and rerunning. If the diffraction intensities vary or if new diffraction peaks arise and others are extinguished, the sample is probably oriented. If the sample is oriented in only one direction, then the sample may have to be again moved 90° perpendicular to the other two directions.

Materials with equal crystal shapes and no cleavages that produce flat samples usually yield random samples using any specimen preparation method. However, if the particles have any crystallographically related shape, the shape will make achieving randomness difficult. As the difference between the maximum and minimum dimensions of the particle increases, it becomes more difficult to make a random sample. Fibrous shapes are generally more difficult to randomize than plate-like shapes. One possible solution to the problem is to mix the sample with a viscous binder and then to mount it into the sample holder. Several authors use a petroleum jelly like Vaseline for this purpose since it is viscous and does not move during the experiment. Vaseline can also protect any hydrated samples from dehydration and usually does not react or recrystallize the original sample. Very dry samples can be dusted directly onto a layer of petroleum jelly, hydrocarbon, or silicon grease. Binding agents can be employed when the sample is *dusty* and/or difficult to pack. Binding agents include acetone, alcohol, water, amyl acetate, oil, and ethers and can even include more viscous materials such as Duco cement. The binder should be used sparingly since the presence of binder reduces the concentration of the sample per unit volume in the X-ray beam. Also, binders either may give a general scattering background if they are noncrystalline or may give rise to their own diffraction pattern, such as that of Vaseline. Moreover, the binder may mask poorly crystalline materials or phases of low concentration. Since some binders are difficult to remove, the use of a binder may prevent sample recovery. Thin binders such as acetone may allow the particles to settle and shaped particles to align.

In some cases, it is desirable to analyze an oriented specimen. In crystalline polymers such as polyethylene or polypropylene, the analyst may want to determine the degree of chain alignment relative to the coordinate system of the sample since this data may relate to the modulus and tensile properties of the material being analyzed. In soil analysis, the detection of clay minerals in the samples can be increased by inducing orientation in the sample. For quantitative analyses by X-ray diffraction, detection limits can often be reduced by using oriented samples. In these cases, standard quantity vs. intensity curves must be prepared. It should be remembered that for ICDD reference standards the intensities were taken on randomized samples, so only known materials should be purposely oriented. Clay minerals rarely show strong diffraction effects from planes other than the (001), so it is highly advantageous to prepare oriented samples.

9.6.1. Use of Standard Mounts

Figure 9.7 shows several of the more common sample holders used in diffractometer analyses. Most of these sample holders use a depression or cavity in which to mount the sample. Cavity mounts are commonly made of aluminum, bronze, Bakelite, glass, or Lucite. In a diffractometer, the X-rays are generally diffracted from the top of the sample. The cavities are usually designed so the surface of the sample is on the diffractometer axis. However, it is not the surface of the sample which is critical; rather, it is the half-depth of penetration of the X-ray beam in the sample that determines the effective diffracting surface and hence the specimen displacement error that shifts the resultant peak position.

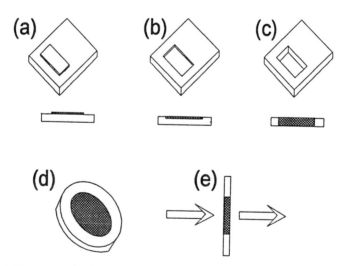

Figure 9.7. Common types of specimen holder: (a) zero background holder; (b) top-loaded; (c) back-loaded; and (d) circular and (e) rectangular pressed mounts.

As mentioned in Section 9.3, the X-ray penetration into the specimen will depend on incident angle, absorption, and elemental composition of the sample. Very low average atomic number compounds, such as most polymers or organics, allow considerable penetration of the X-ray beam, resulting in peak displacement and broadening. In such cases, thick samples in a deep cavity mount may be subject to more displacement errors than thin smears on a glass slide. It is sometimes advantageous to design special specimen holders for a given task; an example of this is the *Kulbiki holder* [11], which was designed to reduce orientation in mineral samples. Frevel [12] designed an adjustable-height sample holder so that the *effective* diffracting surface (i.e., the boundary at the half-depth of penetration) could be made to coincide with the parafocusing surface. The Frevel sample holder consists of a support cylinder with a glass or quartz disk that may be moved by a precise translation screw.

9.6.2. Back and Side Loading

Cavity mounts are most commonly either *side loaded* or *back loaded*. A frosted or serrated glass surface, ceramic or cardboard is placed over the front of the mount (the side exposed to the X-ray beam), and the sample is carefully added through the open back or a side port until the cavity is full. Then the back or side port is covered and the front piece is removed carefully so as not to disturb the surface.

Figure 9.8 shows the steps in the back-loading method. The type of holder employed (a) has a rectangular hole punched through it. The first step (b) is to attach a microscope slide to the top surface using scotch tape or clamps. The holder is then turned over and the powder carefully loaded into the cavity (c). The cavity is filled by applying gentle pressure to the powder as the cavity is filled up. A cover is then placed over the surface of the packed powder. The next step (d) is to turn over the holder again so that the microscope slide is uppermost. When the specimen is to be exposed, the top microscope slide is carefully removed (e).

The advantages of side or back loading can be seen by comparing data from specimens prepared this way with similar data obtained by top loading. As an example, Table 9.5 shows some of the results from a study [13] that compared different specimen preparation techniques for a specimen of molybdenum trioxide (MoO_3). This material is particularly difficult to prepare because, first, it has a rather plate-like habit and tends to orient in the ($0k0$) plane and, second, as was shown in Table 9.2, it allows only a very small X-ray penetration. In order to make comparisons, the data sets in Table 9.5 have each been normalized to the (021) reflection. It will be seen that for the specimen prepared by top loading all ($0k0$) reflections are strongly enhanced. Such a result might be expected when one considers the process of pushing the plate-like particles

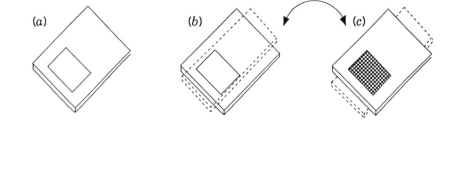

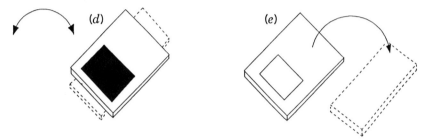

Figure 9.8. Steps in preparing a rear-loaded specimen. (See text for explanation.)

Table 9.5. Data for MoO$_3$ Prepared Using Various Methods

$(hkl) \rightarrow$	(110)	(040)	(021)	(111)	(060)	(200)	(002)
Method			Relative Intensities				
Pressed	141	406	100	27	224	61	122
Side-drifted	79	52	100	27	27	10	16
Spray-dried	72	39	100	26	19	9	15
Calculated	76	38	100	26	21	8	15
PDF 5–508	82	61	100	35	31	13	21

into the cavity of the specimen holder. Conversely the data obtained from the side-drifted specimen agree reasonably well with the calculated data (data are calculated assuming a random orientation). Also shown are data taken from a spray-dried specimen, and here the data agree almost exactly with the calculated data.

The side-loading technique developed by the National Bureau of Standards (see McMurdie et al. [14]), now NIST, is probably the best *packing* method developed to date. This method is similar to the rear-loading method except that the specimen is loaded at the side of the holder. A disadvantage with the side-loading method is that the designs of many of the commercial specimen holders and presentation devices do not allow easy opening up of one side of the holder. An almost equally effective method, which does not have the same difficulty, is that of rear loading [15]. Comparison of the side- and rear-loading methods indicates that, provided adequate care is taken, excellent results can be obtained with either technique [16].

9.6.3. Top Loading

If the holder can only be front mounted, then it is advisable to retouch the surface by lightly cutting grooves with a sharp edge, rolling with a knurled surface, or tamping with a serrated flat. This technique may leave a roughened surface, so it is advisable to adjust the sample height. Pressing the surface smooth with a microscope slide is ill advised as it enhances preferred orientation in the surface layers which contribute heavily to the resultant peak intensities. If preferred orientation is a problem, then the sample may be mixed with a poorly scattering filler. Powdered glass, cork, starch, gelatin, gum arabic, and tragacanth have been used. Starch has a spherical shape and, after picking up particles clinging to its surface, can introduce random packing into the cavity.

9.6.4. The Zero Background Holder Method

The zero background holder (ZBH) sample-mounting method [17] combines the advantages of the external standard method with the accuracy of the internal standard method. A ZBH first proposed by Post [18] is a single crystal that has been cut along a nondiffracting crystallographic direction and then polished to optical flatness. Any X-rays falling on this crystal will be completely extinguished by Bragg extinction. When this method is coupled with use of a diffracted-beam monochromator to remove fluorescent X-rays from the sample and ZBH crystal, the resulting very low background improves the signal-to-background ratio in a powder pattern. The ZBH method entails wiping a thin layer of grease on the crystal holder and then wiping off all but a monolayer. A few milligrams of a sample are manually wet ground (perhaps in acetone) to particle sizes in the range of 1 µm. This powder is dusted onto the grease layer, and any excess is tapped off. Thus, the specimen is prepared as a *monolayer*, so it is possible to see through the loaded sample when viewed normal to the surface of the crystal but not when viewed from a low angle. The total thickness of the sample and grease will be on the order of a few micrometers. It is important to deagglomerate specimens in order that this monolayer condition be met. If exactly the same procedure is used to prepare calibration standards as well as routine samples, the application of the calibration curve will leave only a negligible sample displacement error of a few micrometers uncorrected. There are a number of attractive features to this type of routine sample preparation and calibration:

- Sample displacement and transparency effects are negligible.
- No dilution of the sample is required.
- Fully automatic application of the calibration procedure to all samples can be set as a default condition.
- The grease smear preparation minimizes preferred orientation effects.
- Very small (few-milligram) samples are required.
- The required fine grinding minimizes extinction and optimizes particle counting statistics for the small sample.
- The low intensity due to the small sample is enhanced by the extremely low background of the monocrystal holder produced by Bragg extinction.

The only disadvantage of this procedure is that the overall lower intensity produced by the small sample affects the detectability of weak lines and the particle-counting statistics. This disadvantage does not affect phase identification but clearly hurts the detection of trace phases. In those few cases where low intensities require a second thick mounting to check for trace phases, all of the strong lines from the ZBH run can be used as a secondary standard to calibrate it, avoiding the need for an internal standard.

Table 9.6 shows the effect of absorption on four calibration materials. Table 9.7 shows the effect of external-standard, internal-standard, and ZBH calibration on these four materials. The average $\Delta 2\theta$ is given as the first number after the parentheses of the F_N figure of merit, to be discussed in Section 11.6.3, and is a measure of the precision. Similarly, the standard deviations of the least square refined lattice parameters, given in parentheses as 1σ in the last digit(s), indicates how these methods relate to physical quantities. Figure 9.9 shows typical calibration curves for side-drifted external and internal calibration, and ZBH calibration.

The $\Delta 2\theta$ values for the ZBH and internal-standard calibration methods are comparable in all cases. The overall average $\Delta 2\theta$ for internal-standard calibration in a recent study [17] was 0.011°, while for the ZBH method it was 0.007°. Note that these values for $\Delta 2\theta$ are three times smaller than the National Bureau of Standards data from PDF sets 1–24 [19,20], which include data from the era before diffractometer automation, showing that both hardware and data analysis techniques have improved with time.

Table 9.6. Cu $K\alpha$ Mass Attenuation Coefficients and Penetration Depths for Four Calibration Materials[a]

Material	μ (cm^{-1})	Depth (μm)
Ag	2289	3
LaB$_6$	644	10
α-Al$_2$O$_3$	124	52
Urea [CO(NH$_2$)$_2$]	10	643

[a] Penetration depth is based on 95% of the diffracted intensity at 50° 2θ.

Table 9.7. ZBH Calibrated Data

Material	Parameter	Second-Derivative Peak Location Uncalibrated	Second-Derivative Peak Location Calibrated	Profile-Fitted Peak Location
Ag	F_N	$F(9)=84$ (0.0119,9)	$F(9)=245$ (0.0041,9)	$F(9)=168$ (0.0060,9)
	a (Å)	4.0858(14)	4.08616(5)	4.08639(7)
Al$_2$O$_3$ (Linde C)	F_N	$F(9)=235$ (0.0029,13)	$F(9)=137$ (0.0050,13)	$F(7)=69$ (0.0079,13)
	a (Å)	4.7582(2)	4.7598(3)	4.7592(8)
	c (Å)	12.9849(7)	12.9895(11)	12.9951(75)
LaB$_6$	F_N	$F(13)=172$ (0.0058,13)	$F(13)=190$ (0.0053,13)	$F(13)=274$ (0.0036,13)
	a (Å)	4.1552(1)	4.1562(1)	4.15635(8)
Urea	F_N	$F(11)=57$ (0.0107,18)	$F(11)=56$ (0.0108,18)	$F(11)=53$ (0.0115,18)
	a (Å)	5.6478(7)	5.6499(7)	5.6493(6)
	c (Å)	4.6972(23)	4.6989(24)	4.6982(22)

9.6.5. Spray-Drying

The only completely general way of overcoming effects of preferred orientation is to spray-dry the sample. Figure 9.10 illustrates this technique in which the solid is wet-ground to reduce the particle size to under 10 μm. The grinding is conveniently carried out in a vibratory mill with a slurry composed of

252   SPECIMEN PREPARATION

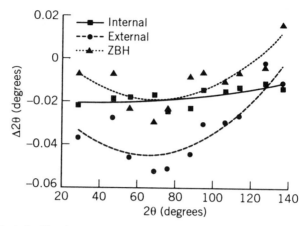

Figure 9.9. Typical calibration curves for side-drifted external and internal calibration, and ZBH calibration.

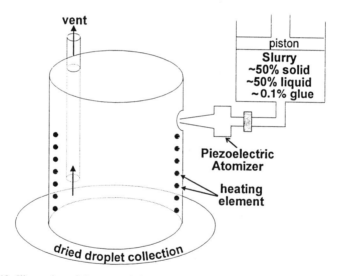

Figure 9.10. Illustration of the spray-drying procedure for preferred-orientation-free specimen preparation.

approximately 50% solids. A small amount of a binder like poly(vinyl alcohol) and a deflocculant like Darvan is added to the slurry, and it is then atomized into a hot chamber so that the droplets dry before striking the walls. Modern piezoelectric-driven atomizers have the ability of producing a mist of droplets with zero velocity, so they may fall by gravity through a lab bench-top drying

SPECIMEN PREPARATION PROCEDURES   253

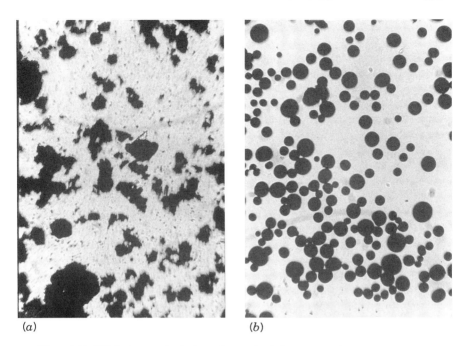

(a)   (b)

Figure 9.11. SEM micrograph of a hematite powder before (a) and after (b) spray-drying.

chamber. Figure 9.11 shows the resulting spherical dried droplets. These spheres, with a typical size of 50 µm, have the crystallites distributed over their isotropic surface, which results in all orientations being equally shown to the X-ray beam.

Although this method of specimen preparation is easy to perform, it requires perhaps 15 min per sample and requires the additional expense of constructing a spray dryer. It has been shown in a number of careful studies [21,22] that the technique is effective in eliminating preferred orientation from all types of samples including plate-like clays and acyclic materials such as wollastonite. Due to the extra work involved with spray-drying, this technique is typically only employed when the relative intensities must be known accurately, as in quantitative phase analysis and in Rietveld structure analysis.

9.6.6. Use of Aerosols

One novel method of obtaining a randomized specimen is to produce an aerosol of the specimen by allowing air to rush into an evacuated chamber containing the powdered specimen. By using this tubular aerosol suspension chamber method, it is possible to randomly deposit small amounts of powder on a filter paper or a fiberglass membrane, which can then be used directly for analysis. Aerosol loads of typically 300–1,000 µg/cm^2 can be obtained using this method. This technique has been successfully employed, for example, for the analysis of a variety of different minerals [23]. Because of the possible different aerodynamic properties of different phases in a mixture, it is important that a complete sample be recovered in the filter to avoid phase separation.

9.7. MEASUREMENT OF THE PREPARED SPECIMEN

Systematic investigations [24,25] of the common errors associated with search/matching procedures for identifying unknowns have shown sample positioning to be the primary cause of failures in the identification process. Correct sample positioning depends on both proper instrumental alignment and on specimen preparation and presentation. It is important to remember that the penetration of the X-ray beam is also very small, particularly at low angles, where it may be as little as a few of micrometers, while it may be more than double that at higher angles. In addition, the irradiation area at the surface of the specimen increases markedly with decrease in 2θ unless the divergence slit is changed. For example, from the data given in Table 9.3, if a scan is started at $10°$ 2θ with a $1°$ divergence slit, the beam splashes off the ends of a typical sample holder. This effect causes the relative intensities of the low-angle diffraction lines to be underestimated.

9.7.1. Specimen Displacement

It will be seen from the specimen displacement error given by Equation 7.6 that at moderate-to-low angles, where $\cos\theta$ approaches unity, the magnitude of the error is roughly proportional to $(-2s/R)$. For a goniometer radius of 17 cm, the error in 2θ is approximately equal to $-0.01°$ 2θ for every 15 µm of specimen displacement. Figure 9.12 illustrates the main causes of specimen displacement. The center of the θ shaft represents the center of the goniometer circle. A reference surface is provided against which the specimen holder can be placed. It is the task of the diffractionist to ensure that the surface of the prepared specimen is on the same plane as the specimen holder and that the specimen holder really does fit correctly against the reference surface. Should there be any misalignment of specimen surface, specimen holder surface, or reference surface, a specimen displacement error will result. If the specimen surface is too high, a positive angular error will occur, and if the specimen surface is too low, a negative angular error will result.

MEASUREMENT OF THE PREPARED SPECIMEN   255

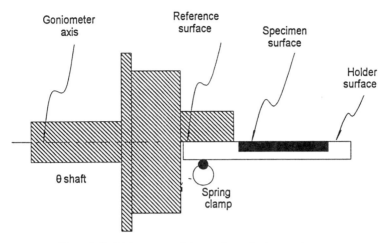

Figure 9.12. Main sources of the specimen displacement error.

The effect of specimen displacement can be minimized by careful preparation and mounting of the thick specimen. With care, this displacement can be kept within the range ± 50 µm, corresponding to a 2θ error on the order of a few hundredths of a degree, and an equivalent d-spacing error in the range of a few parts per thousand. In the ZBH technique, the displacement can typically be held to less than 5 µm and is negligible. Where accurate work is required, for example, in the determination of lattice parameters, it is usual to employ an internal standard or the ZBH technique to correct for this and other errors.

9.7.2. Mechanical Methods for Randomizing

The most common way to reduce oriented peaks is by rotation of the sample during the experiment. Most diffraction equipment comes with some type of optional sample-rotation device. The purpose of sample rotation is to bring more crystallites into a position to diffract for each peak and to average out (randomize) the contribution of the crystallites to each peak in the scan. Figure 9.13 illustrates how the number of crystallites irradiated in a specimen can be increased by spinning the specimen in its own plane during analysis. When the specimen is stationary, the area irradiated is equal to $S_d L_i$, where L_i is the irradiation length (in millimeters), defined as in Equation 9.2, and S_d is the specimen width taken in this instance as 10 mm. The radius r of the rotation circle is given by

$$r^2 = \left(\frac{S_d}{2}\right)^2 + \left(\frac{L_i}{2}\right)^2. \tag{9.14}$$

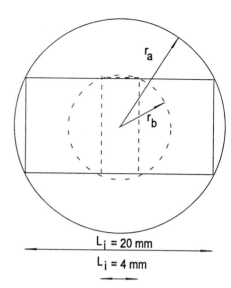

Figure 9.13. Effect of rotating the specimen: L_i is the irradiated length; r_a and r_b are the radii of the sample area and irradiated area, respectively.

Thus, the area A of the rotation circle will be

$$A = \pi\left(\frac{S_d^2}{2} + \frac{L_i^2}{2}\right), \tag{9.15}$$

and the gain G in area (and volume) due to rotation of the specimen is given by

$$G = 0.08\left(\frac{L_i^2 + 100}{L_i}\right). \tag{9.16}$$

Examination of Figure 9.13 and Equation 9.16 show that as the irradiation length L_i is decreased, G increases (see Table 9.8). This gain is minimized at the point at which $L_i = S_d$. There is an equivalent improvement in the particle statistics, the magnitude of which is given by $\sqrt{G}$. This improvement factor is also given in Table 9.8.

The rotator may randomize one or two directions depending upon its mechanical motion. While a sample rotator may provide a means of reducing the results of orientation on the diffraction data, it does not reduce orientation and should not be a substitute for specimen preparation methods necessary to provide a random sample. The sample rotators do not change the crystalline orientation—they only change the position of these crystals relative to the X-ray beam. It will be recognized that where the specimen is strongly preferred in the (001) direction such that all reflecting planes are parallel to the specimen

Table 9.8. Improvement in Particle Statistics Due to Specimen Spinning

2θ	L_i	Area Gain	Statistical Gain
10	18	1.85	1.36
20	9.0	1.58	1.26
30	6.0	1.78	1.33
40	4.6	2.07	1.44
50	3.7	2.41	1.55
60	3.1	2.78	1.67
70	2.7	3.12	1.77
80	2.5	3.34	1.83
90	2.2	3.74	1.94

surface, spinning the specimen will not be effective. An alternative method that is sometimes employed is to rock the specimen about the θ axis during analysis.

9.7.3. Handling of Small Samples

If microsamples are to be analyzed, sample support and position become crucial variables. Single crystals, either cleaved or cut on or off Bragg planes, have been effective sample supports. Fluorite, calcite, and MgO can be easily cleaved. Silicon and quartz can be cut off Bragg planes and chemically polished. Quartz can be polished mechanically by grinding on a flat surface such as a glass plate, with 600 grit SiC followed by a 90 s etch in HF. This treatment effectively eliminates peaks from the damaged surface layers. Once a substrate is chosen, the sample must be placed in the center of the X-ray beam. A binocular microscope is a handy aid in centering any sample. Crystals can be transferred to the support by using glass capillaries, metal fibers, or a set of dental tools (picks of different sizes and curvatures). Once the sample is centered, a binder or lacquer can be used to hold it in place. Double-sided tape can also be used to hold the sample to the center of the support. With small samples, beam scatter can become an important consideration. If possible, the sample chamber and beam path should be evacuated or flushed with helium. In general, the ZBH technique permits the routine analysis of samples as small as 1.0 mg.

9.7.4. Special Samples

For slightly reactive samples, hygroscopic materials, slurries, or liquids, a sealed sample holder may be necessary. The sample can be placed in a depression in the holder, then sealed with a thin film of plastic, or even Scotch cellophane tape. Since the sealing material will give a diffraction pattern, special care must be taken to ensure that the seal pattern does not confuse the measurement of the diffraction pattern of the specimen. It is advisable to run a blank sample so that diffraction peaks from the sample seal can be accounted for. Reactive samples may require complete encapsulation in a special sample holder. Samples for pressure and temperature studies also deserve special consideration. Low-temperature samples crystallized *in situ* often require treatment to produce a good polycrystalline layer. High-temperature, oxidation, and/or reduction experiments generally require special inert holders to prevent unwanted reactions. High-pressure samples are very dependent on apparatus but usually require diluents to allow transmission through the sample.

REFERENCES

1. Bish, D. L., and Reynolds, R. C. Sample preparation for X-ray powder diffraction. In *Modern Powder Diffraction* (D. L. Bish and J. E. Post, eds.), Rev. Mineral, Vol. 20, pp. 72–99. Mineral. Soc. Am. Washington, DC, 1989.
2. Jenkins, R., Fawcett, T. G., Smith, D. K., Visser, J. W., Morris, M. C., and Frevel, L. K. Sample preparation methods in X-ray powder diffraction. In *JCPDS–ICDD Methods and Practices in X-Ray Powder Diffraction*, Sect. 5.2.1. International Centre for Diffraction Data, Newtown Square, PA, 1988.
3. Parrish, W., and Huang, T. C. Accuracy and precision of intensities in X-ray polycrystalline diffraction. *Adv. X-Ray Anal.* **26**, 35–44 (1983).
4. Jenkins, R., and de Vries, J. L. *An Introduction to Powder Diffractometry*, 7000.02.3770.11. Philips, Eindhoven, The Netherlands, 1977.
5. Jenkins, R., and de Vries, J. L. *Worked Examples in X-Ray Analysis*, 2nd ed., pp. 61–63. Springer-Verlag, New York, 1978.
6. Alexander, L. E., and Klug, H. P. Basic aspects of X-ray absorption in quantitative diffraction analysis of powder mixture. *Anal. Chem.* **20**, 886–894 (1948).
7. Smith, D. K. Particle statistics and whole-pattern methods in quantitative X-ray powder diffraction analysis. *Adv. X-Ray Anal.* **35A**, 1–15 (1992).
8. Smith, D. K., and Barrett, C. S. Special handling problems in X-ray diffractometry. *Adv. X-Ray Anal.* **22**, 1–12 (1979).
9. Jenkins, R., Fawcett, T. C., Smith; D. K., Visser, J. W., Morris, M. C., and Frevel, L. K. Sample preparation methods in X-ray powder diffraction. *Powder Diffr.* **1**, (2) 51–63 (1986).
10. de Woolf, P. M., and Visser, J. W. Absolute intensities—outline of a recommended practice. *Powder Diffr.* **3**, 202–204 (1988).
11. Kulbiki, G. L. Dosage des principaux minéraux des roches sédimentaires par diffraction-X. *Rapp. Internes SNPA* (1959).

12. Frevel, L. K. Error analysis of 2θ powder data for cubic or uniaxial phases. *J. Appl. Crystollogr.* **11**, 184–189 (1978).
13. Calvert, L. D., Sirianni, A. F., Gainsford, G. J., and Hubbard, C. R. A comparison of methods for reducing preferred orientation. *Adv. X-Ray Anal.* **26**, 105–117 (1982).
14. McMurdie, H., Morris, M. C., Evans, E. H., Paretzkin, B., and Wong-Ng, W. Methods of producing standard X-ray diffraction patterns. *Powder Diffr.* **1**, 40–43 (1986).
15. McCreery, G. L. Improved mount for powdered specimens used in the Geiger counter X-ray spectrometer. *J. Am. Ceram. Soc.* **32**, 141–146 (1949).
16. Hubbard, C. R., and Smith, D. K. Experimental and calculated standards for quantitative analysis by powder diffraction. *Adv. X-Ray Anal.* **20**, 27–39 (1977).
17. Misture, S. T., Chatfield, L., and Snyder, R. L. Accurate powder diffraction patterns using zero background holders. *Powder Diffr.* **9**(3), 172–179 (1994).
18. Post, B. Laboratory hints for crystallographers. *Norelco Rep.* **20**, 8–11 (1973).
19. Snyder, R. L., Johnson, Q. C., Kahara, E., Smith, G. S., and Nichols, M. C. An analysis of the powder diffraction file. *Lawrence Livermore Lab. [Rep.] UCRL* **UCRL-52505** (1978).
20. Snyder, R. L. The renaissance of X-ray powder diffraction. In *Advances in Material Characterization* (D. R. Rossington, R. A. Condrate, and R. L. Snyder, eds), pp. 449–464. Plenum, New York, 1983.
21. Smith, S. T., Snyder, R. L., and Brownell, W. Minimization of preferred orientation in powders by spray drying. *Adv. X-Ray Anal.* **22**, 77–87 (1979).
22. Cline, J. P., and Snyder, R. L. Sample characteristics affecting quantitative analysis by X-ray powder diffraction. In *Advances in Materials Characterization II* (R. L. Snyder, R. A. Condrate, and P. F. Johnson, eds), pp. 131–144. Plenum, New York, 1985.
23. Davis, B. L., and Johnson, L. R. Sample preparation and methodology for X-ray quantitative analysis of thin aerosol layers deposited on glass fiber and membrane filters. *Adv. X-Ray Anal.* **25**, 295–300 (1982).
24. Schreiner, W. N., and Jenkins, R. Automatically correcting for specimen displacement error during XRD search/match identification. *Adv. X-Ray Anal.* **25**, 231–236 (1982).
25. Huang, T. C., and Parrish, W. A new computer algorithm for qualitative X-ray powder diffraction analysis. *Adv. X-Ray Anal.* **25**, 213–219 (1982).

CHAPTER
10
ACQUISITION OF DIFFRACTION DATA

10.1. INTRODUCTION

Most qualitative X-ray powder diffractometry is carried out using a reduced pattern, in which profiles are expressed as a set of single unique d or 2θ values, along with intensities expressed as a percentage of the strongest line in the pattern. In the examination of errors in the experimental $d-I$ list to be used, it is necessary to consider errors introduced both in the collection of the original (raw) diffractogram and in the production of the reduced pattern. In addition to the geometric systematic aberrations and the sample placement problems discussed in Section 9.7, other influences on the observed d-spacings may be encountered. For example, phases may sometimes form solid solutions, as in the case of sodium and potassium feldspars. When an atom substitutes for another in a given structure, giving a substitutional solid solution, the size difference between the two types of atoms may cause the lattice parameters to increase or decrease. Because the d-value is a direct function of the lattice parameters, solid solution formation will cause significant shifts in the measured d-values. These shifts can be so large as to result in failure to identify a solid solution phase. Although this problem can, in principle, be successfully handled by computer techniques, the quality of the older data in the ICDD[1] Powder Diffraction File (PDF) has inhibited researchers from applying such techniques to any great extent (although some programs are now available that will help in the identification of solid solutions; e.g., see Schreiner et al. [1]).

10.2. STEPS IN DATA ACQUISITION

Table 10.1 lists the typical steps that are involved in the acquisition, treatment, and storage of diffraction data. In this chapter, while the emphasis is on the *acquisition* of data, the effect of the choice of experimental conditions under which the pattern is to be recorded must also be considered. Figure 10.1 shows the relationship between the experimental pattern and the "reduced pattern." The true pattern is a series of unique reflections corresponding to the various

---
[1] International Centre for Diffraction Data, Newtown, Square, PA 19073.

262 ACQUISITION OF DIFFRACTION DATA

Table 10.1. Steps in the Acquisition and Treatment of Powder Diffraction Data

1. Specimen preparation and presentation
2. Selection of instrument variables
   Source: kV, mA, (λ)
   Divergence and receiving slits
3. Data collection
   Range, step size, count time
4. Pattern reduction
   Smooth
   $K\alpha_2$ strip
   Peak locate
   Peak correction/calibration
   Store/report
5. Interpretation

d-spacings allowed by the symmetry and crystal structure of the phase(s) making up the pattern. The experimental pattern samples the true pattern through a filter determined by the experimental conditions. In the illustration, Al_2O_3 has been taken to show the features of the pattern. During the course of the measurement of the pattern, each d-value is represented by a reflection at the appropriate Bragg angle. Each reflection has a finite width, this width depending on the unique instrumental and specimen characteristics. Due to the (generally) polychromatic nature of the source, the lines may appear as doublets, the separation of the doublets being dependent on the angular dispersion of the diffractometer. In addition, extra lines may appear due to, for instance, spectral contamination from the X-ray tube. The diffraction peaks are superimposed on a definite background level that arises mainly from a combination of scatter and fluorescence. The peak and background data are subject to counting statistical fluctuations, the magnitude of such fluctuations being dependent upon the count rates and the process involved in the data acquisition (count times, ratemeter settings, smoothing, etc.). Finally, the range of the pattern is limited due to both instrumental and available analysis time constraints.

In the reduction of the experimental pattern, an attempt is made to obtain a reduced pattern that is as close as possible to the true pattern. The steps in the reduction process include the location of the peak maxima, with or without spectral stripping, the conversion of peak maxima to d-values, data smoothing, removal of background, and normalization of intensities. In these processes, certain decisions have to be made that may markedly affect the number

STEPS IN DATA ACQUISITION   263

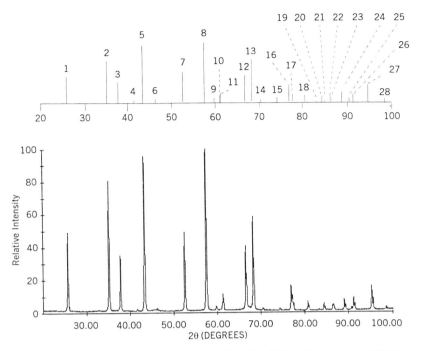

Figure 10.1. The experimental (bottom) and reduced (top) diffraction patterns of Al_2O_3.

of lines found in the pattern. These include loss of small peaks under background, loss or gain of peaks due to noise/smoothing, and/or ineffective peak hunting.

The data in Figure 10.1 represent an actual diffractogram of Al_2O_3 obtained with Cu $K\alpha$ radiation. In this example, the intensity scale is linear in count rate with all peak intensities normalized to the strongest line, which is given the value of 100. The pattern has been recorded between 2° and 100° 2θ, and 28 lines are observed. The experimental intensity data can be displayed in a number of ways including linear, logarithmic, square root, etc. Displaying the intensities in terms of the square roots offers three important advantages over the use of a linear scale:

1. The dynamic count rate range is doubled (1×10^4 rather than 1×10^2).
2. It is easier to see weak lines close to the background level.
3. Since the count error varies as $\sqrt{N}$, the error scale is linear.

Figure 10.2 shows the same corundum pattern as that shown in Figure 10.1,

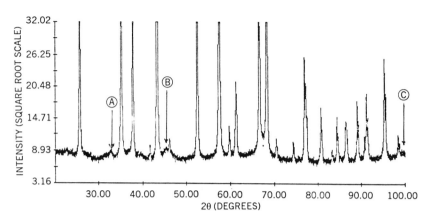

Figure 10.2. Pattern of Al_2O_3 with intensity on a square root scale.

but now with the intensity on a square root scale. Study of the diffraction pattern in the square root intensity format reveals three additional "possible" peaks, labeled in the figure as A, B, and C. As far as data acquisition is concerned, if the small peaks are to be considered as significant (for most qualitative work, the common procedure is to consider lines only down to 1% of the strongest line), it is important that sufficient counts be taken at each step so as to produce statistically significant data. As an example, referring again to Figure 10.2, we note that the average background count is about 64 c/s. With a step time of 1 s, 64 counts are taken at the background with an error (1σ) of 8 c/s. If any signal in excess of the background plus 2.5σ of the background is taken as significant, any value greater than 84 c/s will be statistically significant. On this basis, each of the peaks A, B, and C are significant.

10.3. TYPICAL DATA QUALITY

One of the problems in routine powder diffractometry is the tendency on the part of users to overestimate the accuracy of data. In an ICDD-sponsored round-robin study, Schreiner and Fawcett concluded that "*The major, and not-so-surprising, conclusion from this study is that routine powder diffractometer data is probably 3–5 times worse on the average than most workers are willing to admit*" [2]. A survey taken by the ICDD some years ago revealed that more than half the users polled believed they were routinely able to obtain d-values to better than 1 part in 5000. Results of round-robin tests on actual powder samples indicated that these estimates of accuracy were overstated by about an order of magnitude [3, 4]. The gap between belief and reality is, in

Table 10.2. Average 2θ Precision Obtained with Zinc Oxide (Z) and Calcite (C) Mixtures in a Multilaboratory Round-Robin Test

Line	Average	σ	Line	Average	σ
Z(100)	31.793	0.052	C(012)	23.072	0.050
Z(002)	34.448	0.041	C(104)	29.428	0.051
Z(110)	56.603	0.049	C(113)	39.433	0.051
Z(112)	62.867	0.046	C(202)	43.169	0.053
Z(213)	116.629	0.043	C(116)	48.519	0.046
			C(324)	83.759	0.040

part, fostered by the increasing use of automated diffraction systems. While the automated system clearly offers many advantages, it is also clear that the automation tends to divorce the user from the actual processes involved in collection and analysis of data. Unless the user is prepared to make a conscious and often time-consuming effort to understand the procedures being employed, he or she can easily be misled by such misconceptions. While computer techniques are extremely useful, care must be taken in comparison of computer-treated data and manual treated data. Such is frequently the case in the set of data from a modern automated powder diffractometer (APD) and standard patterns from the PDF. A further complication is that most APD software systems now use a method of peak hunting based on the use of the second differential of the peak. This technique is extremely sensitive to partially resolved peaks and more peaks are generally found than using manual methods.

However, in spite of the numerous sources of error in the measurement of 2θ values and subsequent conversion to d-spacings, the derived d-values are generally sufficient to allow qualitative phase identification. As an example, Table 10.2 shows data from a round-robin test [5] in which participants were asked to measure samples of the binary mixture $CaCO_3/ZnO$. Here the average multiperson/multilaboratory precision is about 0.048°.

10.4. SELECTION OF THE d-SPACING RANGE OF THE PATTERN

It is important that an adequate number of d-values be measured in an experimental pattern, to ensure the complete identification of the material under investigation. A good rule of thumb is that about 50 lines should be recorded for qualitative phase identification, unless the material is of very high symmetry and only gives a few lines. In this instance, it is usual to measure the pattern out to 60° 2θ or so. Most important of all is to ensure that as many as possible of the low-angle reflections are measured since these lines are vital both for correct phase identification and for pattern indexing. The two most important variables that will determine the d-spacing range covered are the 2θ scan range and the experimental wavelength.

10.4.1. Choice of the 2θ Range

The example cited in Section 10.2 illustrated the need to record and treat experimental and reference patterns under similar conditions. One of the first choices to be made in recording an experimental pattern is the angular range over which it should be measured. Of the various factors that come into this choice, the previous experience of the operator is probably the most important. Where such experience does not exist, the following points should be considered. First of all, the largest d-spacings are generally the most useful both in pattern identification and in pattern indexing. Thus, there is a great need to ensure that all of the low lines that could be observed are actually measured. Very large d-spacings, perhaps greater than $80 Å$ or so, are not measurable using conventional powder diffractometers and Cu $K\alpha$ radiation. Large d-spacings occur from large unit cells. In general terms, this will mean organic materials and low symmetry materials. A special problem arises because the conversion of experimental 2θ errors to d-spacing errors includes a cot 2θ term (see Section 11.4). This tends to magnify low-angle errors and requires special attention to be given to the measurement and treatment of the lower 2θ values. This is especially true where a fixed divergence slit is employed (as opposed to a "variable" divergence slit), and the choice of divergence slit aperture can be quite critical below $10° 2\theta$ or so. In general, a diffraction scan using Cu $K\alpha$ radiation should begin at about $5° 2\theta$.

The choice of the maximum value to be measured is also important. For routine phase identification, one would like to have around 50 or so lines to work with. Where the symmetry of the material(s) under investigation is low, this number of lines could be achieved below a maximum angle of $30°$. On the other hand, where the symmetry is high, a phase might give only 10 lines or so up to $120° 2\theta$. Thus, the choice of the maximum scan angle is more difficult to generalize. A good rule of thumb is to routinely scan all *unknowns* from $5°$ to $65° 2\theta$ and if the number of lines is still less than 50 or so, scan to a higher angle.

10.4.2. Choice of Wavelength

Copper $K\alpha$ is by far the most popular wavelength used for powder diffraction work, since it provides a reasonable compromise between dispersion and d-spacing range coverable for most commonly encountered materials. As was explained in Section 6.2, the Cu $K\alpha$ line is actually a doublet and various

SELECTION OF d-SPACING RANGE OF THE PATTERN   267

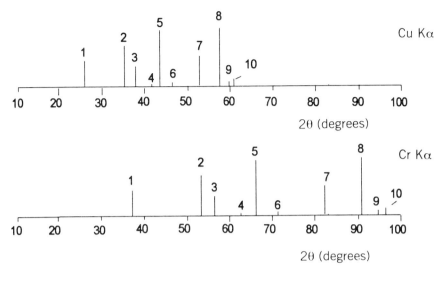

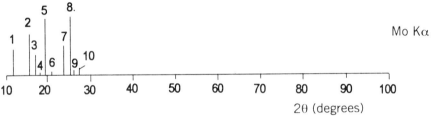

Figure 10.3. Pattern for corundum measured with chromium, copper, and molybdenum radiation.

Table 10.3. Maximum and Minimum d-Values Measurable with Various Wavelengths

Radiation	Wavelength	Maximum d	Minimum d
Cr $K\alpha$	2.291	65.6	1.16
Co $K\alpha$	1.790	51.3	0.91
Cu $K\alpha$	1.542	44.2	0.78
Mo $K\alpha$	0.709	20.3	0.36

measures are available for either removing the $K\alpha_2$ component from the incident beam by primary beam monochromatization or removing the $K\alpha_2$ from the experimental pattern by spectral stripping. The recommended wavelengths for Cu $K\alpha_1$ and Cu $K\alpha_2$ are 1.54060 and 1.54439 Å, respectively. It follows from Bragg's law that an error in d-spacing is linearly related to an

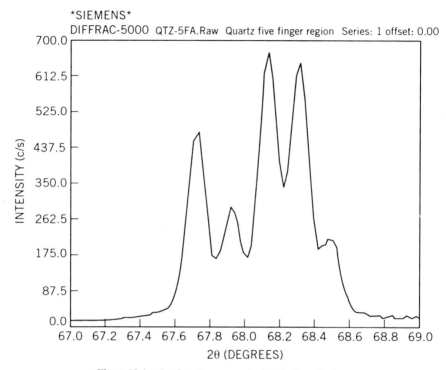

Figure 10.4. (a) α-Quartz measured with Cu Kα radiation.

error in wavelength. For most qualitative phase identification, an accuracy of a few parts per thousand is sufficient for $\Delta d/d$. Provided that the true $K\alpha_1$ emission line is being used (i.e., either the diffracted beam is monochromatic or the contribution from the α_2 is effectively removed from the diffracted line profiles), the wavelength need also be known to about one part per thousand [6]. For the accurate measurement of lattice parameters, it is common practice to utilize an internal standard and, for all intents and purposes, the effect of wavelength errors is calibrated out.

Choice of a wavelength longer than Cu $K\alpha$ increases the dispersion of the pattern and at the same time shifts the pattern to a higher angular range (in turn, reducing the range for the recording of smaller d-values). Use of a wavelength shorter than Cu $K\alpha$ has the opposite effect. An example is given in Figure 10.3, where the first 10 lines of Al_2O_3 are shown as measured with Mo $K\alpha$ (0.709 Å), Cu $K\alpha$ (1.542 Å), and Cr $K\alpha$ (2.291 Å). Table 10.3 shows the full range of d-spacings that are measurable using these three wavelengths in

SELECTION OF d-SPACING RANGE OF THE PATTERN

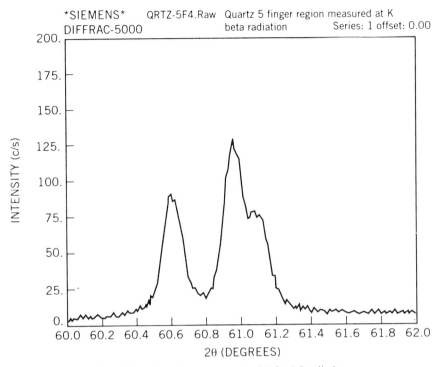

Figure 10.4. (b) α-Quartz measured with Cu $K\beta$ radiation.

the goniometer range of 2–160° 2θ. In those cases where a diffracted beam monochromator is used and where the diffraction pattern is complex with many overlapping lines (especially at higher angles), an advantage may be gained by resetting the monochromator to collect the unresolvable $K\beta$ doublet rather than the resolved $K\alpha$ doublet. Figure 10.4 shows the pattern for the classic "five fingers" of α-SiO$_2$ measured with the α_1, α_2 doublet and the β_1, β_3 doublet. As was described in section 1.4.4, the energy separation of the α_1, α_2 doublet and the β_1, β_3 doublet is due to the spin quantum number. In the case of the α doublet the energy gap due to the spin quantum number is about 12 eV; in the case of the β-doublet it is about 2 eV. Since the absolute energy resolution of the diffractometer is never better than about 2 eV, the β_1, β_3 doublet is never resolved. Even though the absolute intensity is reduced by about a factor of 5 when β rather than α radiation is used, this intensity loss may be justified by the significant simplification of the pattern.

10.5. MANUAL POWDER DIFFRACTOMETERS

10.5.1. Synchronous Scanning

If we now glance ahead at Chapter 12, Figure 12.7 shows a manual chart-recorder tracing of the diffraction pattern of a mixture of Al_2O_3 and ZnO recorded using Cu $K\alpha$ radiation. The scan is essentially a plot of intensity as a function of 2θ. As an additional aid in estimating the correct angles for each peak maximum, a time event marker gives a continuous trace at the edge of the chart with a line that changes its position every $0.5° 2\theta$. The first thing to do in treating such a manual pattern is to number each line and to record the line number, the value of 2θ, and the intensity, as shown in Table 12.9. It may not always be apparent as to what is or is not a true "line," particularly in cases of poor peak-to-background ratios arising from low concentration, poor crystallinity, line broadening, and the like. It is thus necessary that one first define a background threshold (e.g., a peak-to-background ratio of greater than 2:1 could be used). Having listed all line numbers, angles, and absolute intensities, the angle must be converted into d values by use of a chart of 2θ vs. d. Lastly all intensity values must be normalized such that the strongest is equal to 100 units.

Because the detector circuitry gives information in digital form, some system of integration must be used if a continuous scan over an angular range is required. For this purpose, either a ratemeter circuit or a step-scanning system is used. In a typical analog–digital (A/D) counting chain, all pulses from the detector are first passed to the pulse height selector and then allowed to flow in one of two directions. The analog chain makes use of a ratemeter and an x/t recorder.

10.5.2. Use of Ratemeters

In ratemeter integration, the pulses are integrated for a time related to the RC time constant of the ratemeter as described in Section 5.5.1. A ratemeter is essentially a smoothing device that gives a measure of the counting rate, averaged over a given period of time, related to the time constant. In practice, a ratemeter consists of pulse-amplifying and -shaping circuitry that gives pulses of fixed time and voltage dimensions. These pulses are fed into an integrating circuit consisting of a resistor R and a capacitor C. The time to charge the capacitor to 63% of its maximum charge value is called the time constant RC. Integration of the pulses is thus achieved by a succession of charging and discharging stages, the latter utilizing a value of RC selectable by the operator. In practice, care must be taken in matching the time constant with the scanning speed. Too small a value of RC will lead to a noisy recording, whereas too large a value of RC leads to severe distortion of the line profiles. Of these, too large a value of RC is by far the more dangerous. The output of the ratemeter is essentially a voltage level, which is then fed to the x axis of the x/t recorder. This makes the x axis an intensity axis. The time (t) axis of the recorder is synchronized to the angular speed of the goniometer, converting it to a 2θ axis. Because the ratemeter is essentially integrating all X-ray photons impinging on the detector, it is important that the integration time be sufficient to allow a correct measure to be made of photon rate at a given angle, before the receiving slit passes the angle and no longer intercepts the true photon rate. The effect of the use of a ratemeter is to shift the peak location, lower the intensity, and lower the resolution by increasing the full width at half-maximum (FWHM). These distortions increase with increasing value of the ratemeter time constant.

A useful general rule for establishing the value of RC is

$$RC \leqslant \frac{30 \times \text{receiving slit width (degrees)}}{\text{scan speed (degrees/min)}}. \quad (10.1)$$

As an example, with a receiving slit width of 0.2° and a scan speed of 2°/min, the time constant should be $\leqslant 3$. Too small a value of the time constant will lead to large statistical fluctuations in the recording, and too large a value will cause distortion of the line profiles. Because one is severely restricted in flexibility when working with incremental time-constant values and driving the goniometer with mechanical gearing, almost all modern systems employ a goniometer equipped with a pulsed stepping motor and step scanning.

Recording of a count rate r using a ratemeter with a time constant equal to RC is equivalent to collecting the pulses during a time interval $2RC$. It was shown in Section 5.6 that the percentage standard deviation of the count error $\sigma(\%)$ is equal to $100/(rt)^{1/2}$, where r is the counting rate in pulses/second and t the count time in seconds. It follows that the counting error $\sigma(RM)$ using the ratemeter recording is

$$\sigma(RM) = \frac{100}{(r \times 2RC)^{1/2}}. \quad (10.2)$$

A common error in ratemeter scanning is to first choose the goniometer scan speed, then to select the appropriate ratemeter setting. In fact, this sequence is backward. The correct procedure in setting up a ratemeter scan is as follows:

1. Take note of the average count rate of the background.
2. Calculate the statistical error needed to reveal small peaks (Equation 10.2).

3. Calculate the ratemeter setting required to give the needed count error.
4. Allowing for the receiving slit used, calculate the maximum allowable scan speed (Equation 10.1).

Case 1:	
Ratemeter	$= 1$
Peak counting rate	$= 70 \, c/s$
σ(peak)	$= (2 \times 70 \times 1)^{1/2} = \pm 11.8 \, c/s$
Background counting rate	$= 20 \, c/s$
σ(background)	$= (2 \times 20 \times 1)^{1/2} = \pm 6.3 \, c/s$

As an example, if one is using a receiving slit width of 0.2 mm, with a background of 25 c/s, and if a 10% error is considered acceptable to reveal weak peaks, the ratemeter setting can be calculated using Equation 10.2:

$$\sigma(RM) = 10 = \frac{100}{(2 \times 25 \times RC)^{1/2}}, \quad (10.3)$$

from which RC is 2. The maximum scanning speed is now calculated from Equation 10.2:

$$\text{scan speed} = \frac{30 \times 0.2}{2} = 3°/\text{minute}. \quad (10.4)$$

Figure 10.5 shows a pictorial view of the effect of the choice of the ratemeter setting. The left-hand panel shows a line of 50 c/s superimposed on a background of 20 c/s, giving a gross peak height of 70 c/s. In the central panel, a scan has been made over the profile using a ratemeter setting of 1. Equation 10.2 can now be used to calculate the statistical ratemeter errors for the peak and background. The right-hand panel of Figure 10.5 shows the same peak recorded using a ratemeter setting of 8. The statistical ratemeter errors can now be calculated as previously. The reduction of *noise* from the scan by use of the higher ratemeter setting will be clear from the calculations given in Table 10.4.

10.5.3. Step Scanning

Another consideration in system design is the performance degradation over an instrument's useful life. This aspect has become more significant with the advent of automation. For example, the use of stepper motors in place of

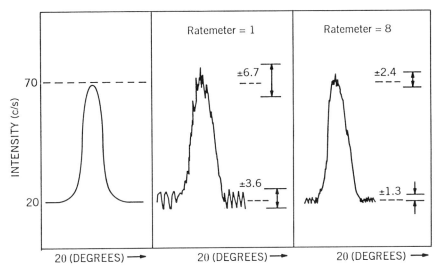

Figure 10.5. Choice of ratemeter settings. (See the text.)

Table 10.4. Calculation of Ratemeter Variation

Case 2:	
Ratemeter	= 8
Peak counting rate	= 70 c/s
σ(peak)	= $(2 \times 70 \times 8)^{1/2}/8 = \pm 4.1$ c/s
Background counting rate	= 20 c/s
σ(background)	= $(2 \times 20 \times 1)^{1/2}/8 = \pm 2.2$ c/s

synchronous motors increases the requirements on goniometer mechanical stability. Stepper-motor-driven goniometers may scan 5–10 times faster than synchronous-motor-driven systems since the computer bypasses the inherently slow ratemeter circuits and directly integrates the digital pulses from the detector. Higher scan speeds mean potentially higher wear rates of a gear train. Furthermore, the stepper motor introduces enormous mechanical stresses and vibrations every time it steps because of its quantized angular rotations. The use of accessories, such as the diffracted beam monochromator and the automated sample changer, adds to the weight of the moving parts and increases wear due to increased pressure on the gear teeth. On the more positive side, the use of computer automation has made it practical to correct for all types of systematic errors, including those which were not significant enough to be of concern when analysis was carried out by hand. This phenomenon has brought the potential for routinely reaching the random error limit much closer to reality.

10.6. AUTOMATED POWDER DIFFRACTOMETERS

More than 90% of all new powder diffractometers now sold in the United States are automated. Many excellent computer programs are provided with these instruments to aid the diffractionist in the reduction of powder data. Unfortunately, there can be pitfalls in the use of some of these programs, especially for the newcomer to the diffraction field. A popular misconception is that automated diffractometers are easier to use than manual systems. What in fact is true is that automated powder diffractometers can give many hours of unattended and reliable data collection and can greatly assist in the tedious and routine tasks of data analysis, provided one is careful to apply the programs in the correct manner.

During the 1960s, as mainframe computers evolved (e.g., the IBM 7094 or CDC 6600), crystallographic computing was performed in a noninteractive environment, where cards were input into a program operating in a batch stream. Due to the very high cost of computers, laboratory automation was nearly nonexistent, so these early developments can be considered as the zeroth generation of laboratory computational software. One of the first automated powder diffractometers was developed by Rex [7] in the mid-1960s. However, it took the microelectronics revolution of the 1970s to initiate the general techniques that have led to today's automated instrumentation. Although the principal thrust in the early 1970s was to develop the hardware interfaces needed to allow a computer to control a diffractometer, this work rapidly gave way to the much more serious problems of devising algorithms for the control of the instrument and the processing of the digital data. The 1970s saw the evolution of the early laboratory computer (e.g., the Digital Equipment Corporation's PDP-8 and PDP-11) and the first generation of process control software. Input to the program was first by paper tape and later by magnetic tape cassette, with programs written in low-level assembler language. Systems of this type were developed at Alfred University, Corning Glass Works, and General Electric. The first commercial automated powder diffractometer (APD) system in the United States was the Philips APD-3500 [8], which developed from involvement in the NASA project for a Lunar Receiving Laboratory. This instrument utilized a 4000 word, 18 bit program logic controller (a forerunner of what was to become the microprocessor). Within the 4KB (kilobytes) of core, routines were available for hardware control, data collection, data manipulation, and sophisticated math—all without hardware arithmetic capability!

The second generation of laboratory software evolved on the next generation of laboratory computers (e.g., the PDP 11/34), with the introduction of floppy disks that permitted convenient use of high-level language compilers. The first of these systems was written in FORTRAN and allowed the user to conduct an interactive dialogue with the computer to set up and execute an automated experiment and analyze the data [9]. The development of the 5 and 10 MB (megabyte) Phoenix and Winchester disk drives allowed the development of the third generation of software involving extensive file structures with help screens on video terminals and default files containing the user's typical settings, rather than interactive dialogue [10,11]. The first commercial system to adopt this approach was the Siemens D500 package introduced in 1983 [12]. A fourth generation of software is currently evolving where all interactions with the user are done through a "point and click" video interface with a mouse [13]. The trend is to eliminate all keyboard activity and convert crystallographic methods, which used to rely on the user evaluating numbers, to visual examination of graphic data.

There are essentially four elements to the automation of a diffractometer:

1. The replacement of the synchronous $\theta:2\theta$ motor with a stepping motor and its associated electronics
2. The replacement of a conventional scaler/timer with one that can be remotely set and read
3. The conversion of the various alarms, limit switches, and shutter controls to computer-readable signals
4. The creation of a computer interface that will allow a computer to control items 1 through 3

These four items are easily obtained today by direct purchase of modules that often plug directly onto the bus of a modern minicomputer, as shown in Figure 10.6.

Today, nearly all laboratory APD systems are controlled by a personal computer. The major impact of computer automation on improving the accuracy of the measurement of diffraction angles has come from the algorithms that bring much more intelligence to the process than has been conventionally used in manual measurements. There are two subject areas that need to be considered here: first, the algorithms that control the collection of data; second, those which reduce the data to d-values and intensities. The first generation of control algorithms were principally nonoptimizing, move-and-count methods. A design for such a program is shown in Figure 10.7. Although various attempts have been made to use the computer's decision-making capability to bring more "intelligence" to the data collection process

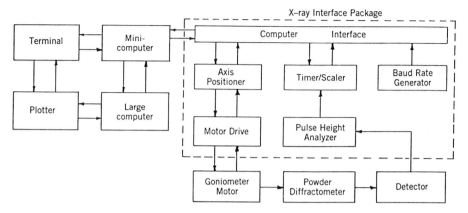

Figure 10.6. Schematic of an automated powder diffractometer (APD) system.

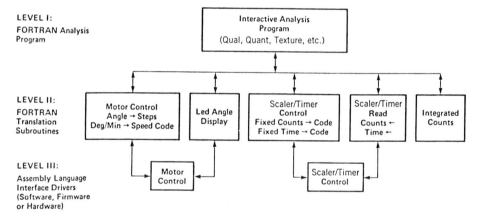

Figure 10.7. Software control scheme for an APD.

[14], the algorithms in use today continue to be nonoptimizing. The reason for this is fundamental, namely, that the determination of the intensity threshold above background limits the system's ability to detect small peaks that could be due to trace phases in the specimen. Thus, in the limit, an intelligent algorithm would spend all of its time counting where the peaks are not present. The key hindrance to the development of intelligent data collection software remains our inability to predict or even accurately describe the diffractometer's background function.

10.6.1. Step Scanning with the Computer

Because of the limitations associated with ratemeter scanning, most modern diffractometers employ a system of step scanning in place of ratemeter scanning. The basis of most step-scanning methods involves the movement of the goniometer in selected fixed angular increments, with the timer/scaler counting for a fixed time increment while the goniometer is stationary. This process allows a certain number of counts to be collected at specific positions of the goniometer where the angular increment is applied by a stepping motor. The step-scanning process is illustrated in Figure 10.8. Here, a five-stage process involving the scaler and timer is used to simulate ratemeter scanning:

1. The scaler contents are displayed as a voltage level on the recorder using an analog–digital (A/D) converter.
2. The timer and scaler are reset to read zero and at the same time the goniometer is incremented by one step.
3. The scaler and timer are started.
4. On reaching the preset time, the timer sends a stop pulse to the scaler.
5. The process repeats back to the first step until the final angle is reached.

The step-scanning output is then a step function as shown in Figure 10.8. Since the recording device is allowed to continue to move in the counting process, the profile appears as a series of *steps*. At the start of the process (step 1), since the scaler has not collected counts, the output is zero. After steps 2 and 3, the scaler has collected counts and redisplay of the scaler contents via the A/D converter gives a sudden sharp increase in the signal. This new level continues to be displayed until a new count cycle is complete.

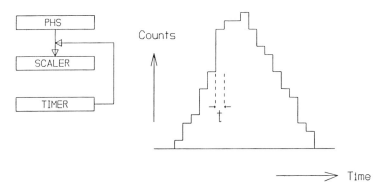

Figure 10.8. Step scanning with the timer/scale (PHS = pulse height selector).

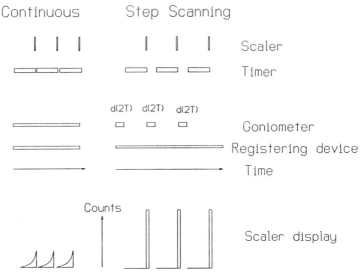

Figure 10.9. Step counting and count/step counting.

It is also possible to use the step-scanning system for continuous scanning. The difference between use of the step-scanning system for *continuous scanning* and *step scanning* is illustrated in Figure 10.9. In the upper part of the figure, the functions of the four components of the step-scanning system—the scaler, the timer, the goniometer, and the registering device—are shown as a function of time. The continuous scanning mode is shown on the left side of the figure. In this instance, the goniometer and the registering device both move at a constant rate. At the same time, the timer counts in fixed time increments, and the contents of the scaler are displayed through the A/D converter at the end of each count cycle. In the step-scanning case, shown on the right side of the figure, only the registering device continues at a constant rate. The goniometer is stepped; then the timer starts and stops the scaler, after which the scaler contents are displayed. The lower portion of the figure shows the way the scaler contents are displayed as a function of time. In the true step-scanning case, for each step/count cycle, the scaler contents are displayed as a single incremental step. This is in contrast to the continuous step-scanning mode where the scaler contents are displayed over a period of time. The advantage of the continuous step-scanning mode is that time can be saved because all three processes—timer/scaler, goniometer, and display—are all taking place at the same time. The disadvantage is that small peak shifts may accrue because the display of count data is *smeared* over a time period, this time period being a function of the count time, the count rate, and the goniometer speed.

10.6.2. Choice of Step Width

Although the use of step scanning offers the diffractionist great flexibility, special care must always be taken in setting up the step-scan conditions, especially in those cases where computer treatment of the raw data is anticipated. For example, too large a step size followed by a high degree of smoothing will lead to marked suppression of the peak intensity and loss of resolution. Conversely, too small a step size and too little smoothing may lead to peak shifts. Figure 10.10 illustrates this problem in a rather simplified way. It sometimes happens that the lines in a pattern may be broad and, as illustrated in the figure, the choice of step size and any subsequent data smoothing is predicated on how much of the profile should be sampled. The ideal step size is one that gives between 10 and 20 individual data points above

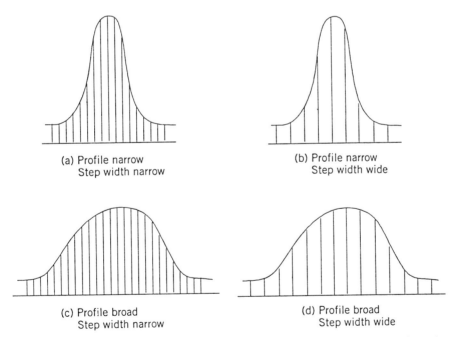

Figure 10.10. Choice of step size as a function of peak width. From R. Jenkins, Experimental procedures. In *Modern Powder Diffraction* (D. L. Bish and J. E. Post, eds.), p. 55, Fig. 5. Mineralogical Society of America, Washington, DC, 1989. Reprinted by permission.

the full width at half-maximum (FWHM) of the peak. For well-crystallized materials the peak width is typically (but not always) about 0.1–0.3° 2θ. Thus, applying the generalization above, one would use a step size of 0.02° 2θ. Clearly, as the peak width broadens due to the influence of particle size and/or strain, one should increase the step width accordingly. Cameron and Armstrong [15] have studied the optimization of step size and concluded that frequently too small a step size is employed. The situation is clearly different where profile-fitting techniques are employed because in that case much more of the profile data are being used.

10.6.3. Open-Loop and Absolute Encoders

The methods of employing encoders can be summarized by characterizing the control mode employed. A goniometer with no encoders is operated as an open-loop system. A goniometer with an encoder on the stepper motor shaft (i.e., on the input shaft) is still an open-loop goniometer system but has closed-loop motor control. A goniometer with an encoder on the output shaft would be a closed-loop goniometer system. A closed-loop goniometer system might also incorporate an encoder on the motor shaft to provide closed-loop motor control. Once a goniometer design employing a precision encoder on the 2θ output shaft is achieved, the importance of backlash and gear accuracy will be greatly diminished because software can be used to drive the 2θ axis directly to the desired angle, independent of perturbations along the way. Use of an encoder on the stepping motor that drives the input worm gear should not be confused with use of a precision encoder on the output shaft. Because most manufacturers employ encoders with 200 or more lines on their stepper motors, many users falsely believe that this determines the accuracy of their goniometers. Such encoders do not compensate for any of the gear or backlash errors in the goniometer. Their purpose is simply to permit closed-loop electronic feedback control of the stepper motor itself and ensure that the motor has taken the desired number of steps.

Software has long been used to reduce errors in goniometers, most commonly to compensate for backlash. The typical approach is to drive the goniometer in one direction (e.g., up-angle) to take up backlash. When the goniometer must be moved down-angle, it is sent 0.2–0.5° below the desired angle and then moved upward to the desired angle. In this manner, the gear teeth always ride on the same face when an angular measurement is made. However, the computers used in today's automated systems are capable of making much more sophisticated corrections than just backlash. Not only is software potentially able to compensate for the geometric and intrinsic errors associated with the parafocusing geometry, it can also be applied to the systematic errors arising from the gearing and/or electronic control circuitry of a goniometer. No two goniometers will ever be exactly alike, yet in principle each goniometer can be calibrated and the measured constants automatically applied to diffraction data. If this approach is employed, the optimum goniometer design will be obtained when the total random error from all mechanical sources is equal to the total random error from all other sources.

10.7. USE OF CALIBRATION STANDARDS

There are various types of standards employed in X-ray powder diffractometry, and many of the more useful of these are available as certified *Standard Reference Materials* (SRMs) from the National Institute of Standards and Technology[1] (NIST; formerly the National Bureau of Standards, NBS). At the present time, NIST has around 1000 SRMs available [16], of which about a dozen or so are specifically designed for X-ray diffraction measurements. As shown in Table 10.5, calibration standards fit into five

Table 10.5. Types of Standard Used in X-ray Powder Diffraction

Type	Use	Material
1. External 2θ Standards		Silicon α-Quartz Gold
2. Internal d-spacing Standards	Primary Primary Secondary	Si (SRM 640b) Fluorophlogopite (SRM 675) W, Ag, quartz, diamond
3. Internal Intensity Standards	Quantitative Intensity Respirable Quartz	Al_2O_3 (SRM 676) α- and β-silicon nitride (SRM 656) Oxide of Al, Ce, Cr, Ti, and Zn (SRM 674a) α-SiO_2 (SRM 1878a) Cristobalite (SRM 1879a)
4. External Sensitivity Standards		Al_2O_3 (SRM 1976)
5. Line profile Standards	Broadening Calibration	Lanthanum hexaboride (SRM 660)

[1] Gaithersburg, MD 20899.

major categories: (1) External 2θ Standards; (2) Internal d-Spacing Standards; (3) Internal Intensity Standards; (4) External Sensitivity Standards; and (5) Line Profile Standards. External 2θ Standards are used in the alignment of the diffractometer and are typically supplied by the instrument vendor when the diffractometer is first purchased.

10.7.1. External 2θ Standards

As was discussed in Section 8.2, because of the various inherent and alignment errors in a given diffractometer, the actual observed experimental 2θ value of a given line may vary from instrument to instrument. As was shown in Figure 8.9, the experimental 2θ value will include both inherent and misalignment errors. One of the main uses for calibration standards is to correct for uncompensated or unsuspected systematic errors in the calibration procedure. As an example, Table 10.6 summarizes the effectiveness of standards in the correction of 2θ errors.

One important use of standards is for the checking of the alignment of a diffractometer. Once an alignment has been established, it is necessary to check first that the alignment is correct, then recheck at frequent intervals to ensure that the integrity of the alignment is maintained. The first of these steps might be quite complex and takes several hours to perform. The second of the steps needs to be based on a rapid, simple procedure that can be completed in a few minutes. In both cases, however, it is necessary that the tests be very reproducible without the introduction of day-to-day errors, for instance, in sample mounting. To aid in this area, most equipment manufacturers supply

Table 10.6. Effectiveness of Standards for the Correction of 2θ Errors

Use of Standard	Type of Standard				
	None	External (2θ)	Internal (2θ)	ZBH (2θ)	External (Intensity)
Instrument misalignment	No	Yes	Yes	Yes	(Yes)
Inherent aberrations	No	Yes	Yes	Yes	No
Specimen transparency	No	No	Yes	Yes	No
Specimen displacement	No	No	Yes	Yes	No
Instrument sensitivity	No	No	No	No	Yes

a mounted, permanent alignment reference with the diffractometer. The standard is typically a specimen of fine-grained α-quartz (novaculite), cemented into a holder and surface ground to ensure that there is no specimen displacement. Other materials that have been used for this purpose include silicon, tungsten, and gold. The type of instrument alignment standard described above can also be used as an external alignment standard to establish a calibration curve for a given instrument. By this means, all data can be corrected to the external standard. One important point to note, however, is that use of the external standard will not correct for the largest of the systematic errors: the specimen displacement error or transparency. It is also not good practice to use an external standard to compensate for poor alignment of the diffractometer.

10.7.2. Internal 2θ and d-Spacing Standards

While the external standard is excellent for the preparation of angular calibration curves to correct for inherent aberrations and misalignments, its use will not correct for specimen displacement and transparency errors. As was discussed in Section 7.5, each of these errors can be quite large and must be corrected for in order to achieve an accurate d-spacing determination. The usual procedure for such a correction is to use an internal standard or the zero background holder method as described in Section 9.6.4. The ideal internal standard [17] should give a good angular coverage, should have a reasonably simple pattern to minimize possible line overlaps, and should be stable, inert, and available in small particle size. As shown in Table 10.4, two primary internal standard materials have been certified by NIST: silicon (SRM 640b) and fluorophlogopite (SRM 675). The silicon SRM is a good general-purpose internal standard with angular coverage from about $24° 2\theta$ upwards. The fluorophlogopite is intended for use at low 2θ values. Being a mica, fluorophlogopite orients strongly in the [00l] direction. The large plate-like crystallites of this standard are intended to be patted on top of the prepared specimen rather than intimately mixed with it. Only the 00l lines will be seen in the diffraction pattern, and these may be used for calibration.

10.7.3. Quantitative Analysis Standards

Quantitative analysis SRMs are powders of high phase purity that exhibit minimal preferred orientation. The NIST SRM 674a consists of five separate standards that cover the range of linear attenuation coefficients from 126 to 2203 cm^{-1} for Cu $K\alpha$ radiation [18]. The phases are Al_2O_3 (corundum structure), ZnO (wurtzite structure), TiO_2 (rutile structure), Cr_2O_3 (corundum structure), and CeO_2 (fluorite structure). The phases are intended as internal standards for quantitative analysis, and as instrument calibration standards for checking the angular intensity response from a given diffractometer. Certified reference intensity ratios (RIRs) are also given for each phase.

Other quantitative analysis SRMs include two SiO_2 materials: SRM 1878, α-quartz, and SRM 1879, cristobalite. These two SiO_2 phases have been certified with respect to amorphous content for the analysis of silica-containing materials in accordance with health and safety regulations. SRM 656 consists of two silicon nitride powders: one is high in α-content; the other is high in β-content. Finally, SRM 676 is an Al_2O_3 powder for the determination of $I/I_{corundum}$ values. A note of warning should be included here on the powder diffraction pattern of Al_2O_3; this material shows a plate-like hexagonal habit and typically shows a few percent orientation in any diffraction mount.

10.7.4. Sensitivity Standards

The instrument sensitivity standard is intended to quantify variations in angular sensitivity between different diffractometers. The material generally used for this purpose is the NIST SRM 1976 *Instrument Sensitivity Standard*. SRM 1976 is a sintered plate of α-alumina, and examples of its use have already been discussed in Section 8.2.

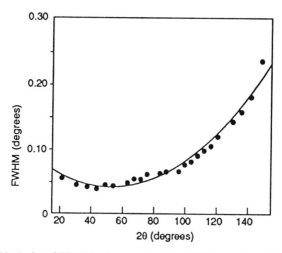

Figure 10.11. A plot of FWHM values as a function of 2θ determined with SRM 660.

10.7.5. Line Profile Standards

With the growth of interest in line profile analysis in the 1980s (see Section 11.3.4), there was a requirement for a suitable line profile standard to define the instrument broadening function. A round-robin test was sponsored by the ICDD [19] to evaluate candidate materials, and it was found that lanthanum hexaboride (LaB_6) was reasonably free from peak broadening due to strain and/or particle size effects; LaB_6 also has a goodly number of reasonably spaced diffraction lines, and these factors combine to make it a good SRM.

Figure 10.11 shows a plot of (FWHM) values as a function of 2θ, determined with SRM 660, which is the NIST LaB_6 material. The FWHM values were obtained by use of a split Pearson VII profile shape function [20]. Accurate fitting of these data (by using second-degree polynomials) indicates proper alignment in the test instrument. Due to its certified lattice parameters, SRM 660 can also be used for determining instrumental parameters through a Rietveld refinement.

REFERENCES

1. Schreiner, W. N., Surduwkowski, C., and Jenkins, R. An approach to the isostructural/isotypical, and solid solution problems in multi-phase X-ray analysis. *J. Appl. Crystallogr.* **15**, 605–610 (1982).
2. Schreiner, W. N., and Fawcett, T. Results of a round robin study of systematic errors found in routine X-ray diffraction raw data. *Adv. X-Ray Anal.* **28**, 310–314 (1985).
3. Jenkins, R., and Schreiner, W. N. Consideration in the design of goniometers for use in X-ray powder diffractometers. *Powder Diffr.* **1**, 305–319 (1986).
4. Jenkins, R., and Hubbard, C. R. A preliminary report on the design and results of the second round robin to evaluate search/match methods for qualitative powder diffractometry. *Adv. X-Ray Anal.* **22**, 133–142 (1979).
5. Jenkins, R., and Schreiner, W. N. Intensity round robin report. *Powder Diffr.* **4**, 74–100 (1989).
6. Jenkins, R. On the selection of the experimental wavelength in powder diffraction measurements. *Adv. X-Ray Anal.* **32**, 551–556 (1989).
7. Rex, R. W. Numerical control X-ray powder diffractometry. *Adv. X-Ray Anal.* **10**, 366–373 (1966).
8. Jenkins, R., Haas, D. J., and Paolini, F. R. A new concept in automated X-ray powder diffractometry. *Norelco Rep.* **18**, 1–16 (1971).
9. Mallory, C. L., and Snyder, R. L. The control and processing of data from an automated powder diffractometer. *Adv. X-Ray Anal.* **22**, 121–132 (1979).

286   ACQUISITION OF DIFFRACTION DATA

10. Snyder, R. L. AUTO: A real-time diffractometer control system, *Nat. Bur. Stand. [Tech. Rep.] NBSIR (U.S.)* **NBSIR-81**–2229 (1981).
11. Snyder, R. L., Hubbard, C. R., and Panagiotopoulos, N. C. A second generation automated powder diffractometer control system. *Adv. X-Ray Anal.* **25**, 245–260 (1982).
12. Snyder, R. L. The renaissance of X-ray powder diffraction. In *Advances in Material Characterization* (D. R. Rossington, R. A. Condrate, and R. L. Snyder, eds), pp. 449–464. Plenum, New York, 1983.
13. Zorn, G. Programs DIFF and SHOW. Siemens AG, Munich (private communication), 1994.
14. Snyder, R. L. Accuracy in angle and intensity measurements in X-ray powder diffraction. *Adv. X-Ray Anal.* **26**, 1–11 (1983).
15. Cameron, D. G., and Armstrong, E. E. Optimization of stepsize in X-ray powder diffractogram collection. *Powder Diffr.* **3**, 32–38 (1988).
16. Dragoo, A. L. Standard reference materials for X-ray diffraction. Part I. Overview of current and future standard reference materials. In *ICDD Methods and Practices Handbook*, Sect. 6.1.1. International Centre for Diffraction Data, Newtown Square, PA, 1988.
17. Wong-Ng, W., and Hubbard, C. R. Standard reference materials for X-ray diffraction. Part II. Calibration using d-spacing standards. In *ICDD Methods and Practices Handbook*, Sect. 6.2.1. International Centre for Diffraction Data, Newtown Square, PA, 1989.
18. Hubbard, C. R. SRM674. Office of Standard Reference Materials, National Institute of Standards and Technology, Gaithersburg, MD, 1983.
19. Fawcett, T. W., et al. Establishing an instrumental peak profile calibration standard for powder diffraction analysis. *Powder Diffr.* **3**, 210–218 (1988).
20. Howard, S. A., and Snyder, R. L. An evaluation of some profile models and the optimization procedures used in profile fitting. *Adv. X-Ray Anal.* **26**, 73–81 (1983).

CHAPTER
11
REDUCTION OF DATA FROM AUTOMATED POWDER DIFFRACTOMETERS

11.1. DATA REDUCTION PROCEDURES

An experimental diffraction pattern is usually recorded in terms of a distribution of absolute intensities as a function of diffraction angle (2θ). In order to perform qualitative phase identification, it is usual to "reduce" the experimental pattern to a table or graph of d-spacings and relative intensities, as was shown in Figure 10.1. The d–I list is referred to as a "reduced" pattern. In this data conversion process, the maximum intensity value of each peak is sought and the corresponding angle at which this maximum occurs is converted to a d-value by use of Bragg's law. This process requires knowledge of the experimental wavelength(s). The experimental absolute intensity values are converted into relative values by calling the strongest line intensity 100%, then scaling each intensity maximum to this value. Either peak height values or peak area values can be used. The advantage in using relative rather than absolute intensity values is that it reduces all intensities to a common scale and avoids problems due to the wide variation in sensitivity exhibited by typical diffractometers.

11.2. RANGE OF EXPERIMENTAL DATA TO BE TREATED

An important decision to be made in setting up a procedure for the reduction of data from an experimental pattern is the angular and intensity range to be treated. The diffractogram shown in Figure 10.1 was obtained by scanning between $20°\ 2\theta$ and $100°\ 2\theta$. The diffractogram is typical of many experimental patterns and contains both strong and weak lines superimposed on top of a significant background. One assumption that has already been made is that the experimental pattern truly represents the material to be analyzed. Care must now be taken to ensure that sufficient information from the experimental pattern is processed to allow complete identification. Table 11.1 shows the results of the data reduction for an angular range from 20–$70°\ 2\theta$, using two different background threshold values. The first selection assumes a background threshold of $130\,c/s$, and 28 peaks result from this first selection. The

Table 11.1. Data Reduction Using Two Different Background Threshold Levels

2θ	d-Value	Background (c/s)	Peak (c/s)	I (%)	1st	2nd
25.622	3.474	316	9820	48.7	Yes	Yes
35.191	2.548	345	16208	80.3	Yes	Yes
37.401	2.403	340	207	1.0	Yes	Yes
37.824	2.377	333	6862	34.0	Yes	Yes
41.697	2.164	287	199	1.0	Yes	No
41.733	2.163	287	193	1.0	Yes	No
43.391	2.084	306	19329	95.8	Yes	Yes
43.450	2.081	307	14937	74.0	Yes	Yes
45.612	1.987	307	152	0.8	Yes	No
46.224	1.962	300	300	1.5	Yes	Yes
46.278	1.960	299	245	1.2	Yes	No
52.570	1.740	291	9722	48.4	Yes	Yes
52.687	1.736	290	6079	30.1	Yes	Yes
57.526	1.601	324	20181	100.0	Yes	Yes
57.662	1.597	324	11521	57.1	Yes	Yes
58.174	1.585	338	140	0.7	Yes	No
59.772	1.546	345	544	2.7	Yes	Yes
59.898	1.543	346	314	1.6	Yes	Yes
60.975	1.518	330	161	0.8	Yes	No
61.175	1.514	330	875	4.3	Yes	Yes
61.332	1.510	336	1970	9.8	Yes	Yes
61.483	1.507	334	989	4.9	Yes	Yes
66.538	1.404	542	7770	38.5	Yes	Yes
66.711	1.401	486	4435	22.0	Yes	Yes

second selection is for the *same* angular range, but this time with the background threshold set at 285 c/s. This second selection yields only 21 peaks.

It can be seen from this example that the number of lines found in a reduced pattern depends not only on the conditions under which the pattern was *recorded* but also on how much of the experimental pattern was *treated*. While it is common practice to treat the whole of the recorded angular range, the selection of the background threshold level and degree of data smoothing is very much an operator judgment.

11.2.1. Computer Reduction of Data

Most modern powder diffractometers utilize computers and software to assist in the data reduction process. The output from an automated powder diffrac-

RANGE OF EXPERIMENTAL DATA TO BE TREATED   289

tometer is a digitized pattern comprising a block of numbers of counts collected at contiguous 2θ increments. In this chapter, both the techniques of automated data reduction and the possible software-introduced errors will be considered. Modern computer methods have much to offer in terms of data smoothing, sensitive peak location methods, and automated calibration. Against this, however, there are many pitfalls to be avoided by the inexperienced user. As an example, the use of computer-automated internal or external standard corrections can almost completely remove instrumental angular errors. While such an idea may sound extremely attractive, when such automated corrections are used to compensate for poorly aligned diffractometers the user can be lulled into a false sense of security, since errors may accrue that are not compensated by use of a simple correction curve. Software-introduced intensity errors are less common and are generally caused by poor choice of step width and/or oversmoothing of raw data. Also, the use of a $K\alpha_2$ stripping program can modify both the number of lines reported and the relative intensities of these lines. As an example, at low 2θ values where the $K\alpha_1, K\alpha_2$ doublet is not resolved, application of an $K\alpha_2$ stripping program will reduce the residual peak height by about one-third.

Table 11.2 shows the output from two commercially available data reduction programs, run on exactly the same data set. Twenty lines are found by each software package. The intensities match within about $\pm 15\%$, and d-spacings match to about ± 2 parts per thousand. The following sections will discuss some of the possible reasons why these variations occur. It should also be appreciated that while computer techniques are extremely useful, care should always be taken in comparing computer-treated data with manually treated data. Such is frequently the case in the use of data from a modern automated powder diffractometer and older reference patterns from the ICDD Powder Diffraction File (PDF). In fact, great care must always be exercised when comparing data treated by one method with data treated by another method. As an example, a frequent major source of confusion is the recognition of peaks by a given automated system on a certain specimen, where data collected by manual and automated methods are being compared. The extreme sensitivity of, for example, the second derivative peak-hunting method may give significantly more lines than would be found manually. Additionally, the number of peaks found is very sensitive to the method of defining a significant signal level above background. Many of these "peaks" may not be true peaks at all, but simply statistical variations in the background or on the shoulders of real peaks. As an example, a recent round-robin study [1] showed that for a well-characterized specimen of Al_2O_3, the number of peaks reported by different users, representing a range of instrument types over the same 2θ range, varied from 25 to 53, whereas the calculated number is 42. The situation is further complicated where α_2 stripping is employed, and in

Table 11.2. Experimentally Reduced Data from Two Different Software Programs

Lines Used		User 1		User 2	
Line No.	(hkl)	d (Å)	I/I^{rel}	d (Å)	I/I^{rel}
1	(012)	3.479	75	3.472	43
2	(104)	2.552	90	2.547	71
3	(110)	2.379	40	2.376	34
4	(006)	2.165	1	2.162	1
5	(113)	2.085	100	2.085	96
6	(202)	1.964	2	1.962	2
7	(024)	1.740	45	1.739	52
8	(116)	1.601	80	1.602	100
9	(211)	1.546	4	1.546	5
10	(122)	1.514	6	1.511	8
11	(018)	1.510	8	1.510	11
12	(214)	1.404	30	1.510	43
13	(300)	1.374	50	1.373	73
14	(125)	1.337	2	1.336	3
15	(208)	1.276	4	1.275	6
16	(10, 10)	1.239	16	1.239	26
17	(119)	1.2343	8	1.235	13
18	(220)	1.1898	8	1.189	13
19	(306)	1.1600	1	1.160	2
20	(223)	1.1470	6	1.147	10

this instance the $d-I$ list obtained following α_2 stripping will certainly differ from a data set similarly obtained using an older manual diffractometer employing a strip-chart recorder. The widths of the diffracted lines along with the efficiency of the peak location process will both be major factors in the number of lines reported by the automated system.

At present some of the most powerful research techniques involving X-ray powder diffraction are embodied in a growing number of computer programs. These include methods for calculating powder patterns from crystal structure information, refining structure models, indexing patterns even of low symmetry, refining lattice parameters by least squares, computing figures of merit, evaluating the quality of powder patterns, carrying out crystallite size/stress and orientation analyses, and automated phase identification. Two principal difficulties hinder the widespread use of these techniques. The first is that the software program usually is developed in an academic environment and often requires a rather high degree of computer literacy. The second is the requirement that the user be conversant with the somewhat complex notation of X-ray crystallography. The development of computer procedures that are more user-friendly and oriented toward the nonexpert is extremely time consuming and hence expensive. However, the development of a broad market of automated diffractometer users desiring to make sophisticated application of their instruments is leading to the evolution in this field of a number of software companies, some associated with instrument manufacturers. The commercial sector has already begun to introduce much more user-friendly software. The use of the computer to translate the powerful tools developed in academe into procedures usable by the noncrystallographer and non–computer expert is currently underway. The near future should bring these tools to a much wider user group.

11.3. STEPS IN DATA TREATMENT

As has been previously discussed, a diffraction pattern is made up of a number of contributions, including diffraction of wanted and unwanted wavelengths, plus scatter and fluorescence from the specimen and specimen support. The goal of the data treatment process is to extract the useful data from the experimental pattern. Figure 11.1 lists the typical steps that may be employed in the data treatment process. Following the collection of the digitized data (step 1), the raw digitized pattern is stored along with the appropriate experimental parameters. In step 2, the data are smoothed to reduce counting statistical fluctuations, as described in Section 11.3.1, below. Next (step 3), the background may or may not be subtracted, depending on the average peak/background ratio across the pattern. The background stripping process is described in detail in Section 11.3.2. Where bichromatic source radiation has been used to record the diffraction pattern, it is sometimes (but not always!) advantageous to remove the $K\alpha_2$ contribution (step 4). This process is described in Section 11.3.3. Next (step 5), the peaks are located, as described in Section 11.3.4. The peak location is typically done by using a second-differential method in which the resulting negative peak is fit to a parabola and the minimum is calculated. A 2θ or d-spacing calibration may then be applied (step 6), based on either an internal standard or an external calibration curve. Where the fully digitized pattern is being stored for future use, it may also be advantageous to correct the whole pattern for the sensitivity of the particular instrument employed. Such a process greatly simplifies the comparison of digitized patterns from different instruments [2]. The various calibration procedures are described in Section 11.5. Finally (step 7), the peak data is converted to a d-spacing and stored to an appropriate significance.

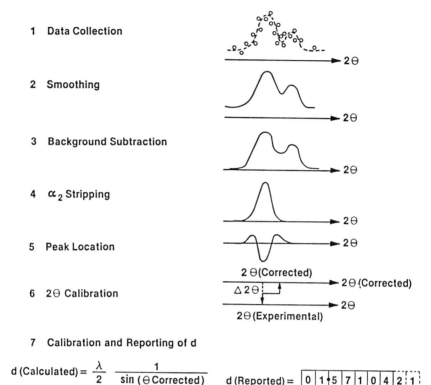

Figure 11.1. Steps in the treatment of diffraction data. From R. Jenkins, Experimental procedures. In *Modern Powder Diffraction* (D. L. Bish and J. E. Post, eds.), p. 59, Fig. 8. Mineralogical Society of America, Washington, DC, 1989. Reprinted by permission.

While different commercial and public-domain data-handling systems exist, most software schemes include each of the aforementioned steps. However, varying approaches to each of the individual steps, when applied in succession, can produce different results, and each of the steps must be considered in the light of all others.

11.3.1. Use of Data Smoothing

Since the production of X-rays is a random process, any count datum will have associated with it a random error, the magnitude of which depends on the number of counts taken (i.e., the product of the count rate and the count time). This counting statistical process introduces random fluctuations in the raw

STEPS IN DATA TREATMENT   293

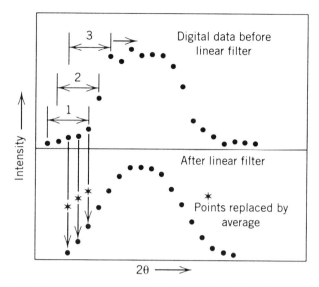

Figure 11.2. Smoothing out statistical noise with a linear digital filter. The five-point-averaged value is indicated with an asterisk.

data. These fluctuations can be partially removed by data smoothing. Figure 11.2 illustrates the fundamental idea behind data smoothing in which an odd number of contiguous data points are averaged and the middle data point is replaced by the average. The starting data point is then incremented by 1 and the process repeated until the whole pattern is smoothed. If three data points are taken at a time, one refers to it as a "three point smooth." If five points are taken at a time, it is called a "five point smooth," etc. In most programs, the number of points to be used in smoothing is a parameter adjustable by the user; however, in some commercial software packages, data smoothing may be applied without user access to control the process. In order to establish how much smoothing has been applied to a given data file, the following expression may be used:

$$N = 2\left(\frac{2\theta_r - 2\theta_s}{\Delta 2\theta}\right) + 1, \tag{11.1}$$

Where $2\theta_r$ is the first angle for which smoothed data is reported; $2\theta_s$ is the angle at which the scan was started; $\Delta 2\theta$ is the step size; and N is the number of points taken by the smoothing routine. For example, if a scan were made starting at $20.00°\ 2\theta_s$ at a step size of $0.02°$ ($\Delta 2\theta$), and the first smoothed data point reported was at $20.06°\ 2\theta_r$, then a seven point smooth would have been applied.

The choice of whether or not to smooth and the degree of smoothing to be used are predicated on a number of factors including the statistical counting error associated with each peak, the sharpness and uniqueness of each peak, and the number of individual data points associated with each peak. Each of these factors must be considered in the light of all others. It is necessary first of all to take a sufficient number of individual data steps to adequately represent the shape of the profile. As was discussed in the previous chapter, if the peak is narrow and the step size is broad, an insufficient number of data points will be accumulated. While smoothing may marginally improve the peak definition, it will also suppress the peak intensity. Conversely, if the peak is broad and the step width small, the large number of data points may not allow the peak maximum of the profile to be defined. In this instance, smoothing helps to reduce the statistical counting error but does little to improve the already poor peak definition. The ideal step size is one that gives between 5 and 15 individual data points above the full width at half-maximum (FWHM) of the peak. For well-crystallized materials, the peak width is typically about 0.1– 0.25° 2θ. Too small a step size and too little smoothing may lead to peak shifts and/or spurious peaks. As the peak width broadens due to the influence of crystallite size and/or strain, one should increase the step width accordingly. The situation is clearly different where profile fitting techniques are employed because in this case much more of the profile data are being used.

In addition to problems of peak definition, statistical fluctuations and the possible presence of noise spikes in the intensity measurements can also lead to the detection of false peaks in the regions above the threshold. So, some smoothing of data is always required. X-ray diffraction profiles step scanned on a parafocusing diffractometer will always be asymmetric, with the intensity falling off more rapidly on the high-angle side, principally due to axial divergence, as explained in Section 7.5.1. The passing of a linear digital filter, as illustrated in Figure 11.2, over such a peak will cause the peak maximum to shift. Thus, in practice, a linear digital filter is never used; rather, a quadratic, cubic, or higher order polynomial of the type

$$2\theta_{calc} = a + b(2\theta_{obs}) + c(2\theta_{obs})^2 + d(2\theta_{obs})^3 + \cdots \quad (11.2)$$

is fit to the odd number of raw data points by using least squares regression to establish the values of the $a, b, c, \ldots$ coefficients for each data interval. The point in the middle of the interval is replaced with the point computed from the interpolating polynomial. Subsequent smoothed data points are produced in a similar manner by applying the procedure to each overlapping successive group of data points. This—perhaps quadratic—digital filter (since it permits a curve to be fit to the data points in the interval) will have much less effect on shifting the position of the peak maximum. As this digital filter slides over the

STEPS IN DATA TREATMENT   295

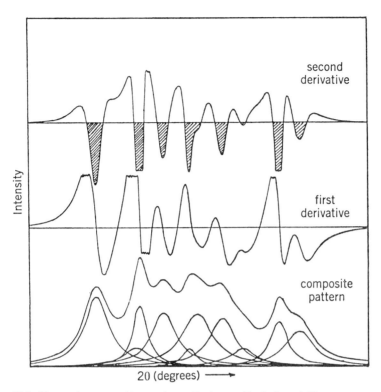

Figure 11.3. Observed pattern with actual contributing profiles indicated. Upper curves are the first and second derivatives produced by the Savitsky–Golay algorithm [3].

data, statistical fluctuations are greatly reduced. However, there is also a corresponding loss in peak resolution that increases with the 2θ step width and the number of points used in the filter. Savitzky and Golay [3] have produced an extremely efficient computational method for applying a quadratic filter to digital data that is illustrated in Figure 11.3. Their procedure is nearly universally used in diffraction data reduction [4, 5], although a cubic spline function has also been successfully applied [6]. Since the interpolating polynomial will be fit to each point, it is clear that the first or second derivative may be easily evaluated at each point and that this may be used to locate the maxima in the data. However, since this is the most common method for locating peaks, some smoothing will be forced on the user in order to compute the polynomial.

Figure 11.4 gives an example of the effect of smoothing. The portion of the diffractogram displayed represents a pair of partially resolved $K\alpha_1$, $K\alpha_2$ doub-

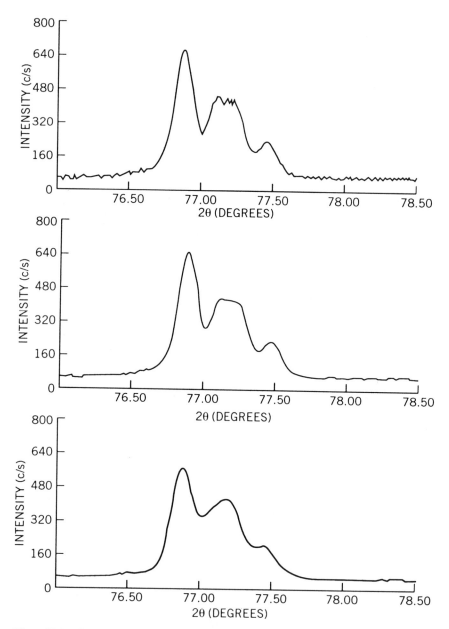

Figure 11.4. Effect of data smoothing on a pair of $K\alpha_1$, $K\alpha_2$ doublets: the $K\alpha_2$ of the lower angle peak is interfering with the $K\alpha_1$ of the higher angle peak. The top plot is the raw data, the middle plot is after an 11-point-smoothing routine, and the bottom plot is after a 25-point-smoothing routine.

lets. The top plot shows the raw data as they were accumulated using a step size of 0.01°. Three lines are immediately obvious, but on closer inspection a partial separation can be seen of the α_2 peak of the left-hand doublet and the α_1 peak of the right-hand doublet. The center plot was obtained by applying an 11-point-smoothing routine to the raw data. Note that the revelation of the center doublet is a little clearer and also that the fluctuations in the background have been almost completely removed. The lower scan was obtained by applying a 25-point-smoothing routine to the raw data. Note the severe loss of resolution in the pattern and the suppression of the peak intensities.

11.3.2. Background Subtraction

If the background in a diffraction pattern were reasonably constant over the 2θ range under examination, there would be no difficulty in simply subtracting an average background value to give net peak intensities. The more usual case of nonlinear background in an X-ray diffraction pattern can, however, cause difficulties in picking out small peaks and defining a true intensity. Variation in the background is mainly due to the following five factors:

1. Scatter from the sample holder (generally seen at low values of 2θ where too wide a divergence slit is chosen)
2. Fluorescence from the specimen (controllable to a certain extent by the use of a diffracted-beam monochromator or by pulse height selection)
3. Presence of significant amounts of amorphous material in the specimen
4. Scatter from the specimen mount substrate [seen in "thin" specimens, but controllable by use of zero background holders (ZBHs)]
5. Air scatter (which has the greatest effect at low 2θ values)

Background values may also be different for fixed and variable divergence slit systems. As would be expected, the fixed divergence slit data show much higher background at low 2θ values due to scatter as the beam exceeds the specimen length.

The operation of differentiating peaks from background noise may be performed in two discrete steps [7]. The first is to linearize the pattern in order to remove the typical low-angle upward curvature and the broad maximum resulting from amorphous scattering. This is performed by selecting and then smoothing the minimum intensity points in each 0.25° 2θ segment of the pattern. Minimum points that are more than $n\sigma$ greater than the adjacent point are assumed to be part of a diffraction peak and so are rejected from the set of minimum intensity values. A least squares polynomial (of the type shown in Equation 11.2) passed through the accepted minima is subtracted from the

298   DATA REDUCTION FROM AUTOMATED DIFFRACTOMETERS

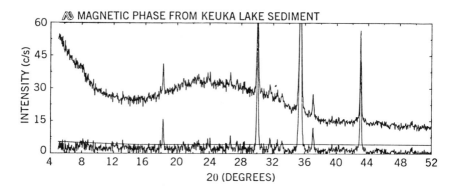

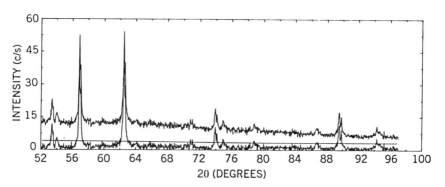

Figure 11.5. Raw data linearization and threshold determination.

observed pattern, producing the linearized pattern shown at the bottom of Figure 11.5. The second step is to determine the threshold of statistically significant data [8]. This is performed in a manner similar to the linearization step, but here the maximum intensity points in each 0.25° pattern segment are collected. Any points in this list that are greater than $n\sigma$ of the average are rejected as probably lying on peaks. The remaining points are fit to a polynomial that represents the threshold above which points are significantly different from the background. Both of these procedures are illustrated in Figure 11.5. This pattern was obtained from a 5 mg sample placed on a glass slide with a small depression etched in it. Due to the very small sample size, the pattern was counted at 20 s per 0.04° step. The raw data shown in the upper curve in each panel exhibit both the effect of the amorphous sample holder and the increased low-angle intensity due to the 1° divergence slit. The lower tracing shows the pattern after linearization. Note that all traces of the distortions have been removed. The smooth line on top of the background

11.3.3. Treatment of the α_2

Most powder diffraction work is carried out using the Cu $K\alpha_1$ $K\alpha_2$ doublet, and one of the greatest experimental inconveniences arising from this choice is the variable angular dispersion of the diffractometer (previously discussed in Section 6.2). Some peak distortion may occur due to partial separation of the Cu $K\alpha_1$, $K\alpha_2$ doublet over some parts of the 2θ range because the angular dispersion of the diffractometer increases with increasing 2θ. Hence, whereas at low 2θ values the α doublet is unresolved, at high 2θ values it is completely separated. In the midangular range, the lines are only partially resolved, leading to some distortion of the diffraction profile and an apparent shift in the maximum of each peak in the doublet toward each other. When manual peak-finding techniques are used, it is common practice to utilize the weighted geometric average of the $K\alpha_1$, $K\alpha_2$ wavelength as the experimental wavelength, at least until a 2θ value is reached where the α_1, α_2 doublet is sufficiently resolved to allow accurate measurement of the α_1 line. Where automated methods are used, α_2 stripping is generally employed over the whole range of the measured pattern.

In manual-type diffractometers, generally no attempt is made to modify the lines arising from the characteristic X-ray tube wavelengths that are diffracted by the specimen, other than to remove or reduce the β-component. Since a bichromatic $\alpha_1:\alpha_2$ *wavelength* is being used, problems occur in establishing the true maximum of an α_1 (or α_2) peak, because the splitting of the α doublet at angles in excess of $30-55°$ 2θ depends on the line width of the profile. The resolving power $(d\lambda/\lambda)$ of a diffractometer is given by $2\tan\theta/\Delta 2\theta$. Cullity [9] has suggested that, for two lines of similar height and width, this expression approximates to $\tan\theta/B$, where B is the peak breadth at half-maximum intensity. Since the α_2 is only one-half the intensity of the α_1, these two lines will start to resolve as soon as their peaks are separated by an angle equivalent to about 1.16 times the line width. At angles lower than this, the two lines will be added with a shift in the observed 2θ maximum. To compensate for this shift, a weighted wavelength α^* of 1.54186 Å is used [i.e., $(2\lambda_{\alpha_1} + \lambda_{\alpha_2})/3$] in Bragg's law to convert the observed angle to a d-value. This is clearly an approximation, and the error involved by assuming that either the α_1 wavelength is always correct or by assuming that the α^* is always correct introduces a maximum angular error of about $0.008°$ per $10°$ 2θ. The corresponding error in $\Delta d/d$ is about 0.7 parts per thousand.

Most modern diffractometers use a graphite, diffracted-beam monochromator that, due to its high mosaicity, allows both the $K\alpha_1$ and $K\alpha_2$ wavelengths to pass. For most work, the $K\alpha$ diffraction peaks can be readily recognized by a computer algorithm based on their location and height. However, when it is desired to completely remove the $K\alpha_2$ peaks from the raw data, computer techniques for $K\alpha_2$ stripping are typically employed. Methods for α_2 elimination are generally based on either the Rachinger technique [10], which is applied to the 2θ's, or on a deconvolution using the Fourier transformation of the powder pattern [11]. The Rachinger method uses our knowledge of the exact wavelengths of the $K\alpha_1$ and $K\alpha_2$ lines and their intensity ratios. The lower illustration of Figure 11.6 shows a simplified schematic of how the method works. The intensity measured at the first point in the profile may be assumed to be entirely due to the $K\alpha_1$. The d_{hkl} causing the diffraction at this 2θ will, of course, also diffract the $K\alpha_2$ at the angle proscribed by Bragg's law, indicated by the dashed line. Since the intensity ratio of the $K\alpha_1:K\alpha_2$ is exactly 2:1, half of the intensity measured at the low-angle 2θ can be substracted from the higher angle raw intensity value where the $K\alpha_2$ diffracts. The negative peak indicated in Figure 11.6 is determined in this manner and on subtracting it from the observed raw data, a pure $K\alpha_1$ profile results.

The Rachinger method is a good first approximation to $K\alpha_2$ deconvolution; however, it assumes that the profile of the $K\alpha_1$ is identical to the $K\alpha_2$. In fact, the $K\alpha_2$ occurs at a higher angle and therefore is more broadened by spectral dispersion than the $K\alpha_1$, and this difference in profile shape gets stronger as 2θ increases. Ladell et al. [12] carefully determined the profiles of the $K\alpha_1$ and $K\alpha_2$ lines of resolved lines like the one shown in the upper portion of Figure 11.6 and determined how their difference in shape varied with 2θ, using Fourier techniques. Based on this analysis, a table was created permitting a portion of the $K\alpha_2$ intensity to be subtracted from a series of adjacent 2θ points. This modified Rachinger method does a very good job of removing the $K\alpha_2$ profiles from an observed pattern. Delhez and Mittemeijer [13] have discussed the need to use an angle-dependent $K\alpha_1$, $K\alpha_2$ separation.

11.3.4. Peak Location Methods

Manual Peak Location. As far as manual methods are concerned, although the human eye is certainly very good at integrating analog data and establishing peak position, position measurements established by manual methods are not always of the best quality. Most of the problems in this area arise from poor technique on the part of the operator. One such problem is the use of too small a value of the 2θ scale on the chart recorder. As an example, with a goniometer speed of $1° 2\theta/\text{min}$ and a chart speed of 1 cm per degree 2θ, in order to measure a peak value to an accuracy of $0.01°$, one must be able to measure the chart to ± 0.1 mm—clearly a very optimistic goal. Another common

STEPS IN DATA TREATMENT 301

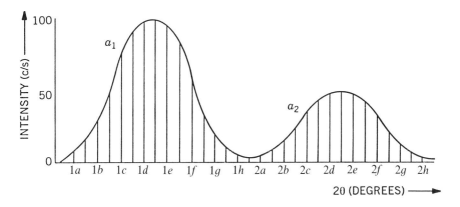

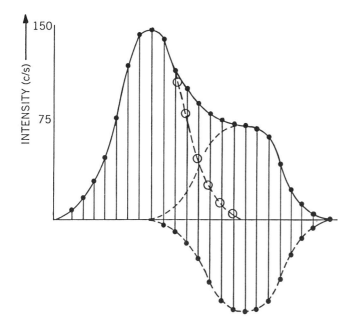

Figure 11.6. Rachinger method for the removal of α_2.

source of systematic error is incorrect alignment of the recorder interval marker pen with the intensity marker pen. In order that data of reasonable quality be obtained, the best solution is to run standard specimens through the whole data collection/treatment process at frequent intervals (at least once a week) and compare the experimental set of $d-I$ values with the theoretical set of values.

Computer Peak Location. The most common peak detection method in use today is based on a combination of smoothing, $K\alpha_2$ removal, and background stripping, followed by a second derivative peak finder. Another procedure gaining in popularity is to fit profiles to the raw pattern, using all of the points in the peak to determine its location. The first technique remains by far the most common, even though the success of derivative methods has a strong dependence on counting statistical noise and the degree of smoothing employed.

As described in Section 11.3.1, the Savitsky–Golay algorithm carries out a least squares fit of a polynomial to each overlapped interval of data points, such that the $a, b, c, \ldots$ coefficients can be used with the 2θ value of the center point of the interval (in Equation 11.2) to compute a "smoothed" value. It is immediately clear that the same coefficients can also be used to evaluate the first and second derivatives at that point in 2θ. Thus, at the same time that the data are smoothed, the derivative of the data is computed. While some algorithms use the first derivative to locate the maxima in the observed data, this procedure will only find resolved peaks. So, most programs use the second derivative, which is sensitive to inflection points and will therefore reveal overlapped peaks. Typically, a least squares parabola is fit to each minimum in the second derivative function to produce an interpolated estimate of the location of the peak. Since the second derivative changes sign at each inflection point, it also produces an estimate of the FWHM of the peak.

The number of peaks found in a procedure will depend on the extent of data smoothing and the signal-to-noise ratio. The number of false peaks found can be minimized by correlating the smoothing parameters with the count time. The accuracy of the peak locations will depend, to some degree, on each of the raw-data-processing steps. However, to achieve absolute accuracy, it is essential that the data be corrected for the aberrations introduced by the instrumental measurement technique. Remember that the sample displacement error described earlier is by far the most serious. Displacements in the tens of micrometers range can cause significant peak shifts of hundredths of a degree.

Profile Fitting. As previously mentioned, there are a variety of factors that determine the shape of a diffracted line profile, and the more important of these include the axial divergence of the X-ray beam, the particle size and/or microstrain of specimen, the monochromaticity of the source, and the degree and type of data smoothing employed in the processing of the raw data. Because of these various influences, the peak shape is typically variable and asymmetric. As an example, the data shown in Table 11.3 were taken [14] on the (020) line of MoO_3 at 12.8° 2θ and show that the peak widths at 20, 50, and 80% of peak maximum vary quite significantly over a range of different

Table 11.3. Data for an Oriented Specimen of MoO$_3$

User	Average Width			Fractional 50% Width	
	20%	50%	80%	20%	80%
1	0.150	0.087	0.050	1.73	0.58
2	0.210	0.110	0.057	1.94	0.52
3	0.228	0.111	0.063	2.06	0.57
4	0.207	0.113	0.060	1.82	0.53
5	0.237	0.120	0.067	1.97	0.56
6	0.253	0.133	0.077	1.90	0.58
7	0.270	1.143	0.083	1.88	0.58
8	0.275	0.145	0.055	1.90	0.38
9	0.270	0.150	0.083	1.80	0.56
10	0.273	0.153	0.077	1.78	0.50
11	0.327	0.153	0.080	2.13	0.52
12	0.340	0.173	0.087	1.96	0.50
Avg	0.254	0.133	0.070	1.91	0.53
Max	0.340	0.173	0.070	2.13	0.58
Min	0.150	0.087	0.050	1.73	0.38

instruments. Table 11.3 also gives the fractional width ratioed to the width at 50%. Even so, the relative *shapes* of the profiles are remarkably consistent. Peak distortion of this type can, in turn, lead to problems in the estimation of peak maxima. As illustrated in Figure 11.7, there are various ways in which a peak maximum can be defined. These include (a) use of the first derivative ($dI/d\theta = 0$); (b) taking the average of inflection points; or (c) use of profile centroids. As long as the peak is reasonably symmetrical, each of these methods should give essentially the same result. However, the asymmetrical peaks from parafocusing diffractometers generally produce different results for each of the different methods.

Peak-hunting methods based on profile-fitting procedures are now becoming popular. These methods are ideally suited for the analysis of closely overlapped peaks, which can be extremely difficult to analyze with the more conventional techniques now in common use. These techniques range all the way from a basic simplex optimization of a symmetric Lorentzian function [4] through total pattern fitting using an internal standard and full deconvolution of the specimens contribution to the profiles [15]. The advantage of profile fitting is immediately apparent: all of the intensity measurements defining the profile are used to determine the best way of fitting an analytical function to the observed envelope. A less apparent advantage is that the systematic

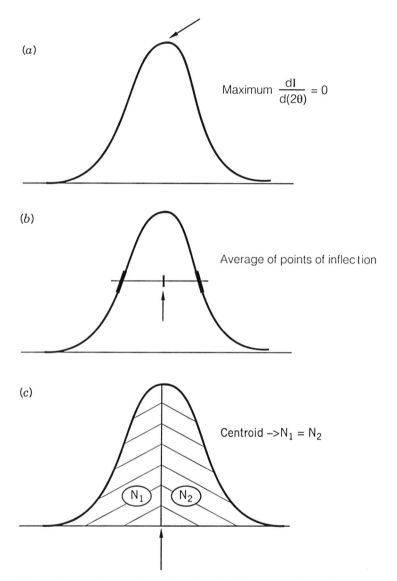

Figure 11.7. Definition of a peak. From R. Jenkins, Experimental procedures. In *Modern Powder Diffraction* (D. L. Bish and J. E. Post, eds.), p. 58, Fig. 7. Mineralogical Society of America, Washington, DC, 1989. Reprinted by permission.

distortions introduced by the peak smoothing and location procedures will be completely removed in this subsequent refinement step. Thus, the estimate of the peak location in 2θ, as well as its FWHM and integrated intensity, will be determined by all measurements rather than just a few. In general, peak locations from profile fit data are twice as precise as those given by the second derivative technique [16].

Early approaches to profile fitting used symmetric functions and ignored the inherent asymmetry of X-ray diffraction profiles, particularly at low angle. Parrish et al. [17] were the first to exactly fit the asymmetrical profile by summing seven independent Lorentzian functions under each profile in order to define the inherent angle-dependent profiles produced by a particular diffractometer, using standards with no specimen broadening. Any further broadening displayed by profiles of a particular specimen are due to size and strain effects, which can then be determined. A more convenient method of fitting asymmetric profiles is to fit different Lorentzian functions to each side of the peak [18]. This split Pearson VII function has also been used to deconvolute the specimen broadening contribution to an observed profile FWHM [19]. A full description of these procedures may be found in a recent review by Snyder [20]. Whole-pattern fitting of the calculated powder diffraction pattern to the observed pattern is becoming a very popular technique for refining structural parameters as well as for determining the phase content of polyphase samples. This Rietveld technique is well described in the recent book just referred to above [20].

11.4. CONVERSION ERRORS

With the growing need for higher quality X-ray powder diffraction data, a problem that is becoming increasingly more difficult to handle is that of the selection of the "practical" wavelength in a powder diffraction experiment. Errors can occur for several reasons. As an example, because of the (generally) polychromatic nature of the diffracted beam, it is sometimes difficult to manually assess where the maximum of a peak occurs, especially in the range of angles from $30°$ to $60°$ 2θ. As an example, Figure 11.8 shows a plot for the angular dispersion of the $\alpha_1 \alpha_2$ doublet, over a range of 2θ values for line widths of $0.10°$, $0.2°$, and $0.3°$ 2θ. The angle at which the $K\alpha_1$, $K\alpha_2$ doublet is resolved (at FWHM) is indicated in each example. Where computer searching for peaks is employed, such methods are invariably more sensitive than manual techniques, meaning that a doublet is recognized as such at a lower angle. This in turn can lead to difficulties during phase identification because all the observed lines may not match the correct standard pattern selection. This problem can be made even worse when α_2 stripping or profile-fitting tech-

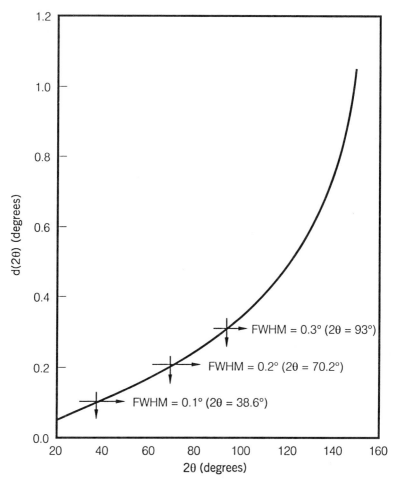

Figure 11.8. Dispersion of the $K\alpha_1, K\alpha_2$ doublet as a function of 2θ for different widths. From R. Jenkins, Experimental procedures. In *Modern Powder Diffraction* (D. L. Bish and J. E. Post, eds.), p. 61, Fig. 9. Mineralogical Society of America, Washington, DC, 1989. Reprinted by permission.

niques are employed on the experimental pattern and have not been employed in the measurement of the reference pattern(s). Another area of difficulty occurs where computers are used for such tasks as the estimation of peak position, the removal of the α_2 contribution, and profile fitting. While these techniques are extremely useful, care must be taken in comparison of computer and manually treated data. Another warning is needed here: when very efficient position-sensitive detectors and very bright X-ray sources are used, such as those from rotating anodes and synchrotrons, often weak peaks are detected that were not observed when the PDF reference pattern was determined. Users must always be on their guard against the effects of changes in technology that may cause discrepancies between the reference and observed patterns.

The measured parameters in a diffraction experiment are the absolute line intensities and the angles at which lines occur. In qualitative phase identification, the parameters used to characterize a phase are the equivalent interplanar d-spacings and the relative intensities. An important point to consider is the

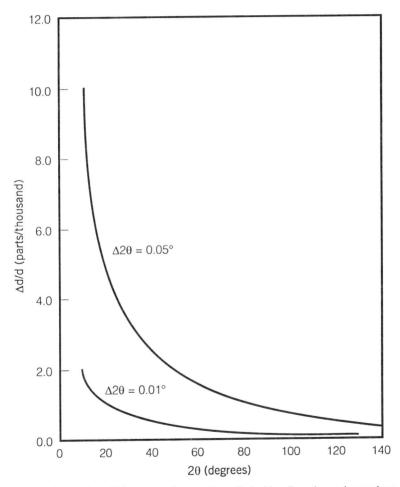

Figure 11.9. Conversion of 2θ errors to d-errors. From R. Jenkins, Experimental procedures. In *Modern Powder Diffraction* (D. L. Bish and J. E. Post, eds.), p. 61, Fig. 10. Mineralogical Society of America, Washington, DC, 1989. Reprinted by permission.

effect of an error in 2θ on the d-spacing. The relationship between the two is of the form

$$\frac{\Delta d}{d} = \Delta\theta \cot \theta. \qquad (11.3)$$

This function is plotted in Figure 11.9 for $\Delta 2\theta$ errors of 0.01° and 0.05°. It will be clear from the plot that a fixed error in 2θ has a greater impact on d-spacing accuracy at low angles than it does at high angles. This last aspect of computer reduction of digitized data has by far produced the most dramatic impact on the precision of peak location. The fitting of analytical profiles to the observed data using some type of optimization procedure (like simplex or least squares) has produced average $\Delta 2\theta$ values lower than 0.005°. In spite of the numerous sources of error in the measurement of 2θ values and subsequent conversion to d-spacings for a typical diffractometer pattern, the derived d-values are generally of sufficient quality to allow qualitative phase identification. However, the improvement in both the quality of the observed pattern and the references in the PDF due to computer automation has made qualitative analysis much more routine and accessible to scientists who are not expert diffractionists.

11.5. CALIBRATION METHODS

In Chapter 10 we discussed the preparation of internal and external calibration curves. It is time now to consider how to use these calibration curves for data correction.

11.5.1. 2θ Correction Using an External Standard

In the use of the external standard, a suitable material is chosen (see Section 10.7) and a series of peaks is measured. Peaks in the standard diagram are then measured and $\Delta 2\theta$ (i.e., $2\theta_{obs} - 2\theta_{calc}$) is plotted against 2θ, as was discussed in Section 9.6.4. The least squares fitting of a polynomial (of the type shown in Equation 11.2 with a $\Delta 2\theta$ dependent variable) to this curve allows the polynomial coefficients to be stored on a disk file, and this curve may then be used routinely and automatically, to correct all experimental patterns for instrumental errors [4]. While the correction line is indeed a curve, for a well-aligned diffractometer the curvature is often not very great and for manual application a good approximation is to assume a straight line. The correction term is now very easy to apply since it simply entails adding (or subtracting) a small angular increment over the appropriate portion of the

Table 11.4. The Effects of Calibration on the Figure of Merit F_N

Method	Arsenic Trioxide ($N = 29$)		Quartz ($N = 30$)	
No correction	9.9	(0.0049,59)	16.4	(0.052,35)
External standard	15.4	(0.026,59)	30.0	(0.028,35)
Internal standard	42.0	(0.012,59)	66.1	(0.013,35)

angular range. As an example, Figure 9.9 shows an external calibration curve. Inspection of the calibration curve shows that the correction curve approximates to a straight line, with a change in 2θ of about $-0.02°$ over the angular range of about $120°$. It is important to remember that the external standard will not normally correct for differences in specimen absorption and for specimen displacement errors, this latter error being typically one of the largest encountered in the measurement of an experimental pattern. However, in the zero background holder (ZBH) method, since the specimen displacement of the external standard and specimen are controlled to be within a few micrometers of each other, the external standard technique will give results equivalent to the internal standard method.

11.5.2. 2θ and d-Spacing Correction Using an Internal Standard

When an internal standard is used, the standard is intimately mixed with the specimen to be analyzed. Diffraction lines from the internal standard phase are used to construct a correction curve. This procedure completely eliminates the effects of specimen transparency (absorption) and specimen displacement. As an example, Table 11.4 lists data for the two phases arsenic trioxide and quartz [16]. In both cases, the data have been collected three times—without correction, with external standard correction, and with internal standards. The data given are the Smith–Snyder figures of merit [21] described below in Section 11.6.1. It will be seen that, in both examples, use of the external standard improves the quality of the average $\Delta 2\theta$ error (the first number in the parentheses) by about a factor of 2. Use of the internal standard gives a further improvement of a factor of 2, giving a $\Delta 2\theta$ of about $0.012°$. In fact, the use of profile fitting will provide another factor of 2 improvement, dropping the average $\Delta 2\theta$ error into the $0.005°$ range.

11.5.3. Sensitivity Correction Using an External Intensity Standard

The purpose of the external intensity standard is to correct for sensitivity (i.e., counts per second per unit concentration) variations across the angular range of the diffractometer. Such variations may be inherent to the design optics of a given instrument or to misalignment problems in a given instrument [14]. Typical contributors to the variation in design optics are placement and aperture size of divergence slits, and configuration of the monochromator. Contributions to misalignment effects on sensitivity include alignment of the divergence slit and alignment of the monochromator. An example of the use of an external intensity standard has already been given in Section 8.2.

11.6. EVALUATION OF DATA QUALITY

One of the greatest problems in all scientific experimentation is to judge the credibility and worth of the results of an experiment. In the fields of X-ray fluorescence and X-ray powder diffractometry, typical judgments to be made include: how well a given instrument is aligned or, indeed, how well it was designed; whether the analysis time available is being used to its best advantage; whether the data obtained are sufficiently precise; whether the experimental data are being treated correctly to provide the desired result; and so on. The tools used for the evaluation purposes range from detailed statistical analysis to simple "gut feel." In most analytical laboratories, analysis time is typically at a premium and it is generally in the analyst's best interests to ensure that the quality of the data is "adequate" in the light of the available time. However, just as one must realize the limitations in the quality of experimental data, one must also be cautious in the interpretation of the experimental results.

The major problem that most diffractionists face in setting up a given diffraction experiment is whether or not their selection of the instrument and/or software variables represent the optimum set of experimental conditions. To this end, the use of a FOM may offer some assistance. The FOM will often provide useful guidelines in the establishment of optimum conditions. Such figures of merit based on counting statistical limitations have long been used in, for example, X-ray fluorescence spectrometry [22]. However, in X-ray fluorescence, the line broadening parameters are determined purely by the fixed collimators and line shapes are easily correlated with diffraction angle. In X-ray diffractometry, the situation is much more complicated, and conventional figures of merit based on peaks and background are not directly applicable.

11.6.1. Use of Figures of Merit

The basic idea of the FOM is to attempt to derive a *quantitative* measure of performance based on measurable parameters. Correctly designed figures of merit can be used to quantify a variety of problems including: the optimization of equipment design; the setting up of specific equipment measurement conditions to assess equipment performance; and to gauge the quality of experimental data and data processing [23]. While there are many different types of figures of merit, most fall into one of two broad categories—*derived* and *intuitive*. Derived figures of merit are based either on measurable parameters such as peak height, peak width, background, and sensitivity or on mathematically derived data such as counting statistics, standard deviations of data sets, and least squares analysis. Most derived figures of merit arise because of the statistical processes involved in the production and processing of X-ray data. Derived figures of merit are used mainly for establishing optimum experimental conditions. Intuitive figures of merit are based on subjective judgment calls of the experimentalist, who draws on previous experience or on personal assessment of the importance of certain parameters. Intuitive figures of merit are widely used in the file-searching process involved with qualitative powder diffraction. Figures of merit can also be based on a combination of measurable and intuitive factors. Table 11.5 [24–30] lists the more commonly employed figures of merit and indicates their basis and their main uses.

Table 11.5. Common Figures of Merit and Their Uses

Authors	Use	Basis	Ref.
Jenkins & de Vries	Setting up equipment	Count statistics: peak and background	[23]
Spielberg & Bradenstein	Trace analysis conditions	Count statistics: sensitivity & background	[24]
Jenkins & Schreiner	Diffractometer evaluation	Count statistics: peak, background, & line width	[1]
Smith & Snyder	Quality of data/ success of indexing	$\Delta 2\theta$	[21]
de Wolff	Success of indexing	$\Delta \mathbf{d}^{*2}$	[25]
Snyder	Evaluation of intensities	Intensity	[16]
Lowe-Ma	Evaluation of intensities	Intensity	[26]
Various	Order of hits	Subjective	[27–30]

312 DATA REDUCTION FROM AUTOMATED DIFFRACTOMETERS

11.6.2. Use of Figures of Merit for Instrument Performance Evaluation

Although intensity is important in powder diffraction measurements, the ease of interpretation of the resultant diffractogram is invariably dependent on the resolution of the pattern. A pattern made up of broad lines super-imposed on a high, variable background is much more difficult to process than one in which the lines are sharp and well resolved and the background is low and flat. In the setting up of instrumental parameters for a given series of experiments, there are many variables under the control of the operator—source conditions, receiving slit width, scan speed, step increment, etc. Jenkins and Schreiner have addressed this problem in their review of data from an intensity round-robin test [1]. Their suggested instrument FOM has been previously discussed in Section 8.3.

In the setting up of powder diffractometers, the peak width is one of several parameters to be considered. As an example, one of the selectable parameters in many diffractometers is the secondary collimator between the specimen and the detector. Removal of the collimator greatly increases the peak-counting rate but also increases the background-counting rate and the line widths. Table 11.6 shows data from a recent round-robin test [1] in which data from 22 different experimentalists, grouped by five different instrument manufacturers, are presented in terms of the improvement in the FOM obtained by removing the secondary Soller slit. In almost all cases, a significant gain is found in the FOM, indicating that for the type of measurement tested, where line overlap was not a severe problem, the optimum instrument configuration is without the use of the secondary Soller collimator.

Note from Equation 8.7 (i.e., $FOM = MW/[W(M + 4B)]^{1/2}$) that the following three conditions apply:

1. As $M \to 0$, then $FOM \to 0$ (i.e., limit of detection).
2. As $M \to \infty$, then $FOM \to M$ (i.e., number of counts).

Table 11.6. Improvement in the FOM, Following the Removal of the Secondary Soller Slit

Instrument Type	Improvement Factor
A (4 users)	1.55
B (3 users)	2.03
C (2 users)	1.85
D (9 users)	1.93
E (2 users)	1.42

Table 11.7. Use of the Jenkins–Schreiner FOM for the Selection of Optimum Instrument Conditions

	A	B	C
M (maximum count rate)	887	2233	8188
B (background rate)	3.8	33	102
W (width of peak)	0.087	0.15	0.12
M/B	235	34.3	79
FOM	87	176	302

3. Note that the FOM varies as a linear function of the full width at half-maximum, W; therefore, for quantitative work, where line overlap is not a problem, one should use conditions to give a large value of W.

An example of the use of the instrument FOM is given in Table 11.7. The table lists three sets of data labeled A, B, and C. These data sets were obtained on the same specimen, but measurements were carried out using different combinations of receiving slit width and specimen irradiation area. The purpose of the experiment was to select the optimum measurement conditions for a quantitative phase determination using the integrated peak and background intensities of a single, nonoverlapped line. It will be seen that the peak/background ratio varies from 34.3 (B) to 79 (C) to 235 (A). If the peak/background ratio were the criterion by which the *best* instrument condition was being selected, clearly data set A would be the one chosen. However, note that the integrated peak intensity for set C is almost an order of magnitude greater than that of set A. This is reflected in the FOM where combination C is clearly the best choice.

11.6.3. Use of Figures of Merit for Data Quality Evaluation

The principal application of X-ray powder diffraction is qualitative phase identification (which will be described in Chapter 12). On comparing patterns in the PDF to an observed pattern, a knowledge of the quality of the reference pattern is of the highest interest. In the past, certain symbols were used in the PDF to supply some indication of quality. These "quality marks" [e.g., the "star" (∗), I, C, and O explained in Section 12.2.1] were assigned by the editors of the PDF using a set of heuristic criteria that changed over the years as experimental data improved due to the influence of computer automation. Smith and Snyder in 1979 [21] introduced an objective FOM that gives a quantitative estimate of the quality of the metric aspects of a powder pattern (i.e., the location of the diffraction maxima not their height) in order to avoid such subjective quality estimates. The American Crystallographic Association

DATA REDUCTION FROM AUTOMATED DIFFRACTOMETERS

Table 11.8. Use of the Smith–Snyder FOM for the Evaluation of the Metric Aspects of a Powder Pattern[a]

No.	$2\theta_{calc}$	I^{rel}	d (Å)	$2\theta_{obs}$	$\Delta 2\theta$
1	6.710	21	13.162	6.790	0.080
2	8.820	48	10.018	8.780	−0.040
3	11.710	12	7.551	11.760	0.050
4	14.320	37	6.180	14.360	0.040
5	17.210	100	5.148	17.200	−0.010
6	18.950	25	4.679	18.970	0.020
7	20.230	95	4.386	20.210	−0.020
8	20.730	45	4.281	20.760	0.030
9	21.819	11	4.070	21.809	−0.010
10	26.263	5	3.391	26.283	0.020
11	31.721	12	2.818	31.727	0.006
12	32.618	—	2.743	—	—
13	34.618	78	2.589	34.602	−0.016
14	38.210	31	2.353	38.221	0.011
15	46.262	3	1.961	46.260	−0.002
16	47.183	—	1.925	—	—
17	47.523	39	1.912	47.517	−0.006
18	48.325	68	1.882	48.318	−0.007
19	49.199	21	1.850	49.200	0.001
20	50.999	4	1.789	51.003	0.004
21	52.503	27	1.741	52.509	0.006
22	56.215	—	1.635	—	—
23	56.973	26	1.615	56.991	0.018
24	58.201	11	1.584	58.200	−0.001
25	59.000	3	1.564	59.012	0.012
26	59.421	2	1.554	59.460	0.039
27	60.772	2	1.523	60.820	0.048
28	61.317	36	1.511	61.295	−0.022
29	63.111	17	1.472	63.126	0.015
30	64.914	2	1.435	64.930	0.016
31	66.782	9	1.400	66.740	−0.008
32	68.011	6	1.377	68.000	−0.011
33	72.771	10	1.298	72.764	−0.007
34	73.182	3	1.292	73.210	0.28
35	73.881	—	1.282	—	—
36	74.082	—	1.279	—	—
37	72.926	4	1.266	74.937	0.011
38	76.102	6	1.250	76.105	0.003
39	77.011	3	1.237	77.020	0.009

EVALUATION OF DATA QUALITY

Table 11.8. (*Cont'd.*)

No.	$2\theta_{calc}$	I^{rel}	d (Å)	$2\theta_{obs}$	$\Delta 2\theta$
40	80.253	5	1.195	80.238	−0.015
41	81.772	9	1.177	81.776	0.004
42	82.025	1	1.174	—	—
43	84.002	1	1.151	—	—
44	84.923	1	1.141	84.916	−0.007
45	85.773	3	1.132	85.770	−0.003
46	8.246	6	1.106	88.249	0.003
47	89.114	2	1.098	89.110	−0.004
48	90.002	1	1.089	—	—
49	90.734	4	1.082	90.720	−0.014
50	91.720	1	1.073	91.726	0.006
Avg $\Delta 2\theta$	0.0066	0.0066	0.0083	0.0104	0.0084
N_{poss}	50	50	40	30	20
N_{obs}	50	42	35	27	18
FOM	151.6	127.4	104.9	86.2	107.3

a Wavelength = 1.54056.

[31], the International Union of Crystallography, and the International Centre for Diffraction Data (ICDD) have all adopted this FOM for the evaluation of the quality of the d_{hkl} measurements.

The figure of merit F_N has the form

$$F_N = \frac{1}{|\Delta 2\theta|} \frac{N}{N_{poss}} \qquad (11.4)$$

and is given as $xx.x(y.yyy, zz)$ where N is the number of experimental lines considered; $xx.x$ is the value of F_N; $y.yyy$ is the average $\Delta 2\theta$ error; and zz is the number of diffraction lines possible within the 2θ range covered by the first N lines. In the computation of N_{poss}, all of the resolvable, space group allowed reflections predicted by Equation 3.10 are counted. Figures of merit around 80–150 are considered high quality, and those less than 20 are rather poor quality.

As an example, Table 11.8 lists 50 possible lines of data from a diffraction pattern. Only 42 of these lines have observable intensity under the conditions of the experiment. The $2\theta_{calc}$ values represent the calculated 2θ values based on the known lattice parameters. The $2\theta_{obs}$ values represent the experimental observed values after calibration corrections have been applied. The $\Delta 2\theta$ values represent the differences between calculated and experimental values. At the bottom of Table 11.8 are listed average $\Delta 2\theta$ values and N_{poss} data for the first 20 lines, the first 30 lines, the first 40 lines, and for all of the lines. Also listed are the respective F_N figures of merit. Note the way that the FOM varies as more and more lines of data are added. The initial FOM for 20 lines is 107.3, but this drops to 86.2 when an additional 10 lines are included. This change is due mainly to the large $\Delta 2\theta$ values for the weak lines at Nos. 26 and 27. Inclusion of another 10 lines improves the FOM to 104.9, and inclusion of all the experimental data improves the FOM further to 127.4. Note, too, that searching through the original experimental pattern to convince oneself that the 8 weak lines predicted from the cell really are "present" would bring about a further improvement in the FOM to 151.6. Such a procedure is, however, extremely dangerous: science provides a number of examples of humans being particularly prone to "seeing" things in background (i.e., in cases of low signal-to-noise ratio) that are not really there. The presence or absence of a peak should be made using the objective criterion that it be 3σ (or certainly not less than 2.5σ) above background. One should note here that, when all of the peaks in a profile are considered, the statistical precision of deciding if a number of sequential points are above background is considerably enhanced. So, profile fitting may be used to help in making difficult decisions concerning the presence or absence of a peak.

11.6.4. Use of Figures of Merit in Indexing of Powder Patterns

The process of assigning *hkl* values to the diffraction lines in the pattern of a new phase is known as *indexing a pattern*. In recent years, very powerful computer procedures have been developed for solving this problem in symmetries as low as triclinic. During the process of indexing, several possible solutions may result from trial unit cells. In order to attempt to quantify these possible solutions, two figures of merit are commonly employed—the de Wolff FOM and the Smith–Snyder FOM. The de Wolff FOM [25] M_{20} is the older of the two and takes the form

$$M_{20} = \frac{\mathbf{d}_{20}^{*2}}{2|\Delta\mathbf{d}^{*2}|} \frac{1}{N_{\text{poss}}}, \qquad (11.5)$$

where $\mathbf{d}_{20}^{*2}$ is the value of $\mathbf{d}^{*2}$ for the 20th line; $|\delta\mathbf{d}^{*2}|$ is the average discrepancy between the $\mathbf{d}^{*2}$ for the observed and calculated patterns for first 20 values; and N_{poss} is the same as used in F_N, above. One of the problems with the de Wolff FOM is that it is difficult to relate the FOM to the actual measured parameter, i.e., the 2θ value. Another problem is that the value of the FOM depends on the cell volume and therefore its scale is different for each crystal class. The F_N FOM avoids these problems and acts as a uniform judge of the correctness of an indexing for all crystal systems.

REFERENCES

1. Jenkins, R., and Schreiner, W. N. Intensity round robin report. *Powder Diffr.* **4**, 74–100 (1989).
2. Smith, D. K., and Jenkins, R. The powder diffraction file: Past, present and future. *Rigaku J.* **6**, 3–14 (1989).
3. Savitsky, A., and Golay, M. J. E. Smoothing and differentiation of data by simplified least squares procedures. *Anal. Chem.* **36**, 1627–1639 (1964).
4. Mallory, C. L., and Snyder, R. L. The control and processing of data from an automated powder diffractometer. *Adv. X-Ray Anal.* **22**, 121–132 (1979).
5. Goehner, R. P., and Hatfield, W. T. A microcomputer controlled diffractometer. *Adv. X-Ray Anal.* **22**, 165–167 (1979).
6. Jobst, B. A., and Göbel, H. E. IDENT: A versatile microfile-based system for fast interactive XRPD analysis. *Adv. X-Ray Anal.* **25**, 273–282 (1982).
7. Snyder, R. L. The renaissance of X-ray powder diffraction. In *Advances in Material Characterization* (D. R. Rossington, R. A. Condrate, and R. L. Snyder, ed.), pp. 449–464. Plenum, New York, 1983.
8. Mallory, C. L., and Snyder, R. L. Threshold level determinations from digital X-ray powder diffraction patterns. *NBS Spec. Publ. (U.S.)* **567**, 93 (1980).
9. Cullity, B. D. *Elements of X-Ray Diffraction*, 2nd ed. Addison-Wesley, reading, MA, 1978.
10. Rachinger, W. A. (1948) A correction for the $\alpha_1:\alpha_2$ doublet in the measurement of widths of X-ray diffraction lines. *J. Sci. Instrum.* **25**, 254–259 (1948).
11. Gangulee, A. Separation of the $\alpha_1:\alpha_2$ doublet in X-ray diffraction profiles. *J. Appl. Crystallogr.* **3**, 272–277 (1970).
12. Ladell, J., Zagofsky, A., and Pearlman, S. Cu $K\alpha_2$ elimination. *J. Appl. Crystallogr.* **8**, 499–506 (1975).
13. Delhez, R., and Mittemeijer, E. J. Improved α_2 elimination. *J. Appl. Crystallogr.* **8**, 609–611 (1975).
14. Schreiner, W. N., and Jenkins, R. Results of the JCPDS–ICDD intensity round robin. *Adv. X-Ray Anal.* **32**, 557–560 (1988).
15. Huang, T. C., Parrish, W., Masciocchi, N., and Wang, P. W. Derivation of d-values from digitized X-ray and synchrotron diffraction data. *Adv. X-Ray Anal.* **33**, 295–304 (1989).
16. Snyder, R. L. Accuracy in angle and intensity measurements in X-ray powder diffraction. *Adv. X-Ray Anal.* **26**, 1–11 (1983).
17. Parrish, W., Huang, T. C., and Ayers, G. L. Profile fitting: A powerful method of

computer X-ray instrumentation and analysis. *Am. Crystallogr. Assoc. Monog.* **12**, 5–73·(1976).

18. Howard, S. A., and Snyder, R. L. An evaluation of some profile models and the optimization procedures used in profile fitting. *Adv. X-Ray Anal.* **26**, 73–80 (1982).
19. Howard, S. A., and Snyder, R. L. The use of direct convolution products in profile and pattern fitting algorithms. I. Development of the algorithms. *J. Appl. Crystallogr.* **22**, 238–243 (1989).
20. Snyder, R. L. Analytical profile fitting of X-ray powder diffraction profiles in Rietveld analysis. In *The Rietveld Method* (R. A. Young, ed.), Chapter 7, pp. 111–131. Oxford Univ. Press, Oxford, 1993.
21. Smith, G. S., and Snyder, R. L. F_N, a criterion for rating powder diffraction patterns and evaluating the reliability of powder indexing. *J. Appl. Crystallogr.* **12**, 60–65 (1979).
22. Jenkins, R., and Gilfrich, J. V. Figures-of-merit: Their philosophy, design and use. *X-Ray Spectrom* **21**, 263–269 (1992).
23. Jenkins, R., and de Vries, J. L. *Practical X-Ray Spectrometry*, 2nd ed., Sect. 5.8 and 5.9. Springer-Verlag, New York, 1977.
24. Spielberg, N., and Bradenstein, M. Instrumental factors and FOM in the detection of low concentrations for X-ray spectrochemical analysis. *Appl. Spectrosc.* **17**, 6 (1963).
25. de Wolff, P. M. A simplified criterion for the reliability of powder diffraction patterns... *J. Appl. Crystallogr.* **1**, 108–113 (1968).
26. Lowe-Ma, C. K. Powder diffraction data for two energetic materials and a proposed intensity figure-of-merit. *Powder Diffr.* **6**, 31–35 (1991).
27. Johnson, G. G., Jr. The Johnson-Vand search/match algorithm. *Norelco Rep.* **26**, 15–18 (1979).
28. Marquart, R. G., Kapsnelson, I., Milne, P. W. A., Heller, S. R., Johnson, G. G., Jr., and Jenkins, R. A search-match system for X-ray powder diffraction data. *J. Appl. Crystallogr.* **12**, 629–634 (1979).
29. Snyder, R. L. A Hanawalt type phase identification procedure for a minicomputer. *Adv. X-Ray Anal.* **24**, 83–90 (1981).
30. Huang, T. C., and Parrish, W. A new computer algorithm for qualitative X-ray powder diffraction analysis. *Adv. X-Ray Anal.* **25**, 213–219 (1981).
31. Calvert, L. D., Flippen-Anderson, J. L., Hubbard, C. R., Johnson, Q. C., Lenhert, P. G., Nichols, M. C., Parrish, W., Smith, D. K., Smith, G. S., Snyder, R. L., and Young, R. A. Standards for the publication of powder patterns: The American Crystallographic Association Subcommittee Final Report. *NBS Spec. Publ. (U.S.)* **567**, 513–536 (1980).

CHAPTER
12
QUALITATIVE ANALYSIS

12.1. PHASE IDENTIFICATION BY X-RAY DIFFRACTION

Since, at least in principle, every crystalline material gives a unique X-ray diffraction pattern, study of diffraction patterns from unknown phases offers a powerful means of qualitative identification. The problem is one of pattern recognition between the unknown and a database of single-phase reference patterns. Although the potential for qualitative phase identification was certainly recognized from the very early days of X-ray diffraction, the first attempts to list standard diffraction patterns were not published in detail until the mid-1930s [1]. Table 12.1 lists major landmarks in the use of powder diffraction as a qualitative analysis tool. Probably the most significant of these milestones is the work of Hanawalt, Rinn, and Frevel [2] of the Dow Chemical Company, who in 1938 published a file of about 1000 diffraction patterns with an indexing system based on the use of the three d-spacings giving the strongest intensities. Today, the file is produced in annual updates by the International Centre for Diffraction Data (ICDD), which is a nonprofit corporation relying on the volunteer efforts of many of the world's powder diffractionists.

The means of archiving and retrieval of patterns proposed in the mid-1930s [1] still provides the basis of many search/match methods in use today. The traditional method for the storage of data is to reduce the experimental pattern to a table of $d-I$ values, often referred to as a *reduced* pattern because the process of data treatment reduces the large volume of data in the raw scan to a concise digital form. Unfortunately, during the data reduction process much information concerning the line shape and intensity distribution is lost. Although it may be more useful in some cases to utilize the full diffractogram, until recently storage limitations have inhibited the development of a pattern reference file of fully digitized patterns.

Problems due to inadequate data treatment in finding the angular positions of the peak maxima, plus uncertainties in the value of the experimental wavelength(s), all conspire to add errors to the experimental d-values [3]. The experimental intensities may be similarly distorted due to problems of preferred orientation, poor crystallinity, partially resolved diffraction wavelength multiplets, and line broadening due to particle size and/or strain consider-

Table 12.1. Major Landmarks in the Use of Powder Diffraction for Qualitative Analysis

1917–1919	P. J. W. Debye and P. Scherrer in Europe and A. W. Hull in the United States point out the potential advantages of powder diffraction as a tool for qualitative analysis.
1927	A. N. Winchell publishes first private collection of diffraction patterns.
1935	A. W. Waldo publishes patterns of 51 copper ores.
1938	J. D. Hanawalt, H. W. Rinn, and L. Frevel publish a file of 1000 patterns with an indexing and search system.
1938	The Institute of Mines in Leningrad tabulates powder data for 142 minerals.
1941	Patterns produced on 3 × 5 cards by the National Research Council (NRC) and Committee American Society for Testing and Materials E4 of the (ASTM).
1941–1945	Other societies join the powder committee of ASTM.
1969	The Joint Committee on Powder Diffraction Standards (JCPDS) is incorporated as an independent nonprofit organization.
1977	JCPDS changes its name to International Centre for Diffraction Data (ICDD).
1994	The ICDD Powder Diffraction File (PDF) grows to 60,000 patterns.

ations. Many of the problems of data reduction may be solved in the future, and indeed great improvements have already occurred during the last several years. As an example, the use of *profile-fitting* techniques [4] has done much to give a better measure of integrated peak intensities and profile maxima.

Experimental errors can blur the match between an observed and reference pattern to a point that the user may fail to identify the presence of a phase. The ability to recognize a reference pattern in an unknown strongly depends on the quality of the d's and I's in both the reference material and the unknown sample. One of the principal problems in the identification of materials by comparison of an experimental pattern with reference patterns is the *variability* in the quality of the data. The experimental technique used to measure the pattern is one of the first quality indications to a user of a reference database. For Debye–Scherrer camera data one should assume an error window of $\pm \Delta 2\theta = 0.1°$; for normal diffractometer data one typically assumes a $\pm \Delta 2\theta = 0.05°$; and for internal standard corrected diffractometer or Guinier camera data a $\Delta 2\theta$ window of $0.01°$ may be assumed. If the peaks have been profile fit to an analytical profile shape function, such as a Lorentzian, the

33-1161 ★

SiO$_2$

Silicon Oxide

Quartz, syn

Rad. CuKα_1	λ 1.540598	Filter Mono.	d-sp Diff.
Cut off	Int. Diffractometer	I/I$_{cor}$ 3.6	
Ref. Natl. Bur. Stand. (U.S.) Monogr. 25, 18 61 (1981)			

Sys. Hexagonal		S.G. P3$_2$21 (154)	
a 4.9133(2)	b	c 5.4053(4)	A C 1.1001
α	β	γ	Z 3 mp
Ref. Ibid.			

D$_x$ 2.65	D$_m$ 2.66	SS/FOM F$_{30}$=77(.013,31)
$\epsilon\alpha$	n$\omega\beta$ 1.544	$\epsilon\gamma$ 1.553 Sign + 2V
Ref. Swanson, Fuyat, Natl. Bur. Stand. (U.S.), Circ. 539, 3 24 (1954)		

Color Colorless
Pattern taken at 25 C. Sample from the Glass Section at NBS, Gaithersburg, Maryland, USA, ground single-crystals of optical quality. Pattern reviewed by Holzer, J., McCarthy, G., North Dakota State University, Fargo, North Dakota, USA, ICDD Grant-in-Aid (1990). Agrees well with experimental and calculated patterns. O$_2$Si type. Quartz group. Also called: silica. Also called: low quartz. Silicon used as internal standard. PSC: hP9. To replace 5-490 and validated by calculated pattern. Plus 6 additional reflections to 0.9089.

dÅ	Int	hkℓ	dÅ	Int	hkℓ
4.257	22	100	1.1532	1	311
3.342	100	101	1.1405	<1	204
2.457	8	110	1.1143	<1	303
2.282	8	102	1.0813	2	312
2.237	4	111	1.0635	<1	400
2.127	6	200	1.0476	1	105
1.9792	4	201	1.0438	<1	401
1.8179	14	112	1.0347	<1	214
1.8021	<1	003	1.0150	1	223
1.6719	4	202	0.9898	1	402
1.6591	2	103	0.9873	1	313
1.6082	<1	210	0.9783	1	304
1.5418	9	211	0.9762	1	320
1.4536	1	113	0.9636	<1	205
1.4189	<1	300			
1.3820	6	212			
1.3752	7	203			
1.3718	8	301			
1.2880	2	104			
1.2558	2	302			
1.2285	1	220			
1.1999	2	213			
1.1978	1	221			
1.1843	3	114			
1.1804	3	310			

42-1849 I

C$_{24}$H$_{36}$O$_5$

Lovastatin

Rad. CuKα	λ 1.54178	Filter Mono.	d-sp Diff.
Cut off 32.7	Int. Diffractometer	I/I$_{cor}$	
Ref. Bernstein, J., Zevin, L., Ben-Gurion University of the Negev, Beer-sheva, Israel, ICDD Grant-in-Aid, (1991)			

Sys. Orthorhombic		S.G. P2$_1$2$_1$2$_1$ (19)	
a 22.154	b 17.321	c 5.968	A C
α	β	γ	Z mp
Ref. Ibid.			

D$_x$	D$_m$	SS/FOM F$_{30}$=18(.027,61)

Sample from Merck Sharp and Dohme Research Lab. CAS#: 75330-75-5. C.D. Cell: a=17.321, b=22.154, c=5.968, a/b=0.7818, c/b=0.2694. Mica used as internal standard. PSC: oP?.

dÅ	Int	hkℓ	dÅ	Int	hkℓ
11.06	14	200	3.331	5	241,431
9.32	100	210	3.289	2	521
8.66	33	020	3.137	13	601,350
8.05	37	120	3.095	7	540,611
6.78	3	310	3.029	1	531
5.76	26	101	2.986	1	002
5.61	27	320,130	2.956	6	102,621
5.46	11	111	2.898	1	251
5.27	30	410,201	2.865	5	160
5.11	19	230	2.834	3	212,022
5.02	85	211	2.804	2	122
4.91	57	021			
4.79	3	121			
4.66	39	420,301			
4.55	6	330			
4.49	29	221,311			
4.29	1	510			
4.25	3	140			
4.079	8	321,131			
3.951	27	411,520			
3.887	42	231			
3.616	5	331,610			
3.511	4	530,041			
3.461	6	141			
3.405	8	440,620			

43-283 ★

Ba$_2$Cu$_3$F$_{0.4}$O$_{6.3}$

Barium Copper Oxide Fluoride

Rad. CuKα	λ 1.5418	Filter	d-sp
Cut off	Int.	I/I$_{cor}$	
Ref. Nakayama, H. et al., Physica C: Superconductivity, 153 936 (1988)			

Sys. Orthorhombic		S.G.	
a 3.8262(8)	b 3.8960(10)	c 11.673(2)	A C
α	β	γ	Z mp
Ref. Ibid.			

D$_x$	D$_m$	SS/FOM F$_{13}$=5(.025,105)

Prepared from oxides and CuF$_2$ by solid state reaction. Zero resistance at 77.4 K. Cell computed from d$_{obs}$. Cell parameters generated by least squares refinement. Calculated intensities are also reported in reference. Reference reports: a=3.822, b=3.892, c=11.673. C.D. Cell: a=3.896, b=11.673, c=3.826, a/b=0.3338, c/b=0.3278.

dÅ	Int	hkℓ	dÅ	Int	hkℓ
2.751	58	013			
2.729	100	110,103			
2.336	23	005			
2.234	17	113			
1.947	33	020,006			
1.912	9	200			
1.585	32	123,116			
1.572	11	213			
1.3763	5	026			
1.3647	13	220,206			
1.2301	6	019,130			
1.2118	4	310,303			
1.1178	7	226,01$\underline{10}$			

Figure 12.1. Examples of PDF images.

Table 12.2. Errors in d-Values Resulting from Fixed 2θ Errors

d (Å)	2θ (degrees)	$\pm \Delta 2\theta$ (degrees)	$\pm \Delta d$ (Å)	$\pm \Delta 2\theta$ (degrees)	$\pm \Delta d$ (Å)
5	17.73	0.1	0.04	0.05	0.014
4	22.20	0.1	0.02	0.05	0.008
3	29.76	0.1	0.01	0.05	0.005
2	45.30	0.1	0.004	0.05	0.002
1.5	61.80	0.1	0.002	0.05	0.0011
1.0	100.76	0.1	0.0007	0.05	0.0004

average $\Delta 2\theta$ will be about 0.005°. The other point to be considered is that although the experimentally measured parameter is generally the 2θ value, the search/match parameter is invariably the d-value. As shown in Figure 12.1 and in Table 12.2, the error relationship between 2θ and d is nonlinear. It is unfortunate that, while the most useful lines for phase identification are the low-angle lines, these are also the lines subject to the largest error in d.

12.1.1. Quality of Experimental Data

The success of qualitative analysis of phases in an unknown material depends both on the accuracy of the measurement of the unknown pattern and the accuracy of the patterns in the ICDD Powder Diffraction File (PDF). As the quality of both of these patterns increases, the problem of pattern recognition becomes easier. There has been a major effort over the past several years to meet the ever-increasing demand for the higher quality data needed because of improved instrumentation and better techniques [5]. The introduction of the computer for data collection, treatment and processing has improved the quality of measured d-spacings, leading to an ongoing need for improvement in the quality of reference patterns. As an example, the modern automated powder diffractometer offers the user the possibility of producing d-spacing accuracies of about 1 part per thousand for all but the larger d-values. This quality of data corresponds to an average angular error of 0.01–0.03° 2θ. An analysis of the data in the PDF up to the mid-1970s [6] showed that the average indexable pattern had a $\Delta 2\theta$ of 0.10° while only the highest quality patterns, determined at the National Bureau of Standards (which, at that time, represented about 15% of the entries in the File), met the desired accuracy for phase identification. The F_N figure of merit (described in Section 11.6.3) was developed in conjunction with this study as a quantitative function for evaluating powder patterns. It is being used as a tool in the continuing process of reviewing the quality of the reference data in the PDF and, where necessary, replacing existing patterns with new data to meet current requirements in the diffraction field. In the early 1980s, the ICDD's editorial system was automated to allow detailed reviews to be made of all new patterns entering the PDF. Also, to assure the quality of existing data in the PDF (data added to the file previous to this time), the ICDD initiated a critical review of all numerical data in the PDF for Sets 1–32 [7].

12.2. DATABASES

Databases are (generally) large compilations of similar data sets. As a simple example, a telephone directory is a database. In order for a database to be useful, it must be organized in such a way that it is easily searchable and so that individual data are easily retrievable. The telephone directory is sorted by name and/or organization, since this is generally the form in which telephone numbers are sought. A telephone directory is frequently subdivided into conveniently sized sections. Thus, by grouping together the A's, the B's, etc., the directory is made easier to employ. A database generally requires an index to allow convenient access to the data. An example might be a series of volumes in an encyclopedia, which are not just subdivided into an A–C section, a D–E section, etc., but are also provided with a comprehensive alphabetic and key-word index. Several different indices might be employed to allow the database to be used in different ways.

As shown in Table 12.3, there are a number of databases available for X-ray crystallographic work. The majority of these databases are designed and maintained for the single-crystal community rather than for the powder community. Nevertheless, much cross-fertilization can and does take place. For example, many of the patterns in the PDF are calculated from single-crystal data of the type contained in the databases listed in Table 12.3. The databases designed to support the powder diffraction community are shown in Table 12.4. The development of powerful personal computers have even permitted Boolean searching for articles containing and combination of key words.

Since the number of patterns in the PDF is large, special ways of organizing the d's and I's into subfiles have been devised, and the more important of these are listed in Table 12.5. In addition, various search indices have been devised to simplify search procedures. These search indices will be discussed in the following subsections. The process of qualitative phase identification is generally referred to as *Search/Match*. The *Search* part of the process is done with an *Index*, and the *Match* part is done with a *File*. It will be seen later in this chapter that the whole process would be better referred to as *Search/Match/Identify* since for every potential *match* found a decision must

Table 12.3. Databases of Crystallographic and Structural Information

Name	Content	Center
Cambridge Structural Database (CSD)	Organic, organometallic	Cambridge, England
Inorganic Crystal Structure Database (ICSD)	Inorganic materials	Karlsruhe, Germany
NRCC Metals Data File (CRYSTMET)	Metals and alloys	Ottawa, Canada
Protein Data Bank (PDB)	Structure of macromolecules	Brookhaven, New York
NIST Crystal Data [NBS(CDF)]	Inorganic and organic unit cells	Gaithersburg, Maryland

Table 12.4. Databases for X-ray Powder Diffraction

Name	Content
Master DB	Master ICDD Database—all known powder data on a single phase, plus editorial marks and comments
PDF-2	User version of the Master DB (does not contain special editorial comments)
PDF-1	Subset of PDF-2, contains d's, I's and names (designed for automated search systems)
PDF-3	Contains raw data as a digitized pattern
CDF	The Crystal Data File (contains cell data, names, and references)
EISI	The Elemental and Interplanar Spacing Index (designed for electron diffraction)

be taken as to whether the result agrees with other factors—sample history, sample chemistry, intuition, prior knowledge, experience, etc.

12.2.1. The Powder Diffraction File

The PDF is a collection of single-phase X-ray powder diffraction patterns in the form of tables of the interplanar spacings (d) and relative intensities (I^{rel}) characteristic of the compound. The PDF has been used for almost five decades [8], and the ICDD maintains the PDF by continually adding new and updated diffraction patterns to the file. Currently 2000 such patterns are added each year, comprising 1500 inorganic patterns and 500 organic patterns. There

Table 12.5. Subfiles of the PDF

Subfile	Entries	Where Available
Inorganic	43,308	Book and computer
Organic	17,661	Book and computer
Metals & Alloys	11,630	Book and computer
Minerals	3,954	Book and computer
Forensic Materials	3,612	Book and computer
Common Phases	3,202	Computer readable
Zeolites	626	Book and computer
Explosives	149	Computer readable
Polymers	248	Computer readable
Cement	360	Computer readable
Superconductors	139	Computer readable
Dyes & Pigments	101	Computer readable
Total as of set 44	59,847	

is a continuing effort by the ICDD to ensure that new patterns being added to the PDF contain a significant proportion of phases that represent current needs and trends in industry and research. The effort is implemented, in part, by sponsoring grants-in-aid for the production of new patterns of phases of current interest or preparation of the phases themselves. The master database of powder patterns is continually undergoing revision and updating, but in order to ensure that all database users have the opportunity to work with the same version, a *frozen* version of the master database is produced each year and is supplied as the *PDF-2* file. A smaller version of the database called *PDF-1* was produced mainly for search programs on minicomputers with limited disk storage. As of Set 44, the PDF-2, database requires about 200 MB (megabytes) of storage and PDF-1 about 25 MB.

The PDF-2 contains a series of individual data sets, and Figure 12.1 shows the layout of three typical PDF-2 images. As is shown in the figure, each individual data set in the file contains, as a minimum, a list of $d–I$ pairs, the chemical formula, the name, a unique identification (PDF) number, and a reference to the primary source. In addition to this information, supplemental data may be added, where available, including Miller indices for all lines, unit cell and space group data, physical constants, experimental details and other comments. For convenience, these pieces of information are arranged in *boxes*. The four boxes on the left-hand side of the card image are for Name and Formula; Experimental Data; Crystallographic and Physical Data; and Comments. The right-hand side of the card image contains lists of d-spacings and relative intensities, plus Miller indices, where available. The top-left-hand corner of the card image has a unique (PDF) number of the form *yy-nnnn*, where *yy* represents the year. For data sets after set 5, the actual year of publication can be derived by adding 1950 to the value of *yy*. As an example, set 42 was published in the year $1950 + 42 = 1992$. The characters *nnnn* represent the number of the pattern published in the appropriate year. For instance, pattern 41-0001 would be the first pattern published in 1991.

Figure 12.1 gives three examples of PDF data set images (traditionally called *cards*). As indicated in the upper-right-hand corner of each card, the top and bottom patterns are of "star" (∗) quality. This means that the data are of the best quality, with an average $\Delta 2\theta$ of $<0.03°$, and that all lines in the pattern have been indexed and the intensities measured quantitatively. The middle pattern is an "I" quality pattern; I patterns are *indexed* with no more than two lines being unaccounted for. The average $\Delta 2\theta$ is $<0.06°$, and again the intensities have been measured quantitatively. Patterns may also be *calculated* and given a "C" quality mark. This means that the *d*-values have been calculated from the unit cell by using Equation 3.10. The intensities, calculated from crystal structure information by using Equation 3.28, may or may not agree with experimental data due mainly to problems of preferred orientation. Two other quality marks are used: the "O" pattern indicates that the data have *low precision* and are *poorly characterized*, and there are *no unit cell data*. The possibility also exists that an O pattern might be a mixture; a "Blank" generally indicates a pattern that does not meet the criteria for a star, an I, or a C. Since there are no unit cell data for such a pattern, it is impossible to assess the accuracy of individual lines in the pattern. When unit cell data are available, permitting computation of the FOM, F_N is also included in the database.

12.2.2. The Crystal Data File

The ICDD also publishes the *Crystal Data File* (*CDF*), which has been built up over the years at the National Institute of Standards and Technology (NIST). This database contains the unit cell information on many materials for which a report has been published. As of 1995, the CDF contains references to about 200,000 unit cells. The CDF includes crystallographic unit cell parameters from both powder and single-crystal sources [9]. Each entry consists of reduced cell and volume; crystal system; space group symbol and number; chemical name and formula; and literature reference. The primary use of the CDF is to match cells determined from experimental data with those previously published. The ICDD supplies a powerful Boolean search program which allows searching on such parameters as Authors' Name, Journal CODEN and Year, Space Group Symbol, Chemical Elements, and Density. Section 12.6

gives examples of the use of a similar Boolean search program for use with the PDF. In addition to the Boolean search software, other software programs are available which perform searches by alternative means. The most powerful of these is the NBS*LATTICE program [10], which allows the identification of unknown materials, calculation of the reduced cell of the lattice, and the calculation and reduction of specified derivative supercells and/or subcells. This program has been available for a number of years, running in a mainframe environment, and has recently been ported down to a personal computer (PC) such that it will work with CD–ROM-supported PC systems.

The most powerful application of the CDF database is to find the literature references to the crystal structure information for a phase and, when the structure is unknown, to find potential isostructural materials that allow one to determine the crystal structure. A good example of the use of this database is a published analysis by Smith et al. of the crystal structure of BeH_2 [11]. Due to the low symmetry and extremely low X-ray scattering ability of BeH_2, this structure defied analysis for many years. Using the very high-resolution diffractometer on beamline X7A at the BNL/NSLS[1] synchrotron source, Smith et al. obtained a very-good-quality pattern for this material. Computer indexing [12] of this pattern indicated an orthorhombic unit cell. A search of the CDF database produced a reference to a high-pressure form of ice (i.e., OH_2). The very-well-resolved peaks allowed isolation of integrated intensities and the solution of the crystal structure based on this model.

12.2.3. The Elemental and Interplanar Spacing Index (EISI)

Electron diffraction methods can give useful diffraction patterns from areas or particles with sizes from hundreds of micrometers down to 10 nm (nanometers) or less. The principal difference between an X-ray and an electron diffraction pattern is the intensities. The atomic scattering factors that go into the intensity equation (3.28) are very different for electron scattering than for X-ray scattering. However, due to preferred orientation in many materials, various search schemes have been developed that rely heavily or even exclusively on the d-values. Indeed, this fact was one of the key factors in the decision to develop an alternative search procedure to the Hanawalt method. This alternative method—the *Fink Search/Match*—is described later in Section 12.4.3. since the d-values are not affected by the scattering mechanism, the PDF may be effectively used to identify phases from their electron diffraction patterns. In addition to the different scattering factors, there are some other important experimental differences between the two methods. Transmission electron microscopes require very thin specimens, and

---

[1] BNL/NSLS: Brookhaven National Laboratory/National Synchrotron Light Source.

the preparation of a thin specimen invariably results in a highly oriented specimen. In fact, most electron diffraction patterns are of single-crystal zones. a second important difference between most electron diffraction and X-ray diffraction instruments is that the majority of modern electron diffraction instruments have energy dispersive spectrometers attached to them; thus, elemental data are almost *always* available to aid in interpretation. A third, and perhaps the most important, difference between the two techniques is the fact that, since electron microscopes are able to focus on small areas (and hence on individual single-phase particles), phase analysis by electron diffraction often involves the identification of rather pure phases. These facts have led to the development of indices that have been specially designed for the electron diffraction community [13, 14].

Since the intensities are less important in electron diffraction than in X-ray diffraction, use can be made of the fact that d-spacings can be readily calculated from unit cell data for those phases where powder data are not available but unit cell data are (generally from a structure determination using single-crystal methods). In order to produce a non-intensity-dependent phase identification system, powder patterns have been computed from the unit cell information for all of the materials in the CDF. These calculated patterns have had the space group systematic-extinct reflections removed, leaving a few d-values in the calculated pattern that might not be in the observed pattern. These occasionally accidentally absent reflections are due to a particular structure factor having a low value, causing the intensity to be below the observational threshold. The new computed d-value patterns were added to the existing observed patterns in the PDF, producing a database with over 200,000 patterns. Because this database, based only on high d-values, is particularly suited to identifying patterns from electron diffraction, it has been called the Electron Diffraction Database (EDD), or the Max-d Index. When the entries in the EDD are arranged by the elements present in each phase (as they would be determined by X-ray fluorescence), the resulting search manual is called the Elemental and Interplanar Spacings Index (EISI) and is also available from the ICDD. This file has been successfully used to identify phases not in the PDF.

12.2.4. The Metals and Alloys Index

The Metals & Alloys Index has been developed [15,16] especially for materials scientists who are dealing with combinations of metals, metal oxides, and related materials. The search manual contains four separate indices which may be used independently, or in conjunction with the PDF. Two of the indices contain data for all materials in the Metals & Alloys PDF, and two contain supporting data. The permuted-sort Alphabetic Formula Index brings together all entries containing a given element being sorted to simplify the reading of the index. With binary and ternary phase diagrams, phases are listed in compositional order to facilitate direct use in phase diagram research. The Pearson Symbol Code Index has entries arranged in order of the Pearson Symbol Code [17]. The Pearson Code was originally based on a recommendation from Committee E-4 of the American Society for Testing and Materials (ASTM); the recommendation proposed symbols for classifying or describing alloy phases. Each symbol comprises the number of atoms in the conventional crystallographic cell, plus one of 14 arbitrary capital letters to designate the Bravais lattices. This system was changed at a later date using mnemonic combinations to replace the arbitrary letters. A full description of the system currently employed for the Pearson Codes is found in the latest edition of the Metals & Alloys Index published by the ICDD. In addition to the Pearson Code for each entry, the index also contains a prototype structure (if one has been assigned), example formula, space group, lattice parameters, PDF number, and quality mark. The Common Names Index permits cross-referencing of common metallurgical names such as *austenite* or *cementite* to the appropriate PDF data. The index also contains a cross-reference for all mineral names. The *Strukturbericht* Symbol Index provides cross-referencing between *Strukturbericht* symbols [18] and the corresponding prototype structures.

12.3. MEDIA ON WHICH ICDD DATABASES ARE SUPPLIED

12.3.1. Historical Evolution of Database Media

Table 12.6 summarizes the methods of data storage and presentation that have been used for the storage of powder diffraction data. In the search for alternative approaches, the basic need is for a low-cost, high-density data storage system, with the following four additional factors to be considered:

1. The physical size of the file
2. The ease and speed of retrieval
3. The display capability
4. The ease of updating

Until the mid-1960s, the traditional method of supplying the PDF was to provide annual data sets on individual cards. However, the continuing growth in the number of file entries over the years has prompted the ICDD to seek alternative means of data storage and display. In the early 1970s the microfiche system was introduced as an alternative, with a dramatic reduction in the

Table 12.6. Characteristics of Media on Which the PDF Has Been Supplied

Category	Cards	Fiche	Tape	Disk	CD-ROM[a]
Retrieval time	Slow		Moderate	Very fast	Fast
File Stored	All			Part/all	All
File access	Random		Sequential	Random	
Updating	Not retroactive		Retroactive		
Size (cm^2)	200,000	3,000	4,500		20

[a]Compact disk–read-only memory.

physical size of the file. PDF-1 became available on computer tape in the early 1960s, supplementing the cards and microfiche. With growing interest in the use of Automated Powder Diffraction (APD) systems during the 1970s, the PDF was supported on other computer-readable media, including floppy and fixed disks. The latest addition to this list is the CD–ROM. The three major advantages offered by the tape, disk, and CD–ROM-based files are that they allow rapid retrieval, are relatively small in physical volume, and are easily updatable. This last point is important because the PDF is under continuous review and the only way to update cards and microfiche would be to issue a completely updated full set each year.

The development of the full digital version of the PDF and its production on a CD–ROM has dramatically improved flexibility in phase identification and characterization. The latest version of the PDF contains nearly 60,000 phases and, due to the high-density format of the CD-ROM, is readily accessible on a PC. The digital version is PDF-2, which contains all of the unit cell, indices, experimental conditions, etc., contained in the full database. The latest generation of commercial third-generation search/match programs use this database and perform identification in less than 60 s. Access to this database is essential to speed up data analysis, in that the instrumental developments (mentioned earlier) have led to the ability to produce data much faster than they can be analyzed using second- or third-generation software or the even slower manual methods.

12.3.2. Computer-Readable Products

The PDF became available on computer tape in the early 1960s, supplementing the cards and microfiche. However, the limited disk storage available on most commercial APD systems prevented its use in that environment. On the order of 200 MB are required for the storage of the full PDF-2 file. Most APD systems developed in the late 1970s and early 1980s were equipped with two 5 MB disk drives, of which about 8 MB were available for the storage of data files. In order to provide users with a useful subset of the full database, in 1977 the ICDD introduced the PDF-1 database comprising the $d-I$ lists along with the PDF number, the quality mark, the chemical formula name, the mineral name, and the $I/I_{corundum}$ value. This database, which even now requires only about 25 MB, has provided, and continues to provide, an extremely useful product, both for individual users and for third-party suppliers of custom-made databases for proprietary search/match computer programs. At this time, a number of different organizations, ranging from automated powder diffractometer suppliers to software entrepreneurs, offer software for searching custom versions of the PDF-1 database, usually with later look-up access to PDF-2 on CD-ROM.

12.3.3. The CD-ROM System

The CD-ROM is probably the most important breakthrough [19] in computer-readable data storage since the magnetic disk. As illustrated in Figure 12.2, the CD-ROM stores data as small etch *pits*, which do not reflect light, and the flat spaces (called *lands*) between the etch pits, which do reflect light. The etch pits are covered with a reflective coating and are protected by a thick surface layer of plastic on top and a lacquer layer on bottom. A low-power infrared laser focuses through the plastic onto the active surface. On playback, the reader reads the reflected light of the laser beam on the active surface of the disk. A binary "1" is represented by a land-pit or pit-land transition, and the path length between transitions represents the number "0." Because the laser focuses through the plastic surface coating, anything on the surface such as scratches or dust is *out of focus* and therefore has minimal effect on the playback quality. Another benefit of the use of reflected light is that laser light can travel a significant distance and still be detected without loss of signal. This, in turn, allows the disk head to be maintained at a distance of more than a millimeter from the surface of the substrate, thus, eliminating the possibility of head crashes. The most important aspect of laser disk technology is the fact that, because the laser operates at wavelengths around that of visible light, as opposed to the smallest size achievable by ceramic technology for tiny ferrite cores, it allows a very high storage density with a concomitant high linear data density. Current state-of-the-art systems allow 1.66 data bits/µm, or 42 kilobits/inch. With a track pitch of 1.6 µm, these figures correspond to 10^6 bits/mm² — about an order of magnitude higher than traditional magnetic materials. Today, a standard 12 cm diameter CD-ROM has a capacity of 680 MB.

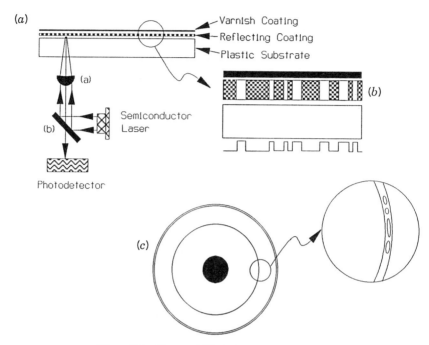

Figure 12.2. Picture of CD-ROM track distribution.

12.4. MANUAL SEARCH/MATCHING METHODS

As was discussed earlier, all databases require some type of index system to allow access to information contained within the database. As far as paper products are concerned, it is common practice to supply a combination *Index* plus *Search Manual* for each of the main subsets of the PDF. A number of manual searching methods have been developed over the last 40 years. There are three methods in common use today: the *Alphabetic method*, the *Hanawalt method*, and the *Fink method*. Each year an alphabetic listing and a Hanawalt search manual are published. At nonregular intervals Fink search manuals and Common Phase manuals have been published. The Common Phase is a list of about 2500 of the most frequently encountered materials. The Common Phase search manual contains three sections: an alphabetic listing, a Hanawalt index, and a Fink index. Table 12.7 illustrates the various entry methods employed in the common indexes. It will be seen that the Alphabetic index is a chemistry-based index, using only elemental information. The Hanawalt index is an intensity-driven index since it employs only the *strongest* lines for searching. The Fink index, on the other hand, is a *d*-spacing-driven

MANUAL SEARCH/MATCHING METHODS   333

Table 12.7. Types of PDF Data-Searching Indexes

Index	Entry Method	Search Parameters
Alphabetic	Chemistry	Permuted elemental symbols
Hanawalt	I/d	Three strongest lines
Fink	d/I	First eight lines
EISI[a]	Chemistry/d	Low high Z elements; d-spacing
Boolean	Various	d-Spacings, chemistry, strong lines, CODEN, physical properties, functional groups, etc.

[a]Elemental and Interplanar Spacing Index.

index since it employs mainly the *largest* d-values. The EISI index employs both chemistry and d-spacing information. Finally, the Boolean indices use many search parameters in different combinations.

12.4.1. The Alphabetic Method

The Alphabetic index is designed to permit a rapid systematic search for all patterns with a specified chemical content. The index lists alphabetically the names of substances in the PDF. Each name is followed by the chemical formula, the d-values of the three strongest intensity lines of the diffraction pattern, the PDF number of the corresponding data card, and occasionally the reference intensity ratio (I/I_c). The I/I_c value is a measure of the diffraction intensity of the phase with respect to the intensity of the corundum pattern and can be used in quantitative analysis (see Section 13.6.1). Each d-value has a subscript denoting the relative intensity: X stands for 100, 9 for 90, 8 for 80, etc. The intensities are rounded off from the more exact intensity values that are listed on the card of the complete pattern. Figure 12.3 shows a sample entry from the Alphabetic manual.

There are very specific rules of nomenclature and rotation of duplicate entries applied to all entries in an Alphabetic index. The user should be aware of these rules if searches are to be completed successfully. A list of the rules is always included in the *front matter* of a search manual. As shown in Figure 12.4, the search procedure is based on a series of preconceived notions as to what the specimen being analyzed might contain. Such ideas might stem from prior knowledge of the source of the specimen, chemical analysis, etc. A given phase may appear several times in the index; this is achieved by rotation of the components of the name. The proper name end is indicated by a colon(:). The rotation components of the name retains the requirements of

*	Sodium Chromium Oxide :	β-Na$_2$Cr$_2$O$_7$	4.67$_x$	4.46$_9$	3.24$_9$	30–1178	
*	Sodium Chromium Oxide :	Na$_2$CrO$_4$	2.90$_x$	2.73$_7$	4.07$_7$	22–1365	1.20
	Sodium Chromium Oxide :	Na$_2$CrO$_4$	2.68$_x$	4.37$_9$	3.78$_4$	29–1199	
*	Sodium Chromium Oxide :	NaCrO$_2$	2.16$_x$	5.32$_7$	2.45$_x$	25– 819	2.60
*	Sodium Cyanide :	NaCN	2.94$_x$	2.08$_3$	1.70$_1$	37–1490	
*	Sodium Fluoride :/Villiaumite	NaF	2.32$_x$	1.64$_2$	1.34$_1$	36–1455	
i	Sodium Germanium Oxide :	Na$_4$GeO$_4$	6.89$_x$	3.88$_6$	5.11$_8$	36– 62	
*	Sodium Hydrogen Carbonate :/Nahcolite, syn	NaHCO$_3$	2.94$_x$	2.60$_4$	2.96$_7$	15– 700	0.30
	Sodium Hydrogen Phosphate :	Na$_3$H(PO$_3$)$_2$	5.03$_x$	3.11$_9$	3.21$_8$	9– 101	
i	Sodium Hydrogen Phosphate :	NaH$_2$PO$_4$	3.20$_x$	3.94$_3$	3.30$_6$	11– 659	
i	Sodium Hydrogen Phosphate :	Na$_2$H$_2$P$_2$O$_7$	2.93$_x$	3.09$_6$	3.43$_6$	10– 192	
i	Sodium Hydrogen Sulfate :	Na$_3$H(SO$_4$)$_2$	3.95$_x$	2.89$_4$	2.73$_x$	32–1090	
*	Sodium Hydrogen Sulfate :	β-NaHSO$_4$	3.59$_x$	3.74$_9$	2.94$_6$	26– 960	
*	Sodium Hydrogen Sulfate :	NaHSO$_4$	3.39$_x$	4.48$_3$	3.43$_3$	25– 833	
*	Sodium Hydroxide :	Na(OH)	2.35$_x$	2.85$_3$	1.70$_2$	35–1009	
*	Sodium Iodate :	NaIO$_3$	2.98$_x$	4.28$_9$	3.20$_x$	8– 474	
*	Sodium Iodide :	NaI	3.24$_x$	3.74$_6$	2.29$_7$	6– 302	1.20
	Sodium Iron Cyanide :	Na$_4$Fe(CN)$_6$	2.75$_x$	1.94$_3$	5.70$_1$	1–1026	
o	Sodium Magnesium Aluminum Silicate Hydroxide :/Montmorillonite-14A	Na$_{0.3}$(Al,Mg)$_2$Si$_4$O$_{10}$(OH)$_2$·xH$_2$O	13.6$_x$	4.47$_3$	3.34$_1$	13– 259	
	Sodium Magnesium Aluminum Silicate Hydrox Hyd :/Montmorillonite-21A	Na$_{0.3}$(Al,Mg)$_2$Si$_4$O$_{10}$(OH)$_2$·xH$_2$O	21.5$_x$	4.45$_6$	3.15$_4$	29–1499	

*	Iron Carbonate :/Siderite	FeCO$_3$	2.80$_x$	1.73$_3$	1.74$_3$	29– 696	
	Iron Chloride :/Molysite, syn	FeCl$_3$	2.68$_x$	2.08$_5$	5.90$_3$	1–1059	
	Iron Chloride :/Lawrencite, syn	FeCl$_2$	2.54$_x$	5.90$_6$	1.80$_6$	1–1106	
i	Iron Chromium :\434-L Stainless Steel	Fe-Cr	2.04$_x$	1.17$_5$	1.44$_3$	34– 396	0.59
	Iron Chromium Oxide :	FeCr$_2$O$_4$	2.55$_x$	2.94$_5$	2.48$_3$	24– 511	
	Iron Cyanide : Sodium	Na$_4$Fe(CN)$_6$	2.75$_x$	1.94$_3$	5.70$_1$	1–1026	
*	Iron Fluoride :	FeF$_3$	3.73$_x$	1.87$_3$	1.69$_2$	33– 647	
	Iron Hydroxide :	Fe(OH)$_2$	4.60$_x$	2.40$_4$	2.82$_8$	13– 89	
c	Iron Hydroxide : Calcium	Ca$_3$Fe$_2$(OH)$_{12}$	4.50$_x$	1.70$_5$	5.19$_6$	32– 166	
i	Iron Magnesium Aluminum Silicate Hydroxide :/Clinochlore-IIb, ferroan	(Mg,Fe)$_6$(Si,Al)$_4$O$_{10}$(OH)$_8$	7.07$_x$	3.54$_6$	14.1$_4$	29– 701	
	Iron Magnesium Aluminum Silicate Hydroxide :/Chamosite-IIb	(Fe,Al,Mg)$_6$(Si,Al)$_4$O$_{10}$(OH)$_8$	7.05$_x$	3.52$_6$	2.60$_9$	21–1227	
	Iron Magnesium Aluminum Silicate Hydroxide :/Chamosite-IIb	(Fe,Al,Mg)$_6$(Si,Al)$_4$O$_{10}$(OH)$_8$	7.05$_x$	3.52$_6$	2.60$_9$	21–1227	
i	Iron Magnesium Silicate :/Hypersthene	(Fe,Mg)SiO$_3$	3.18$_x$	2.88$_5$	2.56$_6$	31– 634	
i	Iron Magnesium Silicate :/Fayalite, magnesian	(Fe,Mg)$_2$SiO$_4$	2.81$_x$	2.49$_7$	2.55$_6$	31– 633	
c	Iron Magnesium Silicate :/Forsterite, ferroan	(Mg$_{0.54}$Fe$_{0.46}$)$_2$SiO$_4$	2.48$_x$	2.79$_5$	2.54$_7$	33– 657	

o	Copper Tin :	ε-Cu$_3$Sn	2.08$_x$	2.16$_4$	1.24$_4$	1–1240	
	Copper Zinc :	CuZn	2.14$_x$	2.09$_7$	2.39$_1$	35–1152	
*	Cyanamide : Calcium	CaCN$_2$	2.94$_x$	1.85$_3$	2.42$_3$	32– 161	
i	Cyanate : Potassium	KOCN	2.72$_x$	3.05$_5$	2.54$_3$	8– 471	
i	Cyanide : Gold	AuCN	5.08$_x$	2.94$_x$	2.54$_7$	11– 307	
*	Cyanide : Potassium	KCN	3.27$_x$	2.31$_5$	3.77$_2$	37–1491	
*	Cyanide : Silver	AgCN	3.00$_x$	3.70$_5$	2.35$_1$	23–1404	4.00
*	Cyanide : Sodium	NaCN	2.94$_x$	2.08$_3$	1.70$_1$	37–1490	
	Cyanide : Sodium Iron	Na$_4$Fe(CN)$_6$	2.75$_x$	1.94$_3$	5.70$_1$	1–1026	
i	Dysprosium Chloride :	γ-DyCl$_3$	2.73$_x$	1.88$_5$	5.95$_8$	22– 257	
*	Dysprosium Oxide :	Dy$_2$O$_3$	3.08$_x$	1.88$_3$	2.67$_4$	22– 612	
	Erbium Chloride :	ErCl$_3$	6.09$_x$	2.74$_9$	1.88$_7$	24– 398	
*	Erbium Oxide :	Er$_2$O$_3$	3.05$_x$	1.86$_5$	2.64$_5$	8– 50	
i	Europium Chloride :	EuCl$_3$	2.53$_x$	2.08$_5$	3.47$_8$	12– 387	
*	Europium Oxide :	Eu$_2$O$_3$	3.14$_x$	1.92$_4$	2.72$_3$	34– 392	

Figure 12.3. Sample entry from the Alphabetic Common Phase manual. The quality mark is followed by the chemical/mineral name and then the formula. Next are the d-values with intensity codes as subscripts of the three strongest lines, followed by the PDF number and $I/I_{CORUNDUM}$.

the rules of nomenclature. For example, a search for Sodium Iron Cyanide (PDF No. 1-1026) would find it under the following entries: Sodium Iron Cyanide, Iron Cyanide: Sodium, and Cyanide: Iron Sodium. However, it could not be located as Sodium Cyanide Iron, Iron Sodium Cyanide, or Cyanide Sodium Iron.

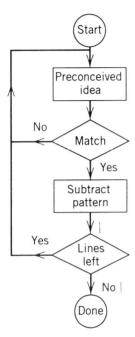

Figure 12.4. Schematic search procedure using the Alphabetic index.

Clearly, the greatest use of the Alphabetic search manual is to check the patterns of unknowns for the presence of a particular phase. For example, if $BaCO_3$ has been fired with TiO_2 and the reaction product is being checked, one certainly does not have to treat the pattern as a complete unknown. Chemical information should always be used in an identification. A simple schematic search procedure is shown in Figure 12.4. The first step in the example given is to look up the pattern of barium titanate. If this phase is indeed present, the lines from the identified pattern are subtracted from the experimental pattern; then, if there are unaccounted for lines, it is necessary to look for unreacted starting materials. If unidentified lines still remain that do not fit any preconceived notion, it may be best to take these lines onto one of the other search methods described below.

12.4.2. The Hanawalt Search Method

The Hanawalt method involves sorting the patterns in the PDF according to the d-value of the 100% intensity line. This list is then broken into small d-intervals called *Hanawalt groups*: at this time there are 40 Hanawalt groups in the Inorganic Search manual, and 24 Hanawalt groups in the Organic

Search manual. A small (± 0.02) overlap in d between intervals is employed to reduce the chance of lines being missed due to the analyst overestimating the accuracy of the experimental data. Each interval is sorted on the d-value of the second most intense line. Subsequent lines are listed in order of decreasing intensity. The eight most intense lines are listed for each phase. Entries (permutations) are made based on the following rules:

- All patterns appear at least once as (1, 2), where 1 is the most intense and 2 is the next most intense d-spacing.
- Patterns appear twice, (1, 2) and (2, 1), when $I_2/I_1 > 0.75$ and $I_3/I_1 \geqslant 0.75$
- Patterns appear three times (1, 2), (2, 1), and (3, 1), when $I_3/I_1 > 0.75$ and $I_4/I_1 \leqslant 0.75$
- Patterns appear four times, (1, 2), (2, 1), (3, 1), and (4, 1), when $I_4/I_1 > 0.75$

The reason for the multiple entries is to attempt to minimize problems of preferred orientation. A sample of the Hanawalt index is shown in Figure 12.5, and the boldface characters at the top give the Hanawalt group, i.e., 3.39 – 3.32 $\pm$ 0.02 (in other words, embracing the strongest line between 3.41 and 3.30 Å). There are 12 columns of figures. The first column shows the quality mark. Lack of an indication in this column indicates that the pattern is a "blank" quality mark pattern. The Columns 2 through 4 contain the three strongest lines, the second of which is sorted in terms of decreasing d-value. Columns 5 through 9 contain the next five strongest lines. Column 10 shows the formula (and mineral name where appropriate), and column 11 gives the PDF number. finally, column 12 gives the I/I_c value when this is available.

A flowchart of the search strategy is given in Figure 12.6. Using the data in Table 12.8 as an example, we find that the first eight lines in the pattern are sorted in terms of decreasing intensity; then the index is entered using the appropriate Hanawalt group, in this case 3.39–3.32. The second column of the group is now scanned to give a match to the second strongest line of 4.25 Å, within reasonable experimental error. This process will probably reveal several potential candidates, so a match is sought using the third strongest line at 1.815 Å. This yields pattern number 33-1161 (plus perhaps 10-423). A full check (the *Match*) is now made using the whole pattern.

The ease of interpretation of diffraction data obviously decreases with the complexity of the pattern. For example, in mixtures, the problem will always be to decide which lines belong to which patterns. In cases where no other evidence is available (previous history, chemical analysis, etc.), the only solution is to start with the strongest line, combined in turn with each other strong line in the pattern. A match is sought using different line combinations,

3.39 – 3.32 (± .02)

										Formula/Name	File No.	I/Ic
i	3.38_9	8.58_x	3.04_8	4.11_8	3.18_8	1.69_7	2.65_6	1.88_5	(Mg,Fe)$_2$Al$_4$Si$_5$O$_{18}$/Cordierite, ferroan	9– 472	
	3.33_x	6.72_x	3.19_8	8.09_7	3.28_7	5.18_4	3.10_4	4.30_4	C$_{10}$H$_{18}$N$_2$O$_6$	29–1716	0.20
i	3.31_8	6.40_x	6.10_8	3.85_6	2.77_5	6.70_4	3.48_4	2.64_4	C$_{12}$H$_8$Cl$_6$	17–1054	
i	3.38_9	6.13_x	8.66_9	3.20_4	3.29_5	9.70_3	4.57_3	3.46_3	C$_{17}$H$_{11}$N$_5$·HCl	28–1749	
i	3.34_x	5.93_x	5.19_1	3.77_1	3.65_1	3.51_1	2.94_1	1.67_1	C$_4$H$_8$N$_2$O$_2$	26–1863	3.30
	3.37_x	5.85_8	3.86_8	3.72_7	3.52_7	3.03_7	2.70_7	7.72_6	C$_6$H$_9$N$_3$O$_2$·HCl	5– 459	
*	3.31_8	5.73_x	3.43_7	3.59_6	3.19_5	4.36_4	4.19_4	3.27_2	C$_6$H$_5$NO$_2$	30–1845	1.00
	3.30_x	5.44_7	5.63_5	3.24_4	4.97_3	6.58_3	4.58_3	3.15_2	(NH$_4$)$_4$P$_2$O$_7$	20– 102	
	3.38_x	5.30_x	3.49_6	5.90_5	3.67_5	3.26_5	3.18_5	2.99_5	KH$_3$P$_2$O$_7$	15– 509	
	3.35_x	5.21_8	4.86_8	4.33_8	4.04_8	3.90_8	3.55_8	2.73_8	β-C$_9$H$_{11}$NO$_2$	22–1874	
i	3.40_x	5.01_x	3.09_7	4.10_4	3.00_4	4.03_3	6.74_2	3.45_2	C$_3$H$_6$N$_6$	24–1654	1.10
*	3.30_x	4.76_6	4.18_6	5.73_5	2.93_3	3.98_3	2.38_2	3.35_2	C$_6$H$_6$O$_4$	37–1919	
	3.31_x	4.71_8	3.50_8	5.56_3	3.84_3	3.03_3	7.02_2	2.30_2	C$_6$H$_5$NO$_2$·HCl	29–1827	
*	3.39_x	4.48_5	3.43_5	3.01_5	4.09_4	2.98_4	2.78_4	3.18_3	NaHSO$_4$	25– 833	
	3.34_9	4.42_x	10.1_9	1.48_9	2.56_8	1.68_8	1.28_7	1.23_7	Al$_2$Si$_2$O$_5$(OH)$_4$·2H$_2$O/Halloysite-10A	9– 451	
i	3.40_9	4.38_x	2.88_7	5.76_4	2.61_4	4.09_4	2.76_4	1.76_3	Y$_2$O$_5$/Shcherbinaite, syn	9– 387	1.60
i	3.33_x	4.30_5	2.82_5	6.08_4	4.72_4	1.71_4	3.52_1	2.15_1	(NH$_4$)$_2$Ca$_2$(SO$_4$)$_3$	22–1037	2.30
	3.37_x	4.28_3	1.84_3	1.55_1	2.47_1	2.31_1	1.39_1	2.14_1	AlPO$_4$/Berlinite, syn	10– 423	
o	3.34_x	4.26_2	1.82_1	1.54_1	2.46_1	2.28_1	1.37_1	1.38_1	SiO$_2$/Quartz, low, syn	33–1161	3.60
	3.36_x	4.23_x	3.57_7	5.27_4	3.72_4	4.04_3	3.97_3	7.19_2	C$_6$H$_8$N$_2$O$_2$S	30–1944	
*	3.35_x	4.22_x	3.25_7	4.43_6	3.67_5	6.22_4	2.89_4	3.55_3	C$_6$H$_4$N$_4$O$_4$	37–1915	
*	3.32_x	4.22_x	5.28_8	4.97_8	3.52_8	3.48_7	2.65_8	2.48_7	C$_6$H$_4$N$_4$O$_2$·HCl	25–1541	
i	3.35_8	3.88_x	3.73_6	3.54_8	2.91_7	2.52_7	2.32_7	1.92_5	C$_5$H$_5$NO$_4$·HCl	25–1925	
*	3.39_x	3.87_9	3.29_8	2.36_6	3.01_5	2.04_4	1.99_4	2.41_3	HgSO$_4$	31– 867	
*	3.41_9	3.84_x	3.52_8	3.26_9	3.87_7	3.03_7	2.74_3	2.37_3	KHSO$_4$/Mercallite, syn	11– 649	
*	3.31_x	3.77_8	4.22_8	3.24_7	3.29_6	2.99_5	3.47_5	2.90_3	KAlSi$_3$O$_8$/Orthoclase	31– 966	
*	3.36_x	3.52_x	7.69_x	6.16_5	3.84_4	3.14_4	3.79_1	3.09_1	C$_6$H$_4$(CO)$_2$C$_6$H$_4$/Hoelite, syn	28–2002	
*	3.35_x	3.50_x	5.04_6	3.56_6	4.00_5	3.15_5	5.58_4	2.48_3	(NH$_4$)$_2$S$_2$O$_8$	31– 69	
*	3.36_x	3.47_7	6.52_6	2.59_6	3.02_6	3.28_5	3.56_5	2.61_4	BaAl$_2$Si$_2$O$_8$/Celsian, syn	38–1450	
*	3.36_x	3.47_7	6.52_6	2.59_6	3.02_6	3.28_5	3.56_5	2.61_4	BaAl$_2$Si$_2$O$_8$/Celsian, syn	38–1450	
	3.30_x	3.47_9	3.66_9	4.87_5	3.06_5	3.03_3	2.88_3	2.86_3	K$_2$Cr$_2$O$_7$/Lopezite, syn	27– 380	0.63
*	3.33_x	3.46_6	3.79_5	3.26_5	3.01_5	2.58_5	2.91_3	2.77_3	(K,Ba)(Si,Al)$_4$O$_8$/Orthoclase, barian	19– 3	
	3.38_x	3.45_7	3.44_7	2.28_6	6.30_4	2.23_4	2.75_3	5.07_3	KPO$_3$/Potassium metaphosphate	35– 819	
*	3.39_x	3.43_5	2.21_6	5.39_5	2.54_5	2.69_4	1.52_4	2.12_3	Al$_6$Si$_2$O$_{13}$/Mullite, syn	15– 776	
i	3.39_x	3.41_9	2.31_4	2.84_3	3.82_2	2.14_2	2.03_2	2.00_2	NaBF$_4$/Ferruccite, syn	11– 671	
i	3.41_9	3.39_x	2.31_4	2.84_3	3.82_2	2.14_2	2.03_2	2.00_2	NaBF$_4$/Ferruccite, syn	11– 671	
i	3.38_x	3.39_8	2.53_8	3.11_7	2.29_6	3.57_4	2.41_4	2.37_4	Gd$_5$S$_3$	20–1056	
i	3.39_8	3.38_x	2.53_8	3.11_7	2.29_6	3.57_4	2.41_4	2.37_4	Gd$_5$S$_3$	20–1056	
i	3.30_x	3.29_x	4.76_6	4.18_6	5.73_5	2.93_3	3.98_3	2.38_2	C$_6$H$_6$O$_4$	37–1919	
*	3.33_x	3.28_8	2.97_7	3.83_4	2.36_4	2.34_1	2.35_1	2.13_1	CdSO$_4$	14– 352	

Figure 12.5. Sample of the Hanawalt search manual. The quality mark is followed by the d and I code, as a subscript, for the eight most intense lines, followed by the formula, PDF number, and I/I_{corundum}.

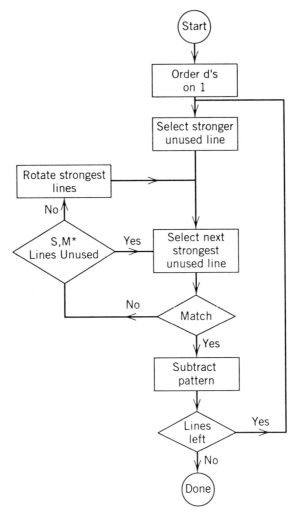

Figure 12.6. Flowchart of the Hanawalt search method.

and, once a phase is identified, the whole pattern from this phase is eliminated from the *unknown* pattern and the process restarted.

As an example, Figure 12.7 shows the diffraction pattern of a mixture of two compounds. As before, each line is numbered and the 2θ and absolute intensity values are listed. The d-values and relative intensities are then calculated. As will be seen from Table 12.9, line 7 is the strongest and is found in Hanawalt group 2.49–2.45 Å. Lines 3 and 9 are the second and the third strongest lines in

Table 12.8. Experimental Data for α-Quartz

I^{rel}	$d(Å)$	I-Order	I^{rel}	$d(Å)$	I-Order
25	4.25	2	2	1.658	—
100	3.34	1	10	1.538	4
7	2.455	7	2	1.451	—
9	2.278	5	6	1.381	—
5	2.235	—	7	1.374	8
6	2.2125	—	8	1.371	6
4	1.978	—	3	1.286	—
17	1.815	3	2	1.254	—
6	1.670	—			

the pattern but these do not give a match. However, lines 3 and 5 do match the pattern of ZnO. Elimination of the ZnO pattern from the mixture leaves lines 9, 6, and 13 as the three strongest of the remaining lines. Line 9 is found in the Hanawalt group 2.00–2.05 Å and proves to be due to Al_2O_3. A check of both the ZnO and the Al_2O_3 patterns from the index against the unknown gives an excellent d match and quite a fair intensity match. It is thus apparent that the crystalline portion of the given mixture is made up of ZnO and Al_2O_3.

From the foregoing it will be clear that the identification of complex mixtures purely on the evidence of their diffraction patterns can prove to be a time-consuming task. However, when backed, for example, with quantitative or semiquantitative spectroscopic analysis, the task can be greatly simplified. Where elemental data is not available, a few simple tests can often give useful additional information. Visual examination with a pocket magnifier will often disclose the presence of different phases. Color, hardness to the touch, smell, water solubility, etc. are all simple tests that may yield key information. In difficult cases, for example, where complex hydrates are present, drying or even ignition, followed by re-recording of the diffraction pattern, will often give a clue as to the composition of the original material.

12.4.3. The Fink Search Method

One of the major problems with the Hanawalt search/match system is that, since patterns are indexed based on their strongest lines, complications arise when preferred orientation and superposition of lines in complex mixtures occur. This is the principal reason that difficulty is found in using the Hanawalt method for data obtained from highly oriented samples or data taken on transmission electron microscopes. Of the various alternatives proposed to the Hanawalt system, the Fink system is probably the most useful.

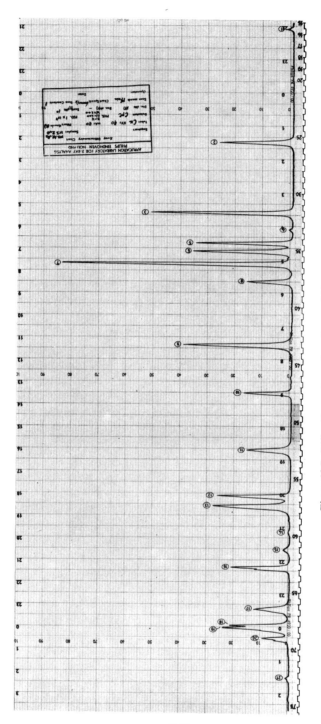

Figure 12.7. Diffraction pattern of a mixture of Al_2O_3 and ZnO.

Table 12.9. Interpretation of the Diffraction Pattern of Al_2O_3 and ZnO

Line No.	$d(\text{Å})$	I^{rel}	ZnO $d(\text{Å})$	I^{rel}	Al_2O_3 $d(\text{Å})$	I^{rel}
1	5.68	2.5	—	—	—	—
2	3.48	31	—	—	3.48	70
3	2.814	62	2.814	57	—	—
4	2.698	2	—	—	—	—
5	2.612	42	2.603	44	—	—
6	2.557	43	—	—	2.551	97
7	2.482	100	2.476	100	—	—
8	2.384	19	—	—	2.379	42
9	2.089	47	—	—	2.085	100
10	1.914	20	1.911	23	—	—
11	1.743	19	—	—	1.740	42
12	1.629	31	1.625	32	—	—
13	1.605	33	—	—	1.601	82
14	1.549	1	—	—	1.546	2
15	1.514	3	—	—	1.511	7
16	1.478	25	1.477	29	—	—
17	1.408	15	1.407	4	1.405	30
18	1.382	26	1.378	23	—	—
19	1.376	30	—	—	1.374	45
20	1.360	11	1.358	11	—	—
21	1.340	1	—	—	1.336	1

This method was introduced by the ICDD in the early 1960s. The Fink system was designed by Bigelow and Smith [20] and was named after William Fink, the long-time chairman of the JCPDS (formerly Committee E4 of ASTM). The original Fink system indexed a pattern based on the d-values of its eight strongest lines, and eight separate entries were made using a cyclic permutation of d-values. Because of the great size of the index, due in turn to the large number of permutations, more recent versions of the Fink index have been based on the use of the eight strongest lines in the pattern, of which only the four strongest are permuted. As an example, consider a pattern composed of the following ten lines, ordered on d:

$$d_2, \ d_9, \ d_x, \ d_3, \ d_1, \ d_8, \ d_7, \ d_4, \ d_6, \ d_5$$

The subscripts are the relative intensities of the lines. The eight most intense lines exclude d_2 and d_1, and these are chosen to be listed in the index. The

7.99 – 7.00

								Formula	File No.	I/Ic
*	7.75₅	7.50₃	6.25₂	5.75₅	4.85ₓ	4.56₈	4.42₂	2.81₆	$C_7H_{11}Cl_2N_3O_4S_2$	31–1786
	7.45₆	7.41₈	6.65ₓ	4.02₅	3.64ₓ	3.61₁	3.43₁	3.19₃	$C_{17}H_{10}O_2$	37–1923
	7.80₁	7.31₂	5.00₁	4.44₁	3.65₁	3.56₁	3.33₁	14.6ₓ	$C_3H_4NNaO_3S \cdot 2H_2O$	29–1902
-	7.70₅	7.28ₓ	4.36₃	3.64ₓ	3.32₃	3.22₃	2.76₄	2.33₃	$C_2H_8N_2 \cdot HCl$	29–1743
	7.70₄	7.14₃	5.88₃	5.21₅	4.03₄	3.50₃	3.29₃	10.5₂	$C_{18}H_{21}NO_4 \setminus Oxycodone$	6– 100
	7.50ₓ	7.12₁	3.76₃	3.70₁	3.42₁	3.38₇	3.30₂	3.15₂	$C_8H_{10}N_2O_2$	29–1594
	7.82₄	6.97₆	6.24₃	5.70₅	5.01ₓ	4.28₇	3.72₁	3.41₅	$C_{20}H_{24}N_2O_2$	9– 651
*	7.56₅	6.97ₓ	6.24₈	5.13₈	4.50ₓ	4.18₆	3.62₈	3.51₈	$Ce(NO_3)_3$	16– 196
-	7.58₇	6.94₁	6.73₆	4.71ₓ	4.52₈	3.59₃	3.53₃	2.35₁	$C_{12}H_{22}O_{11}$	24–1977
	7.63₃	6.93₃	5.23₃	4.53₆	4.07₉	3.34₈	2.40₃	8.67₃	$C_{19}H_{16}ClNO_4$	31–1733
-	7.38₄	6.91₁	6.17₃	5.58₆	4.17₂	3.66₃	3.10₂	10.2₁	$C_{20}H_{34}O_{13}$	29–1848
	7.33₄	6.91₈	5.06ₓ	4.32₈	4.00₃	3.62₈	3.45₂	2.54₂	$C_{15}H_{21}NO_3 \cdot HCl$	6– 128
*	7.73₂	6.78₂	3.87₃	3.79₂	3.58₈	3.20₈	2.83₁	2.49₁	$C_8H_8O_2$	30–1930
o	7.00₈	6.75₅	5.60₅	5.15₇	4.85₇	4.40ₓ	3.80₄	3.60₅	$C_7H_{20}N_2S \cdot HCl$	28–1766
-	7.26₃	6.73₂	5.21₁	4.18₁	3.68₁	3.26₁	11.0₁	8.33ₓ	$C_8H_{10}N_4O_2 \cdot H_2O$	31–1570
	7.71₆	6.66₂	6.07₁	4.79₁	3.83₂	3.33₁	3.30₄	10.0₂	$C_7H_8N_2O_2 \cdot H_2O$	26–1893
-	7.41₆	6.65ₓ	4.02₅	3.64₆	3.61₁	3.43₅	3.19₃	7.45₆	$C_{17}H_{10}O_2$	37–1923
*	7.50₇	6.61₅	5.31₈	4.72₃	3.85₅	3.37ₓ	3.22₂	8.34₁	$C_{20}H_{12}O_3$	9– 817
	7.42₆	6.59₈	6.23₄	5.98₈	5.05₆	4.17₆	3.46₆	10.1₃	$C_{17}H_{19}NO_3 \cdot H_2O$	31–1805
o	7.37₅	6.59₃	6.25₂	6.00ₓ	5.09₈	4.16₃	3.42₃	10.1₂	$C_{17}H_{19}NO_3$	31–1804
c	7.21ₓ	6.59₈	6.33₃	5.45₃	4.87₁	3.75₈	3.31₂	7.58₁	$C_{19}H_{21}NO_4 \cdot HCl \cdot 2H_2O$	30–1821
*	7.78₂	6.57₁	5.41₈	5.09ₓ	4.71₁	3.10₁	2.64₁	13.4₁	$C_8H_{12}N_2O_3 \setminus barbital$	31–1546
	7.04₆	6.54₁	5.41₄	4.90ₓ	4.54₈	9.70₈	9.02₈	7.75₁	$C_{20}H_{24}N_2O_2 \cdot H_2O$	5– 251
	7.00₈	6.54₅	5.29ₓ	5.08₅	4.13₅	4.02₅	3.50₅	8.10₃	$C_{21}H_{23}NO_5 \setminus Heroin$	6– 118
-	7.50₄	6.48₃	4.72₃	4.57ₓ	4.34₃	3.20₃	3.10₃	12.4₃	$C_{21}H_{27}NO \cdot HCl$	31–1782

	I/Ic
(row 5)	0.70
(row 12)	1.91
(row 14)	1.50
(row 16)	0.44
(row 21)	1.13

Figure 12.8. A section from a Fink index. Values in boldface are the four most intense lines, which determine the positions of entry into the index.

342

d-values of the four most intense lines, d_9, d_x, d_8, d_7, are chosen and the phase is entered in the Fink index in the following four ways:

$$\begin{array}{cccccccc}
d_9, & d_x, & d_3, & d_8, & d_7, & d_4, & d_6, & d_5 \\
d_x, & d_3, & d_8, & d_7, & d_4, & d_6, & d_5, & d_9 \\
d_8, & d_7, & d_4, & d_6, & d_5, & d_9, & d_x, & d_3 \\
d_7, & d_4, & d_6, & d_5, & d_9, & d_x, & d_3, & d_8
\end{array}$$

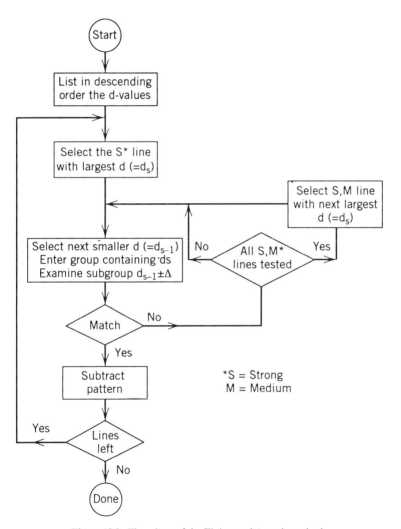

Figure 12.9. Flowchart of the Fink search/match method.

Figure 12.8 shows an example of a portion of the Fink index. As was the case with the Hanawalt index, there are 12 columns of data. The only difference with the Fink index is that the four lines used as index entries are indicated in bold text. Forty-five nonoverlapping d-spacing groups are used in a manner similar to the Hanawalt system.

A flowchart for the Fink search/match system is shown in Figure 12.9. The technique for using the Fink method is to list all lines in an unknown according to decreasing d-value, then to look into the index for each of the highest d-values in the unknown. Although the goal of this technique is to base the search on d-values rather than intensity, the probability that weak high d-spacing lines may not be observed forces the compromise of selecting the index entries based on intensity. A computer version of the Fink maximum-d search is not restricted by this assumption and more closely achieves the goal of a pure d-value search [21]. Nonetheless, the deemphasis on the intensity allows materials with strong preferred orientation, or with intensity distortions due to the use of a radiation other than Cu $K\alpha$, to be identified. In fact, this is the method of choice for identifying electron diffraction patterns or patterns obtained by energy dispersive techniques.

12.5. LIMITATIONS WITH THE USE OF PAPER SEARCH MANUALS

One of the problems that is constantly under scrutiny by the ICDD is the problem of the physical size of search manuals and indexes. The number of entries in a search manual (N) is essentially the product of three variables: the number of unique phases (U), the number of permutations and combinations (C), and the overlap factor (F). Thus,

$$N = UCF. \qquad (12.1)$$

For example, a Hanawalt file containing 50,000 unique phases, based on an average of two rotations, with an overlap factor of 10%, would contain $50{,}000 \times 2 \times 1.1 = 110{,}000$ entries. The number of pages in a search manual (P) is roughly proportional to the number of entries divided by twice the number of entries/page (E). Finally the thickness of a manual (T) is the number of pages divided by the number of pages/cm (S). Typical values currently employed are 90 entries/page and 250 pages/cm. In practice, this figure must be increased by a further factor of about 1.25 to allow for additional data—alphabetic entry search, front matter, etc. Thus, by combining these factors one can roughly predict the thickness of a search manual as follows:

$$\text{thickness} = 3 \times 10^{-5}(UCF) \quad \text{cm}. \qquad (12.2)$$

For example, the Hanawalt manual alluded to earlier would be roughly 3.6 cm thick. A last point to be mentioned about the design of search manuals is that it is very difficult to bind a manual much above a thickness of 6 cm (1500 pages).

A few simple calculations soon reveal that a Fink search manual of 60,000 phases with eight permutations would be impossible to bind as a single book (it would be about 15 cm thick). Mainly because of this problem, there has been no book-form Fink manual since the early 1980s. A CD–ROM version is under consideration by the ICDD.

12.6 BOOLEAN SEARCH METHODS

The development of the full computer-readable version of the PDF and its production on a CD–ROM has dramatically improved flexibility in phase identification and characterization [22]. As of set 44, the PDF-2 contains nearly 60,000 phases and, due to the high-density format of the CD–ROM, it is easily storable and thus also readily accessible on a personal computer. A Boolean search program, PC-PDF, is supplied by the ICDD with the database. The search program works via a series of index files associated with the PDF-2 database and uses a Boolean parsing expression consisting of a key range specification, optional white space, parentheses, wild cards (*), and certain Boolean operators. These operators include AND (&), IOR(—) and ANDNOT (−). Retrieval from the PDF-2 data base is possible on 18 index keys allowing searches on items including Chemistry, Strong Lines, Journal CODEN, Author and Date, Physical Properties, etc. Following a search, the output list can be displayed in terms of d, 2θ, $\sin^2\theta$, or $1/d^2$, with intensities for either fixed or variable divergence slits. Since the database format of the PDF and CDF were designed to be completely compatible [23], programs and techniques designed for searching the PDF are almost directly applicable to the CDF. The Boolean search program is no exception to this general rule. Search criteria are applied using an interactive software program. When the search criteria are initialized, the program builds a Boolean search tree that is used to apply the criteria to the index files to establish candidate matches. The operator can then elect to display the matched patterns one at a time. If a display is requested, the index file pointer is used to identify the appropriate database record, which is read from the CD–ROM. A card image is generated from the NBS*AIDS83 data format and displayed on the screen. The process is repeated for each pattern meeting the selected set of search criteria.

A schematic of this powerful Boolean-type searching is shown in Figure 12.10. Here, a search is being made for minerals containing silicon with

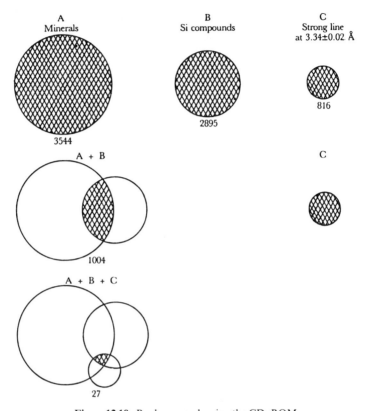

Figure 12.10. Boolean search using the CD–ROM.

a strong line at 3.34 Å. At the time of the test, the full file contained 47,847 entries. Figure 12.10 indicates the size of the PDF subsets based first on minerals only (3544 entries); secondly, on compounds containing silicon (2895 entries); and, thirdly, on patterns with one of its three strongest lines at 3.34 ± 0.02 Å (816 entries). By intersecting the first of these two files (i.e., taking minerals that contain silicon), 1004 matches are found. By intersecting the third subset as well, we now find that there are only 27 potential matches for minerals containing silicon that give a diffraction pattern with its strongest line at 3.34 Å.

12.7. FULLY AUTOMATED SEARCH METHODS

12.7.1. First-Generation Programs

The first two attempts to use computers to aid in the search/match procedure were carried out by Johnson and Vand [24] and by Nichols [25]. Both of these algorithms were designed to run on the large mainframe computers of their mid-1960s era. This first generation of computer search/match algorithms was forced to assume very wide error windows on both the d's and I's of a powder pattern, due both to the poor quality of the data in the PDF at that time and to the relatively poor quality of the typically observed pattern. Because of the limited availability of computer storage, both procedures converted the floating point d-values into integers by dividing them into 1000 and rounding off. Thus, both the d's and I's were stored as integers and the search was carried out in fast integer arithmetic.

The Johnson–Vand approach was to use the computer's speed to carry out an exhaustive search, comparing each reference pattern in the PDF to the unknown pattern and computing a FOM for each match. The absolute value of the FOM was not important since it was computed for all reference patterns and the best 50 matches listed for evaluation. If the unknown pattern contained a large number of lines, then it became quite likely that some of the reference patterns with only a few lines would match well. Thus, these patterns tended to rise to the top of the list. For this reason, the principal quantities to examine in deciding a candidate for a match were the number of lines that matched, followed by the FOM. Even when a reference matched a large number of lines and had a high FOM it still could contain chemical components known not to be present, so chemical constraints on the search were strongly recommended.

The Nichols algorithm took a Hanawalt-type approach to searching, relying on the intensities of the lines in the observed pattern. This strategy used an *inverted file* that was sorted on the d-value of the most intense line in each pattern. This approach required the processing of much less information than the Johnson–Vand method and therefore ran more rapidly. However, like the Johnson–Vand approach, this early algorithm was forced to assume wide error windows on the d's and I's in the reference patterns due to the uncertain quality of the file and experimental data.

Although some further developments of the first-generation mainframe algorithms occurred in the 1970s [26,27], today these programs should be viewed as aids in identification rather than final solutions.

348 QUALITATIVE ANALYSIS

12.7.2. Second-Generation Search/Match Algorithms

Major strides were made in the 1970s in the evolution of automated powder diffractometer (APD) systems. Several groups worked independently on the development of minicomputer-based systems for general diffractometer control along with associated data collection and reduction [28–30]. These developments led to the production of several very successful commercial products (e.g., see Jenkins et al. [31] and Snyder et al. [32]) that permitted routine external and internal standard calibration, precision alignment checks, and new levels of accuracy and precision in data collection and analysis. The publication and distribution of the first public FORTRAN-based algorithms [29] stimulated other manufacturers to develop commercial systems, and these began to appear in the early 1980s. Both the new level of precision available and the ability to create complex algorithms in FORTRAN designed to run on the laboratory minicomputer permitted the evolution of several second-generation search/match algorithms [33–38]. Many of these programs offered great versatility. As an example, the system developed at Alfred University [33] implemented a precision Hanawalt-type strategy in addition to a Johnson–Vand or Fink-like exhaustive approach with a hierarchical use of databases in order to speed up operation on a minicomputer with as little as 0.5 MB of floppy disk storage.

The development of second-generation search/match codes greatly improved the success rate at identifying phases in multiphase unknowns. The increased precision available in automated data permitted a more exact quantitative subtraction of each identified phase from the observed pattern. Thus, new types of search strategies could be implemented. With the improvement over first-generation codes, one typically finds that most one- and two-phase unknowns are readily identified. However, samples containing three, four, or more unknown phases are seldom fully automatically analyzed. To improve performance, chemical constraints are commonly employed. Today all manufacturers have at least a second-generation search/match algorithm as part of their analysis software.

12.7.3. Commercial Search/Match Programs

Since second-generation algorithms are still in wide use today, a few examples of their use and some of the different strategies among them will aid the current user of search/match procedures. One of these codes [33] was developed for Siemens and ultimately grew into their D500 and D5000 systems. Its design, in 1979, included operation on minicomputers with only 5 MB of storage on two floppy disks. On such primitive machines speed of searching and efficient use of space was critical, so this algorithm used a Hanawalt strategy with an inverted file that could be rapidly accessed randomly. The search was carried out using a hierarchy of databases, starting with the 300 phases that accounted for more than 90% of all phases identified at Dow Chemical Corporation over 40 years [26]. Any phases positively identified were quantitatively scaled and subtracted from the observed pattern, and the search continued with the residual pattern. Next, a database containing the 2500 patterns in the ICDD frequently encountered phase list was searched, and finally, if any residual pattern remained, the full PDF was searched. As computer speed and disk capacity increased, this and other algorithms worked so rapidly against the full PDF file that the hierarchical strategy was no longer needed; the current version of this program searches the full PDF.

The key to the improved success rate of second-generation codes lay in the precision of separating out identified phases and generating a reliable residual pattern for further searching. In addition to a much improved success rate [21], this strategy generated a list of the exact phases identified in the unknown, rather than a list of the top 50 matches, which still required extensive manual user evaluation. Another second-generation strategy was taken by Philips Electronics (See Schreiner et al. [36]) in developing a method of *probability searching* whereby a least squares analysis can be applied to sets of lines and the standard deviation of the *mismatch* between line pairs included within the FOM. In this program, the mismatch from the least squares fitting is expressed as *equivalent displacement error* and used in the calculation of the FOM.

12.7.4. Third-Generation Search/Match Algorithms

In the second generation of computer search/match algorithms, a strong reliance was placed on the location and intensities of the diffraction peaks. The success rate of these codes was (and remains) limited by the precision and accuracy of the reference and experimental patterns. Preferred orientation and limited particle statistics distort intensities to a few percent in most materials and to factors of more than 10 in some, limiting the reliability of a residual pattern computed by subtracting phases away from the observed $d-I$ list. A new strategy has recently been introduced [39] that has dramatically improved the success rate of the search/match process. The new idea is to search the whole observed pattern with its background (not just the $d-I$ list) and to *add* candidate phases together to compose, rather than decompose, an observed multiphase pattern. At first glance, this change in strategy does not look so dramatic; however, the results are now heavily influenced by where peaks do not occur instead of only by where they are present. This extra information radically changes the search/match process, permitting searches of the full file without chemistry in just a few seconds. The success rate of this procedure is often 100% even for four-phase unknowns with significant amounts of preferred orientation. The result of this new strategy, which is finding wide acceptance, has been so dramatic as to be considered a new generation of computer software.

12.8. EFFECTIVENESS OF SEARCH/MATCHING USING THE COMPUTER

The effectiveness of computer *search/matching* of the PDF is an especially difficult task to quantify due to the variable quality of experimental and reference data. Earlier search methods were based mainly on experience gained with traditional manual searching techniques and relied on a tally of *hits* and *misses* between lines in the experimental pattern and lines in a potential hit (match) pattern. This situation persists today even in the most modern computer search/match methods.

All generations of computer search/match algorithms use a FOM to evaluate the quality of a match. A typical FOM is

$$\text{FOM} = d_R I_R^2 d_U, \qquad (12.3)$$

where d_R is the percentage of reference lines that match the unknown; I_R is the percentage of the reference intensity matched; and d_U is the percentage of the unknown lines matched. The FOM calculation thus takes into account all the necessary factors by which one is able to judge how close the match is. The relative weight of the above factors is set so that the goodness of a match can be truly represented by the FOM. The higher the FOM, the higher the probability of a match. The FOM of a perfect match is commonly scaled to 100, which would rarely be attained in practice.

Table 12.10. Use of FOMs for PDF Search/Matching

Lines	Result	Reason
Found	Correct	Good technique
	Incorrect	Poor smoothing/peak hunting
	Incorrect	Contaminant lines
	Incorrect	Other phases
Not found	Correct	Outside experimental range
	Correct	Outside reference range
	Correct	Under background
	Correct	Not there

While search/match FOMs may vary in their method of application, the basic principle is as illustrated in Table 12.10. Each matching line contributes a positive number to the *match score*, and each unmatched line contributes a negative number. In either case, a weighting factor is applied by the software program designer to vary the influence of the indicated factors. One of the major problems with the use of FOMs in computer search/match procedures is that the method of weighting the various factors is unique (and often undisclosed) for each program. This will clearly affect the performance of each program in the solving of different types of problems.

REFERENCES

1. Hanawalt, J. D., and Rinn, H. W. Identification of crystalline materials. *Ind. Eng. Chem., Anal. Ed.* **8**, 244–247 (1936).
2. Hanawalt, J. D., Rinn, H. W., and Frevel, L. Chemical analysis by X-ray diffraction—Classification and use of X-ray diffraction patterns. *Ind. Eng. Chem., Anal. Ed.* **10**, 457–512 (1938).
3. Jenkins, R. Profile data acquisition for the JCPDS–ICDD database. *Aust. J. Phys.* **41**, 145–153 (1988).
4. Snyder, R. L. Analytical profile fitting of X-ray powder diffraction profiles in Rietveld analysis. In *The Rietveld Method* (R. A. Young, ed.), Chapter 7, pp. 111–131. Oxford Univ. Press, Oxford 1993.
5. Jenkins, R., Holomany, M., and Wong-Ng, W. On the need for users of the Powder Diffraction File to keep current. *Powder Diffr.* **2**, 84–87 (1987).
6. Snyder, R. L., Johnson, C., Kahara, E., Smith, G. S., and Nichols, M. C. An analysis of the Powder Diffraction File. *Lawrence Livermore Lab.* [*Rep.*] **UCRL-52505** (1978).
7. Wong-Ng, W., Holomany, M., McClune, F., and Hubbard, C. R. The JCPDS data base—Present and future. *Adv. X-Ray Anal.* **26**, 87–88 (1982).
8. Hanawalt, J. D. History of the Powder Diffraction File. In *Crystallography in North America—Apparatus & Methods*, (D. McLachlan and J. P. Glusker, eds.), Chapter 2, pp. 215–219. American Crystallographic Association, New York, 1983.
9. NBS Crystal Data File. (1982) *A Magnetic Tape of Crystallographic and Chemical Data Compiled and Evaluated by the NBS Crystal Data Center*. National Bureau of Standards, Gaithersburg, MD, 1982 (available from JCPDS-International Centre for Diffraction Data, Newtown Square, PA).
10. Himes, V. L., and Mighell, A. D. NBS*LATTICE—*A Program to Analyze Lattice Relationships*. NBS Crystal Data Center, National Bureau of Standards, Gaithersburg, MD, 1985.
11. Smith, G. S., Johnson, Q. C., Smith, D. K., Cox, D. E., Snyder, R. L. Zhou, R. S., and Zalkin, A. The crystal and molecular structure of beryllium dihydride. *Solid State Commun.* **67**, 491–494 (1988).

12. Smith, G. S., Johnson, Q. C., Cox, D. E., Snyder, R. L., Smith, D. K., and Zalkin, A. Synchrotron radiation applied to computer indexing. *Adv. X-Ray Anal.* **30**, 383–388 (1987).
13. Anderson, R., and Johnson, G. G., Jr. The Max-*d* alphabetical index to the JCPDS database: A new tool for electron diffraction analysis. *Proc.—37th Annu. Meet., Electron Microsc. Soc. Am., San Antonio, Texas* (1979).
14. Carr, M. J., Chambers, W. F., and Melgaard, D. A search/match procedure for electron diffraction data based on pattern matching with binary bit maps. *Powder Diffr.* **1**, 226–234 (1986).
15. Mueller, M. H., Wallace, P. L., Huang, T. C., and Dann, J. N. *A Phase Diagram Research Tool, Proc. Symp. Mater. Sci. Div.*, ASM., Minerals, Metals and Materials Society, Washington, DC, 1993.
16. Wallace, P. W., Weissman, S., Mueller, M. H., Calvert, L. D., and Jenkins, R. The new ICDD Metals & Alloys Indexes, usefulness and potentialities. *Powder Diffr.* **9**, 239–245 (1994).
17. Pearson, W. B. *Handbook of Lattice Spacings and Structures of Metals*, Vol. 2, pp. 1–3. Pergamon, Oxford, 1967.
18. *Strukturbericht*, Vols. I–VII. Akad. Verlagsges., Leipzig, 1931–1939 (originally published as supplements to the *Zeitschrift für Kristallographie*).
19. Thomas, G., and Ophey, W., Optical recording. *Phys. World* **3**(12), 36–41 (1990).
20. Bigelow, W., and Smith, J. V. *ASTM Spec Tech. Publ.* **STP 372**, 54–89 (1965).
21. Cherukuri, S., Snyder, R. L., and Beard, D. Comparison of the Hanawalt and Johnson—Vand computer search match strategies. *Adv. X-Ray Anal.* **26**, 99–105 (1983).
22. Jenkins, R., and Holomany, M. PC-PDF—A search/display system utilizing the CD-ROM and the complete Powder Diffraction File. *Powder Diffr.* **4**, 215–219 (1987).
23. Mighell, A. D., Hubbard, C. R. and Stalick, J. K. NBS*AIDS80—A FORTRAN Program for crystallographic data evaluation. *NBS Tech. Note (U.S.)* **1141** (1981).
24. Johnson, G. G., Jr., and Vand, V. A computerized powder diffraction identification system. *Ind. Eng. Chem.* **59**, 19 (1965).
25. Nichols, M. C. A FORTRAN II program for the identification of X-ray powder diffraction patterns. *Lawrence Livermore Lab.* [*Rep.*] *UCRL* **UCRL-70078** (1966).
26. Frevel, L. K., Adams, C. E., and Ruhberg, L. R. A fast search program for powder diffraction analysis. *J. Appl. Crystallogr.* **9**, 300–305 (1976).
27. Marquart, R. G. A search–match system for X-ray powder diffraction data. *J. Appl. Crystallogr.* **12**, 629–634 (1979).
28. Jenkins, R., Haas, D. J., and Paolini, F. R. A new concept in automated X-ray powder diffractometry. *Norelco Rep.* **18**, 1–16 (1971).
29. Mallory, C. L., and Snyder, R. L. The control and processing of data from an automated powder diffractometer. *Adv. X-Ray Anal.* **22**, 121–132 (1979).

30. Goehner, R. P., and Hatfield, W. T. A microcomputer controlled diffractometer. *Adv. X-Ray Anal.* **22**, 165–167 (1979).
31. Jenkins, R., Hahm, Y., Pearlman, S., and Schreiner, W. N. The APD3600—A new dimension in qualitative and quantitative X-ray powder diffractometry. *Norelco Rep.* **26**, 1–15 (1979).
32. Snyder, R. L., Mallory, C. L., Smith, S. T., Osgood, B. C., and Howard, S. A. The rebirth of X-ray powder diffraction. *N.Y. State Coll. of Ceram. Rep.* **3** (5), 17 (1979).
33. Snyder, R. L. A Hanawalt type phase identification procedure for a minicomputer. *Adv. X-Ray Anal.* **24**, 83–90 (1982).
34. Jobst, B. A., and Göbel, H. E. A versatile microfile-based system for fast interactive XRPD phase analysis. *Adv. X-Ray Anal.* **25**, 273–282 (1981).
35. Huang, T. C., and Parrish, W. A new computer algorithm for qualitative X-ray powder diffraction analysis. *Adv. X-Ray Anal.* **25**, 212–219 (1981).
36. Schreiner, W. N., Surdukowski, C., and Jenkins, R. A new minicomputer search/match/identify program for qualitative phase analysis with the powder diffractometer. *J. Appl. Crystallogr.* **15**, 513–523 (1982).
37. Goehner, R. P., and Garbauskas, M. F. PDIDENT—A set of programs for powder diffraction phase identification. *X-Ray Spectrom.* **13**, 172–179 (1984).
38. Toby, B. H. The POWDER SUITE: Computer programs for searching and accessing the JCPDS–ICDD powder diffraction database. *Powder Diffr.* **5**, 2–7 (1990).
39. Caussin, P., Nusinovici, J., and Beard, D. W. Using digitized X-ray powder diffraction scans as input for a new PC–AT Search/Match program. *Adv. X-Ray Anal.* **31**, 423–430 (1988).

CHAPTER
13
QUANTITATIVE ANALYSIS

13.1. HISTORICAL DEVELOPMENT OF QUANTITATIVE PHASE ANALYSIS

Early examples of quantitative phase analysis by X-ray powder diffraction include work in 1925 by Navias [1], who quantitatively determined the amount of mullite in fired ceramics, and in 1936 when Clark and Reynolds [2] reported an *internal standard method* for mine dust analysis by film techniques. In 1948 Alexander and Klug [3] presented the theoretical background for the absorption effects on diffraction intensities from a flat briquette of powder. Since then there have been numerous methods developed based on their fundamental equations [4]. Methods applicable to a wide range of phases and samples include the *absorption–diffraction method* [3], the *method of standard additions* (also called the *spiking method*) [5], and the *internal standard method* [6]. Formalisms have been established to permit inclusion of overlapping lines and chemical constraints in the analysis [7].

Quantitative analysis using any instrumental method is a difficult undertaking. It requires painstaking calibration of the instrument, using carefully prepared standards, and many repetitions during the setup phase to establish the required technique. In general, quantitative analysis of a new phase system will require a minimum of a few days and often a week of setup time. After the technique and standards have been established, a well-designed computer algorithm (e.g., see Snyder and Colleagues [8,9]) can produce routine analyses in as little as a few minutes to an hour per sample, depending on the accuracy desired.

Perhaps it is the difficult nature of quantitative analysis that has limited the number of carefully performed literature examples. This, in turn, has resulted in a concentration of effort on a specific analysis technique, rather than on understanding the sample and instrument-dependent limitations of quantitative phase analysis by X-ray powder diffraction.

In Section 2.4, the use of Vegard's law was described as a method for the determination of the composition of substitutional solid solutions. This is the only application of diffraction to the problem of quantitative analysis that does not rely on the intensity of the diffraction lines. Since the lattice parameters are related to the composition of a substitutional solid solution, it is the measurement of the diffraction maxima position that can be related to the composition. Normally, quantitative X-ray powder diffraction is taken to mean quantitative *phase* analysis of polyphase mixtures.

13.2. MEASUREMENT OF LINE INTENSITIES

Table 13.1 summarizes the various factors that control the absolute and relative intensities of lines in a powder pattern. The basis of quantitative phase analysis by powder diffraction is to attempt to equate the concentration of a given phase or phases with the intensity of a single line, a number of lines, or

Table 13.1. Factors Affecting X-ray Powder Diffraction in Line Intensities

Factor	Parameter
1. Structure-sensitive	Atomic scattering factor
	Structure factor
	Polarization
	Multiplicity
	Temperature
2. Instrument-sensitive	Source intensity
(a) *Absolute intensities*	Diffractometer efficiency
	Voltage drift
	Takeoff angle of tube
	Receiving slit width
	Axial divergence allowed
(b) *Relative intensities*	Divergence slit aperture
	Detector dead time
3. Sample-sensitive	Microabsorption
	Crystallite size
	Degree of crystallinity
	Residual stress
	Degree of peak overlap
	Particle orientation
4. Measurement-sensitive	Method of peak area measurement
	Degree of peak overlap
	Method of background subtraction
	$K\alpha_2$ stripping or not
	Degree of data smoothing employed

even all of the lines in a pattern. Since each of the factors listed in Table 13.1 affect the line intensity to some degree, they will each be considered in the following subsections.

Structure-Sensitive Parameters. The structure-sensitive factors for phase α are grouped into the $K_{(hkl)\alpha}$ factor in Equation 13.2. One of the few of these parameters that the experimentalist can change to affect the intensity is the wavelength. The atomic scattering factor depends strongly on the wavelength, such that longer wavelengths will sharply increase the intensity. The temperature also increases thermal motion and so will affect the intensity.

Instrument-Sensitive Parameters. These parameters affect the absolute intensity and therefore may change the measurement from one day to the next. Figure 13.1 shows the change in the diffracted intensity of a Si (111) reflection continually measured for 24 h. This effect is due to voltage drift of the incoming 220 V source, which transforms to cause a change in the incident-beam intensity and, in turn, in the diffracted intensities. Accurate voltage stabilization is an essential feature on instruments to be used for quantitative analysis, and it should be checked periodically.

The detector dead time discussed in Section 5.2.2 will cause very intense peaks to be observed at intensities lower than the true intensity. All intensity measurements should be routinely corrected for dead time by using Equation 5.1.

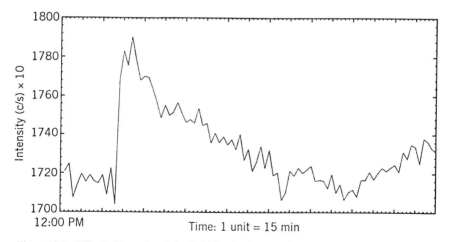

Figure 13.1. Diffracted intensity of the Si (111) reflection as a function of time of measurement.

Sample-Sensitive Parameters. These are the most important class of parameters that can affect both the absolute and relative intensity of diffraction lines. Preferred orientation, described in Section 3.9.1, is the most serious effect and is present to some degree in most specimen mounts. In the Bragg–Brentano geometry, the absorption of the sample is the same at all angles, since there is a constant volume of specimen in the beam. However, microabsorption, described in Section 3.7.1.e, can distort the intensities of all of the lines associated with phases in polycrystalline mixtures. To minimize this effect, one must ensure that the specimen has been ground to less than 10 μm particle size. In fact, there are two other effects that may also severely distort the measured intensities unless the particle size is on the order of 1 μm: crystallite statistics and extinction. Section 9.4.3 described the effect of large particles causing too few diffracting crystallites in a diffraction cone. This effect can produce large intensity fluctuations. Section 3.7.1 described the effect of large-domain perfect crystals causing the intensity of strong reflections to be observed too low. Crystallites as small as 20 μm can cause intensity changes as large as 25% [10]. Depending on the synthesis and sample history, the degree of crystallinity can also cause intensity changes on the order of 25% [11].

Measurement-Sensitive Parameters. The most important parameter in this category is the selection of the 2θ points at which background will be measured and the scan range over which a peak will be integrated. Figure 13.2 shows the diffraction pattern for a 50:50 mixture of Si and Al_2O_3 measured with molybdenum radiation. Since the use of Mo radiation compresses the pattern toward low 2θ, the effect of tails of peaks overlapping neighboring peaks is

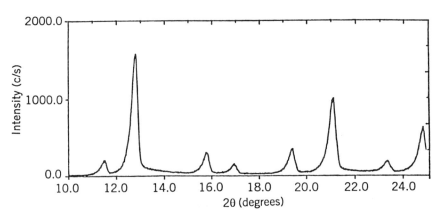

Figure 13.2. Diffraction pattern of a Si/Al_2O_3 mixture using Mo radiation.

emphasized. The selection of correct background points for the peak at 12.5° is impossible because of strong overlap with the peak at 11.5° and some overlap even with the peak near 16°. The best approach to measuring background-corrected integrated intensities in this case is to choose a longer wavelength radiation to improve the peak resolution. However, changing wavelengths is often not a practical solution to the problem of peak overlap.

Even when a peak is fully resolved, as shown in Figure 13.3, the intensity of the tails extend out farther than one might imagine. Figure 13.4 illustrates how

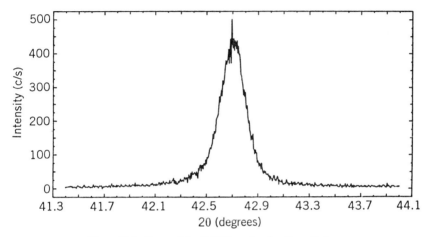

Figure 13.3. Trace of the Si (111) peak using Cr radiation.

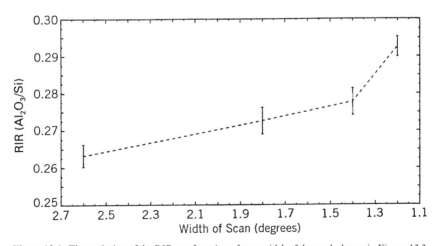

Figure 13.4. The variation of the RIR as a function of scan width of the peak shown in Figure 13.3.

360 QUANTITATIVE ANALYSIS

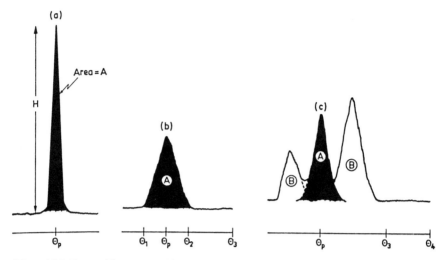

Figure 13.5. Types of line measured in quantitative analysis. From R. Jenkins and J. L. de Vries, *An Introduction to Powder Diffractometry*, p. 35, Fig. 46. Copyright © 1977, N. V. Philips, Eindhoven, The Netherlands.

integrated intensities vary with width of a scan. The figure plots the reference intensity ratio (RIR, to be discussed in Section 13.6.1, below), which is the ratio of the background corrected integrated intensities of the Si (111) peak shown in Figure 13.3 to the 100% line of Al_2O_3. The measurement of the Al_2O_3 peak was held constant, and the scan width of the Si peak was varied from 1.2° to 2.6°. Typically, one must scan a window of at least 2° in order to obtain the integrated intensity of a sharp peak at a 2θ below 50°.

Figure 13.5 illustrates the three basic types of line encountered when attempting quantitative phase analysis:

- *Peak height is proportional to peak area.* In the first case (peak a), the diffraction line is sharp and nonoverlapped. Here, the peak height H (which is easy to measure on a manual system) can be used as an estimate of the intensity. However, in general, one should always use background-corrected integrated intensities for quantitative analysis.
- *Peak height is not proportional to peak area.* The more usual circumstance in the measurement of line intensities is that the diffraction line may be broadened as illustrated in Figure 13.5 (peak b). The peak is distributed over a fairly wide angular range, $2\theta_1$ to $2\theta_2$. Most modern diffractometers are equipped with stepper motors and are easily able to integrate counts between selected angular positions. However, where

a synchronous motor is used, another technique has to be employed. As an example, a scan can be started at $2\theta_1$ and at the same time the scaler/timer is switched on. On reaching $2\theta_2$ the scanning motor stops and sends a stop pulse to the scaler/timer, which then reads the integrated counts collected N_p in time t_p. If a stop pulse is not automatically generated, the operator must manually halt the scaler when the $2\theta_2$ angle is reached. Background measurements are then made at two fixed positions, namely, $2\theta_1$ and $2\theta_2$, using the fixed-time counting mode. Each of these two fixed-time measurements may be made using exactly one-half of the peak and background integration time t_p, or the background and peak measurements will need to be scaled to the same time basis. The sum of these two background counts is then taken as the true number of background counts N_b, and the number of counts in the integrated peak (area A) is given by $N_p - N_b$.

- *Peak area is overlapped by other peaks.* It frequently happens that many of the potentially useful lines that might be used for quantitative analysis are overlapped by diffraction lines from other phases in the analyte specimen. Although there are several mathematical methods for correcting this effect, there is really no substitute for the use of optimum conditions for the proper separation of the peaks. However, the use of profile-fitting methods has proven especially useful in the line overlap situation (see Section 13.8.2, below). Where such methods are not available, recourse will have to be made to the use of the generalized internal standard method described in Section 13.6.5, in which peak groups can be integrated together. A typical situation is illustrated in Figure 13.5 (peak c). In principle, the measurement of the integrated peak intensity is exactly the same as was discussed previously for the measurement of the nonoverlapped peak (b) except that entirely different backgrounds θ_3 and θ_4 have to be chosen. Since the value of N_p for phase A obtained in this instance may be significantly influenced by the unresolved interfering peaks from phase B, it is obvious that an additional error is introduced. Such an error becomes extremely large as the intensity of the interfering line approaches that of the combined line being used.

13.3. FOUNDATION OF QUANTITATIVE PHASE ANALYSIS

Equation 3.28 developed in Section 3.7.2 gives the functional dependence of the intensity of diffraction line *hkl* of a phase α in a polyphase mixture as measured on a flat briquette specimen with a conventional diffractometer. When the volume fraction is converted to the more conventional weight fraction [6], the full-and formidable-intensity equation is

$$I_{(hkl)\alpha} = \left[\frac{I_0 \lambda^3}{64\pi r}\left(\frac{e^2}{m_e c^2}\right)^2\right]\left[\frac{M_{hkl}}{V_\alpha^2}|F_{(hkl)\alpha}|^2\left(\frac{1+\cos^2 2\theta \cos^2 2\theta_m}{\sin^2 \theta \cos \theta}\right)\right]\left[\frac{X_\alpha}{\rho_\alpha (\mu/\rho)_s}\right].$$
(13.1)

Recognizing the terms that are constant for a particular experimental setup as K_e (in the first set of square brackets) and those which are constant for each (hkl) diffraction line of each phase (e.g., α) in a mixture as $K_{(hkl)\alpha}$ (in the second set of square brackets), we can write Equation 13.1 in the more tractable form

$$I_{(hkl)\alpha} = \frac{K_e K_{(hkl)\alpha} X_\alpha}{\rho_\alpha (\mu/\rho)_s},$$
(13.2)

where X_α is the weight fraction of phase α; ρ_α is the density of phase α; $(\mu/\rho)_s$ is the mass attenuation coefficient of the polyphase specimen; and K_e and $K_{(hkl)\alpha}$ were defined in Equations 3.29 and 3.30. Remember that μ is the linear attenuation coefficient of phase α.

The fundamental problem in quantitative analysis of a homogeneous, randomly oriented powder lies in the $(\mu/\rho)_s$ term in the foregoing equations since, to solve for the weight fraction of phase α, it is necessary to compute, or measure, $(\mu/\rho)_s$ via

$$\left(\frac{\mu}{\rho}\right)_s = \sum_j \left(\frac{\mu}{\rho}\right)_j X_j.$$
(13.3)

The total absorption of a polyphase specimen for a given wavelength is simply the sum of the products of individual attenuation coefficients and weight fractions, as shown in Equation 13.3. Solution of this equation requires a knowledge of the weight fractions of each phase. Thus, the single-intensity equation 13.2 contains two unknowns [i.e., X_α and $(\mu/\rho)_s$], and so the problem of quantitative phase analysis is mathematically underdetermined. Each of the following sections describe ways of supplying some required extra information so that the problem becomes solvable.

13.4. THE ABSORPTION–DIFFRACTION METHOD

In the *absorption–diffraction method*, Equation 13.2 is written twice—once for line (hkl) of phase α in the unknown, and again for the same (hkl) line of a pure sample of phase α. The ratio here gives

$$\frac{I_{(hkl)\alpha}}{I^0_{(hkl)\alpha}} = \frac{(\mu/\rho)_\alpha}{(\mu/\rho)_s} X_\alpha, \tag{13.4}$$

where $I^0_{(hkl)\alpha}$ is the intensity of line (hkl) in pure phase α. Equation 13.4 is the basis of the absorption–diffraction method for quantitative analysis. There are several methods for implementing Equation 13.4, and a few cases will be discussed in some detail. Each of these cases is dependent only on the validity of Equation 13.4, and since no assumptions have been made in the derivation, this equation is rigorous. It is sometimes the case that quantitative X-ray diffraction phase analysis is carried out on samples whose chemical composition is known, perhaps from X-ray fluorescence analysis. When this is the case, the mass attenuation coefficients of the pure phase to be analyzed and that of the mixture can be computed from Equation 13.3. Thus, if $I_{(hkl)\alpha}$ and $I^0_{(hkl)\alpha}$ are measured from the unknown and a pure sample of α, respectively, the weight fraction of the analyte (i.e., X_α) may be computed from Equation 13.4.

It can also happen that the mass attenuation coefficient of the mixture and the phase to be analyzed are the same. In this case, Equation 13.4 reduces to

$$\frac{I_{(hkl)\alpha}}{I^0_{(hkl)\alpha}} = X_\alpha, \tag{13.5}$$

since $(\mu/\rho)_\alpha$ exactly the same as $(\mu/\rho)_s$. In cases where there are mixtures of polymorphs, the absorption–diffraction method of analysis becomes very attractive. Common examples of this case are the analysis of cristobalite in a quartz and amorphous SiO_2 matrix, or the analysis of the cubic, tetragonal, and monoclinic forms of ZrO_2 in pure ZrO_2 bodies. Another common example is the determination of anatase in rutile. Titanium dioxide exists in three forms—rutile, anatase, and brookite—all of which analyze as TiO_2. Rutile and anatase have different properties when used as pigments in paint, paper-glazing materials, etc. (Brookite is stable only at certain water content levels and can generally be ignored for analytical purposes.) In the quality control of rutile, careful checks have to be made to ensure that the anatase content is within control limits. Figure 13.6 shows a section of the diffractograms, and it will be seen that the (110) rutile line and (101) anatase line are both sharp and well resolved. Thus, use of simple peak intensities with background correction is quite sufficient to give acceptable intensities. Since both rutile and anatase have the same mass attenuation coefficient and about the same density, and since the diffraction lines are not unduly broadened, there is a very simple relationship between the concentrations of rutile X_r and

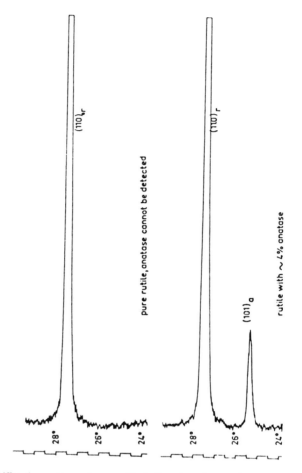

Figure 13.6. Diffraction pattern of pure rutile (left) and a mixture of rutile and anatase (right). From R. Jenkins and J. L. de Vries, *An Introduction to Powder Diffractometry*, p. 35, Fig. 46. Copyright © 1977, N. V. Philips, Eindhoven, The Netherlands.

anatase X_a and their background-corrected peak intensities I_r and I_a. Dividing Equation 13.5, written once for rutile and again for anatase, gives

$$\frac{X_r}{X_a} = \frac{I^0_{(101)a}}{I^0_{(110)r}} \frac{I_r}{I_a}. \qquad (13.6)$$

Note that the term $I^0_{(101)a}/I^0_{(110)r}$ is a measure of the absolute intensities of pure samples of rutile and anatase and will be described below as an RIR. For the moment, it is sufficient to recognize this as a constant that has been measured to be about 1.33 [12]. Thus, for mixtures of rutile and anatase

$$X_r = 1 - X_a,$$

from which it follows that

$$X_a = \frac{1}{1 + 1.33(I_r/I_a)}. \tag{13.7}$$

Using this value for K, as well as count data from the specimens shown in Figure 13.6, one can now easily estimate the concentration of anatase in the example. In this instance, the value for the net peak intensity of the rutile line is 36,176 counts and of the anatase line it is 1675 counts. Substitution of these data in Equation 13.7 gives a value of 3.4% anatase.

Another special case of the absorption–diffraction method is that of a lightly loaded filter where the sample approximates a monolayer and presents a special case of Equation 13.4. As long as all the particles lie alongside each other and do not shade each other from the X-ray beam, all specimen absorption effects are insignificant. There can, of course, be no preferential absorption. Each crystallite of each phase diffracts as if it were a pure phase. Thus, the attenuation coefficient for each phase is the same as for the pure phase, and a plot of $I_{(hkl)\alpha}/I^0_{(hkl)\alpha}$ vs. X_α will be linear (and immediately allows for quantitative analysis). This procedure is valid up to the limit of concentration of particles, where crowding invalidates the assumption that $(\mu/\rho)_\alpha = (\mu/\rho)_s$. This method is commonly used, for example, in the analysis of respirable silica in air-filtered samples collected on silver membranes.

13.4.1. Use of Klug's Equation

The treatment of a binary mixture is a special case in the use of the absorption–diffraction method. In this instance, Equation 13.2 becomes

$$I_{(hkl)\alpha} = \frac{K_e K_{(hkl)\alpha} X_\alpha}{\rho_\alpha [X_\alpha (\mu/\rho)_\alpha + X_\beta (\mu/\rho)_\beta]}. \tag{13.8}$$

Equation 13.2 for the pure phase α may be written as

$$I^0_{(hkl)\alpha} = \frac{K_e K_{(hkl)\alpha}}{\rho_\alpha (\mu/\rho)_\alpha}. \tag{13.9}$$

Dividing Equation 13.8 by Equation 13.9 gives

$$\frac{I_{(hkl)\alpha}}{I^0_{(hkl)\alpha}} = \frac{X_\alpha (\mu/\rho)_\alpha}{X_\alpha (\mu/\rho)_\alpha + X_\beta (\mu/\rho)_\beta}. \tag{13.10}$$

Since, in a binary mixture

$$X_\alpha + X_\beta = 1,$$

one can eliminate X_β from Equation 13.10:

$$\frac{I_{(hkl)\alpha}}{I^0_{(hkl)\alpha}} = \frac{X_\alpha(\mu/\rho)_\alpha}{X_\alpha[(\mu/\rho)_\alpha - (\mu/\rho)_\beta] + (\mu/\rho)_\beta}. \quad (13.11)$$

Equation 13.11 can be used to compute the standard plot of $I_{(hkl)\alpha}/I^0_{(hkl)\alpha}$ vs. X_α for all possible compositions of mixtures of α and β. A further rearrangement of Equation 13.11 allows its use for direct determination of X_α in binary and pseudobinary mixtures:

$$X_\alpha = \frac{(I_{(hkl)\alpha}/I^0_{(hkl)\alpha})(\mu/\rho)_\beta}{(\mu/\rho)_\alpha - (I_{(hkl)\alpha}/I^0_{(hkl)\alpha})[(\mu/\rho)_\alpha - (\mu/\rho)_\beta]}. \quad (13.12)$$

This form of the intensity equation is generally referred to as *Klug's equation*. Klug's equation allows the determination of a single phase in a binary mixture where values of *both* mass attenuation coefficients are known. For example, in the determination of SiO_2 in the $SiO_2/CaSiO_3$ system, the mass attenuation coefficients for Cu $K\alpha$ radiation of SiO_2, $(\mu/\rho)_\alpha$, and $CaSiO_3$, $(\mu/\rho)_\beta$, are 35.9 and 74.1, respectively. If a given specimen gave 4270 counts on a selected SiO_2 line and 10,000 counts for pure SiO_2 on the same line, Equation 13.12 can now be used to calculate the concentration of SiO_2 in the unknown specimen:

$$X_\alpha = \frac{0.427 \times 74.1}{35.9 - (0.427 \times (35.9 - 74.1))} = 60.6\%.$$

In Equation 13.12, the $(\mu/\rho)_\beta$ term can also be used to represent the mass attenuation coefficient of the *remainder* of the specimen (i.e., everything except the analyte phase). In this instance, the system is referred to as a *pseudobinary* system. There is an important difference between the Klug equation (13.12) and the absorption–diffraction method equation (13.4). The $(\mu/\rho)_s$ term in Equation 13.4 represents the *total* mass attenuation coefficient of the specimen, whereas in Equation 13.12 the $(\mu/\rho)_\beta$ term represents the mass attenuation coefficient of the *remainder* of the specimen. As an example, if we take the SiO_2 concentration calculated above for the unknown specimen, Equation 13.3 can now be used to calculate the mass attenuation coefficient of the *total* specimen:

$$(0.606 \times 35.9) \times (0.394 \times 74.1) = 50.96 \text{ cm}^2/\text{g}.$$

Substituting the mass attenuation coefficient and count data into Equation 13.4 gives

$$\frac{4270}{10000} \times \frac{50.96}{35.9} = 60.6\%.$$

13.4.2. Use of Measured Mass Attenuation Coefficients

Leroux et al. [13] were the first to propose that quantitative analysis might be performed on unknowns by experimentally measuring the density and determining the linear attenuation coefficients of the unknowns. This can indeed be done using absorption experiments on specimens of different thickness and employing the mass attenuation law given in Equation 1.10. However, the difficulty with this approach is that the measurement of the attenuation coefficient is somewhat prone to error and the low inherent precision of the measurement limits the accuracy of a quantitative analysis.

A variation on the absorption–diffraction method has been proposed by Norrish and his co-workers [14] in which the *mass thickness* ρx of the specimen is measured. Figure 13.7 illustrates the principle of this technique. A known weight of specimen is carefully pressed into the absorption cell. The cell is simply a piece of thick plastic into which a hole of known diameter has been cut. It will be seen from Equation 1.10 that in order to estimate the value

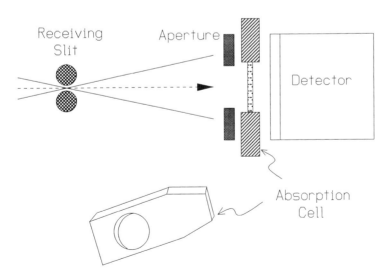

Figure 13.7. Measurement of the mass attenuation coefficient of a sample.

of (μ/ρ) from a measurement of $I_0(\lambda)$ (i.e., beam without absorber) and $I(\lambda)$ (i.e., beam with absorber in path), it is necessary to know both the sample thickness x and the sample density ρ. This product is called the *mass thickness*. However, since

$$\rho = \frac{\text{mass}}{\text{volume}} = \frac{\text{mass}}{\text{area} \times \text{thickness}}, \qquad (13.13)$$

we have

$$\text{mass thickness} = \frac{\text{mass}}{\text{area of hole}}. \qquad (13.14)$$

Hence, by use of a known weight of sample, pressed into a holder of known area, a value for ρx can be estimated. As shown in Figure 13.7, an aperture is placed between the receiving slit and the absorption cell to ensure that the beam passes only through the specimen. It is now necessary to measure the intensity of the beam with and without the absorber in place to allow the mass attenuation coefficient of the specimen to be derived.

13.4.3. Use of Mass Attenuation Coefficients Derived from Elemental Chemistry

It is often the case that the elemental analysis of a sample is known. For example, it commonly happens that chemicals are mixed together in known quantities and then reacted under conditions that do not permit loss of any of the components. In other cases, elemental chemical composition is determined as part of the characterization of samples. In either of these cases, Equation 13.3 permits the computation of $(\mu/\rho)_s$ and thus the general equation for the absorption–diffraction method, 13.4, may be used directly.

As an example, in the late 1970s [15] corings of more than 200 samples of the gas- and oil-bearing Devonian shales from the eastern United States were fully characterized in order to evaluate possible oil recovery strategies. Part of this project entailed the quantitative analysis of the full chemical composition as well as the phase composition. These shales typically contain such phases as quartz, feldspar, pyrite, siderite, and three clays—kaolinite, chlorite, and illite. To analyze such complex mixtures containing clays, which show extreme preferred orientation, the materials had to be spray-dried as described in Section 9.6.5. A particularly important aspect of this analysis was the development of standard specimens for the measurement of I_0. Mineral specimens often vary in solid solution content and crystallinity, causing considerable variation in the intensities of diffraction lines. It is essential that the standard being used in the absorption–diffraction method be carefully compared to the

Table 13.2. Absorption–Diffraction Analysis of a Spray-Dried Prepared Shale Composition

Phase	Measured	Prepared
Illite	45.1% ± 4.1	47.2%
Quartz	31.1% ± 1.9	34.1%
Feldspar	11.6% ± 1.7	11.4%
Chlorite	7.4% ± 0.6	7.3%

phase in the specimen being analyzed. For the shale study, the illite standard had to be isolated from the shale by sedimentation, because of the strong variation in illite's diffraction pattern with geographic location. Table 13.2 shows the results of the absorption–diffraction method, with $(\mu/\rho)_s$ values computed from the known elemental composition of a prepared specimen of typical shale composition.

13.5. METHOD OF STANDARD ADDITIONS

Lennox [5] was one of the first to develop the *spiking method*, which has been widely adopted in X-ray fluorescence spectrometry as the *method of standard additions*. This method, like the internal standard method, is perfectly general, applying to any phase α in a mixture. The only requirement is that the mixture also contain, among other phases, a phase β, which has a diffraction line not overlapped by any line from α and may be used as a reference. Phase β is not analyzed and does not even need to be an identified phase.

In a sample containing α and β, the ratio of the intensities of a line from each phase can be obtained from Equation 13.2 as,

$$\frac{I_{(hkl)\alpha}}{I_{(hkl)'\beta}} = \left(\frac{K_{(hkl)\alpha}}{K_{(hkl)'\beta}}\right)\left(\frac{\rho_\beta}{\rho_\alpha}\right)\left(\frac{X_\alpha}{X_\beta}\right). \tag{13.15}$$

Using the method of standard additions, some of the pure phase α is added to the mixture containing the unknown concentration of α. After adding Y_α grams of the α phase, per gram of the unknown, the ratio $I_{(hkl)\alpha}/I_{(hkl)'\beta}$ becomes,

$$\frac{I_{(hkl)\alpha}}{I_{(hkl)'\beta}} = \frac{K_{(hkl)\alpha}\rho_\beta(X_\alpha + Y_\alpha)}{K_{(hkl)'\beta}\rho_\alpha X_\beta}, \tag{13.16}$$

QUANTITATIVE ANALYSIS

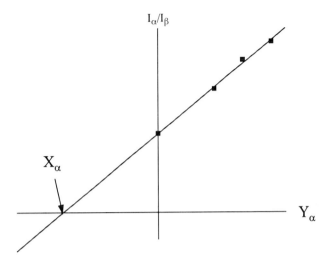

Figure 13.8. Spiking method analysis: plot of the ratio of I_α to the intensity of a reference line as a function of Y_α, the number of grams of α added per gram of sample.

where

- X_α = the initial weight fraction of phase α;
- X_β = the initial weight fraction of phase β;
- Y_α = the number of grams of pure phase α added per gram of the original sample.

In general, the intensity ratio is given by

$$\frac{I_{(hkl)\alpha}}{I_{(hkl)'\beta}} = K(X_\alpha + Y_\alpha), \tag{13.17}$$

where K is the slope of a plot of $I_{(hkl)\alpha}/I_{(hkl)'\beta} \cdot Y_\alpha$, with Y_α in units of grams of α per gram of sample. Multiple additions are made to prepare a plot like the one shown in Figure 13.8, in which the negative x intercept is X_α, the desired concentration of the phase α in the original sample.

13.6. THE INTERNAL STANDARD METHOD OF QUANTITATIVE ANALYSIS

The internal standard method is the most general of any of the methods for quantitative phase analysis by X-ray powder diffraction and by far the most commonly used. The method is based on elimination of the absorption factor $(\mu/\rho)_s$, by dividing two equations of the type shown in Equation 13.2, giving,

$$\frac{I_{(hkl)\alpha}}{I_{(hkl)'\beta}} = k\frac{X_\alpha}{X_\beta}, \qquad (13.18)$$

which is linear in the weight fraction of phase α. The k is the slope of the internal standard calibration curve derived from a plot of $I_{(hkl)\alpha}/I_{(hkl)'\beta}$ vs. X_α/X_β. Equation 13.18 forms the basis of the internal standard method of analysis. The addition of a known amount, X_β, of an internal standard to a mixture of phases (which may include amorphous material) permits quantitative analysis of each of the components of the mixture by first establishing the values of k for each phase from standards of known concentration. Use of a preestablished calibration curve defining the slope k permits the weight fraction of any phase α in the original mixture to be computed. Note that k is a function of (hkl), $(hkl)'$, α, and β for a constant-weight fraction of the internal standard.

Care must be taken in the selection of an internal standard for a given type of analysis. Clearly, the addition of a new phase to the analyte material means that the complexity of the pattern will increase, which may cause peak measurement problems. In general, any pure phase may act as an internal standard; but in practice an internal standard needs to have one or more fully resolved lines not overlapping any lines in the analyte phases, and be free from microabsorption (i.e., be of very small particle size or have an absorption similar to the specimen) and extinction. Usually F-centered cubic materials with small unit cells are chosen, due to their simple patterns. Sometimes the analysis of a single phase in a rock or a mineral may have a rather large particle size and a severe propensity to orient. In this case, a common error is to add an internal standard chosen from the chemical storeroom that has been made by chemical precipitation and has a very fine particle size. On mixing, the finely distributed internal standard coats the outside of the larger particles of the specimen, creating local heterogeneity leading to problems far worse than the original absorption problem. For this and many other reasons, previously discussed, it is essential to reduce the crystallite size of specimens into the 1 μm range.

The internal standard method lends itself most easily to generalization into the RIR (reference intensity ratio) method (see Section 13.6.2) and, even further, can be generalized into a system of linear equations that allows for the use of overlapped lines and chemical analysis constraints (see Section 13.6.5).

13.6.1. $I/I_{corundum}$ and the Reference Intensity Ratio Method

It is clear from Equation 13.18 that a plot of

$$X_\beta \left(\frac{I_{(hkl)\alpha}}{I_{(hkl)'\beta}} \right) \quad \text{vs.} \quad X_\alpha$$

will be a straight line with slope k; therefore k is a measure of the inherent diffracted intensities of the two phases. Visser and de Wolff [16] were the first to propose that k values could be published as material constants by defining the reference phase (β) to be corundum in a 50:50 wt mixture with phase α and using the hkl's of the 100% intensity lines. This $I/I_{corundum}$ (or I/I_c) value has been widely accepted and is now published with patterns for many of the phases in the ICDD Powder Diffraction File.

The simple two-line procedure for measuring I/I_c is quick but suffers from several drawbacks: preferred orientation commonly affects the observed intensities unless careful sample preparation methods are employed; other problems include extinction, inhomogeneity of mixing, and variable crystallinity of the sample due to its synthesis and history. All of these effects conspire to make the published values of I/I_c subject to substantial error. For greater accuracy, multiple lines from both the sample (α) and the reference phase (corundum, in this case) should be used [17]. Such an approach often reveals when preferred orientation is present and provides realistic measurement of the reproducibility in the measurement of I/I_c.

13.6.2. The Generalized Reference Intensity Ratio

The concept of the I/I_c value as a materials constant leads naturally to a broader definition permitting the use of reference phases other than corundum, lines other than just the 100% relative intensity line, and arbitrary concentrations of the two phases α and β. The most general definition of the reference intensity ratio (RIR) [18] can be given as

$$\text{RIR}_{\alpha,\beta} = \left(\frac{I_{(hkl)\alpha}}{I_{(hkl)'\beta}} \right) \left(\frac{I^{rel}_{(hkl)'\beta}}{I^{rel}_{(hkl)\alpha}} \right) \left(\frac{X_\beta}{X_\alpha} \right). \quad (13.19)$$

Equation 13.19 permits the computation of an RIR from any combination of lines in a two-phase mixture of any composition, with hkl (hkl)', and X_β as variables rather than fixed by definition. The RIR of course remains the slope of the calibration curve for phase α with internal standard β but now has been normalized so that it may be computed from any pair of diffraction lines in a calibration mixture. It is also seen that I/I_c values are simply RIR values

THE INTERNAL STANDARD METHOD OF QUANTITATIVE ANALYSIS

where β is corundum. The RIR, when accurately measured, is a true constant and allows comparison of the absolute diffraction line intensities of one material with another [19,20]. It also enables quantitative phase analysis in a number of convenient and useful ways.

13.6.3. Quantitative Analysis with RIRs

The equation for the internal standard method of quantitative analysis, stated in terms of an RIR, is obtained by rearranging Equation 13.19 thus:

$$X_\alpha = \left(\frac{I_{(hkl)\alpha}}{I_{(hkl)'\beta}}\right)\left(\frac{I^{rel}_{(hkl)'\beta}}{I^{rel}_{(hkl)\alpha}}\right)\left(\frac{X_\beta}{RIR_{\alpha,\beta}}\right). \quad (13.20)$$

The RIR value in Equation 13.20 may be obtained through careful calibration, by determining the slope of the internal standard plot, or by derivation from other RIR values via

$$RIR_{\alpha,\beta} = \frac{RIR_{\alpha,\gamma}}{RIR_{\beta,\gamma}}, \quad (13.21)$$

where γ is any common reference phase. When γ is corundum, the RIR values are simply I/I_c values. Hence, one can combine I/I_c values for phases α and β to obtain the RIR for phase α relative to phase β. Taking β to be an internal standard and substituting Equation 13.21 into Equation 13.20 and then rearranging gives

$$X_\alpha = \left(\frac{I_{(hkl)\alpha}}{I_{(hkl)'\beta}}\right)\left(\frac{I^{rel}_{(hkl)'\beta}}{I^{rel}_{(hkl)\alpha}}\right)\left(\frac{RIR_{\beta,c}}{RIR_{\alpha,c}}X_\beta\right). \quad (13.22)$$

This equation is quite general and allows for the analysis of any crystalline phase in an unknown mixture, as long as the RIR is known, by the addition of an internal standard. However, if all four of the required constants ($I^{rel}_{(hkl)\alpha}$, $I^{rel}_{(hkl)'\beta}$, $RIR_{\beta,c}$, and $RIR_{\alpha,c}$) are taken from the literature, the results should be considered as only semiquantitative, since each of them may contain significant error.

Equation 13.22 is valid even for complex mixtures that contain unidentified phases, amorphous phases, or identified phases with unknown RIRs.

13.6.4. The Normalized RIR Method

The ratio of the weight fractions of any two phases whose RIRs are known may always be computed by choosing one as X_α and the other as X_β. That is,

$$\left(\frac{X_\alpha}{X_\beta}\right) = \left(\frac{I_{(hkl)\alpha}}{I_{(hkl)'\beta}}\right)\left(\frac{I^{rel}_{(hkl)'\beta}}{I^{rel}_{(hkl)\alpha}}\right)\left(\frac{RIR_{\beta,c}}{RIR_{\alpha,c}}\right). \quad (13.23)$$

If the RIR values for all n phases in a mixture are known, then $n - 1$ equations of the form shown in Equation 13.23 may be written. It has been pointed out by several authors [19, 21–23] that if no amorphous phases are present, the additional normalization equation holds:

$$\sum_{j=1}^{n} X_j = 1. \quad (13.24)$$

Equation 13.24 permits analysis without addition of any standard to the unknown specimen by allowing us to write a system of n equations to solve for the n weight fractions via

$$X_\alpha = \frac{I_{(hkl)\alpha}}{RIR_\alpha I^{rel}_{(hkl)\alpha}} \left[\frac{1}{\sum_{j=1}^{No.\ of\ phases} (I_{(hkl)'j}/RIR_j I^{rel}_{(hkl)'j})}\right]. \quad (13.25)$$

Chung [22] referred to the use of Equation 13.24 as the matrix flushing or adiabatic principle, but it is best referred to as the normalized RIR *method*. It is important to note that the presence of any amorphous or unidentified crystalline phase invalidates the use of Equation 13.25. Davis and Johnson [24] have discussed the use of additional correction terms in those cases where either there is an amorphous phase present or where the sample is less than "infinitely thick." Of course, any sample containing unidentified phases cannot be analyzed using the normalized RIR method in that the required RIRs will not be known.

When the RIR values are known from another source, for example, published I/I_c values, it may be tempting to use them along with the I^{rel} values from the PDF card to perform what some call a completely "standardless" quantitative analysis using Equation 13.25. These I^{rel} and I/I_c values are seldom accurate enough to be used directly in quantitative analysis. The analyst should accurately determine the relative intensities and RIR values for an analysis by careful calibration measurements. In fact, the word "standardless" is a misnomer. Standards are always required; however, the RIR method allows data to be used from the literature for semiquantitative analysis.

13.6.5. Constrained XRD Phase Analysis: Generalized Internal Standard Method

A quantitative analysis combining both X-ray diffraction and elemental analysis with a knowledge of the composition of the individual phases can yield results of higher precision and accuracy than are generally possible with only one kind of observation. The analysis becomes more complex when several phases in a mixture have similar compositions and/or potential compositional variability, but it is possible, with appropriate constraints during analysis, to place limits on the actual compositions of the constituent phases.

The most general formulation of these ideas has been described by Copeland and Bragg [7]. Their internal standard equations for multicomponent quantitative analysis, with possible line superposition and chemical constraints, take the form of a system of simultaneous linear equations. In terms of RIR values, each equation is of the form

$$\frac{I_n}{I_{(hkl)'\text{std}}} = \left(\frac{I^{\text{rel}}_{(hkl)1}}{I^{\text{rel}}_{(hkl)'\text{std}}} \text{RIR}_{1,\text{std}}\right) \frac{X_1}{X_{\text{std}}} + \left(\frac{I^{\text{rel}}_{(hkl)2}}{I^{\text{rel}}_{(hkl)'\text{std}}} \text{RIR}_{2,\text{std}}\right) \frac{X_2}{X_{\text{std}}}$$
$$+ \cdots + \left(\frac{I^{\text{rel}}_{(hkl)j}}{I^{\text{rel}}_{(hkl)'\text{std}}} \text{RIR}_{j,\text{std}}\right) \frac{X_j}{X_{\text{std}}} + \varepsilon, \quad (13.26)$$

where

- I_n is the intensity of a particular 2θ region, from the mixture of j phases, which may contain contributions from one or more lines from one or more phases in the mixture;
- $I_{(hkl)'\text{std}}$ is the intensity of a *resolved* line of the internal standard;
- X_j is the weight fraction of phase j in the mixture of sample plus the internal standard;
- ε is a least squares error term.

Line overlap is allowed for by having each intensity ratio $(I_n/I_{(hkl)'\text{std}})$ have contributions from multiple lines from multiple phases. As many terms as needed involving contributions to the intensity ratio are included in each linear equation. The Copeland–Bragg analysis results in n equations in j unknowns, which may be solved via least squares as demonstrated in Equation 13.26. Quantitative elemental data obtained from X-ray fluorescence analysis, for example, can be added to the system of equations without increasing the number of unknowns, making it overdetermined. Coupling both the elemental and diffraction data will result in more accurate quantitative analysis of multicomponent mixtures. More importantly, estimates of the standard deviation of the results will be more accurate. Both the line overlap and chemical analysis features of the Copeland–Bragg formalism may be incorporated into an automated quantitative system [8, 9].

Table 13.3. Constrained RIR Analysis of a Prepared Fly Ash with and without the Use of Overlapped Lines[a]

Phase	Theory X	Number Using Lines Resolved	X		Number Using All Lines	X	
Mullite	0.50	1	0.530	(20)	6	0.506	(22)
Quartz	0.15	1	0.171	(10)	4	0.166	(7)
Hematite	0.10	2	0.101	(3)	5	0.104	(5)
Glass[b]	0.25	—	0.198	(23)	—	0.224	(24)

[a] Numbers are weight fractions with the standard deviation of the least significant digit in parentheses.

The generalized RIR method with chemical constraints and line overlap was employed in an analysis of industrial fly ash conducted by Hubbard et al. [25]. Table 13.3 shows the results of the analysis of a prepared sample. The samples and standards were spray-dried to eliminate preferred orientation, and multiple lines were used from each phase. Silicon was used as the internal standard. The first analysis reported in Table 13.3 was determined only from the fully resolved lines. The values at the right were determined using a Copland–Bragg analysis with overlapped lines. This study is the most exacting one reported in the literature, with all aspects of the specimen preparation and measurement procedures being conducted well beyond that of a typical analysis. Therefore, the numbers reported in Table 13.3 [25] represent the best precision and accuracy one is likely to attain using quantitative X-ray powder diffraction without utilizing the full pattern.

13.7. QUANTITATIVE PHASE ANALYSIS USING CRYSTAL STRUCTURE CONSTRAINTS

Equation 13.1 permits the direct computation of diffraction intensities from phases with known crystal structures. The ability to calculate intensities has been used in the determination of retained austenite in steel. Austenite is a solid solution of α-Fe, having a cubic structure, containing carbon and possibly nickel, chromium, cobalt, and/or manganese. In the preparation of steel of relatively low carbon content, cooling of the melt yields austenite (γ-phase), which in turn transforms to the required martensite (α-phase). However, some of the austenite may be retained, depending on the cooling rate and initial melt composition, and this austenite may have detrimental effects on the steel produced.

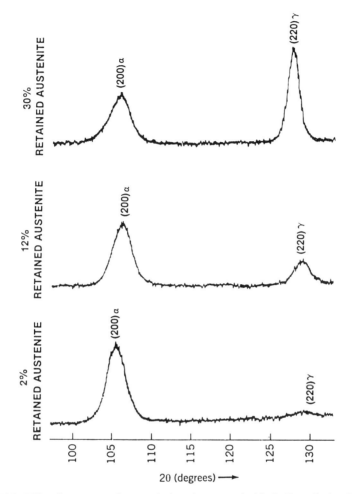

Figure 13.9. Diffraction pattern of a quenched steel, measured with Cr $K\alpha$ radiation. From R. Jenkins and J. L. de Vries, *An Introduction to Powder Diffractometry*, p. 36, Fig. 48. Copyright © 1977, N. V. Philips, Eindhoven, The Netherlands.

Figure 13.9 shows the diffraction pattern of a quenched 52100 steel, obtained using Cr $K\alpha$ radiation [26]. Here it can be seen that the (200) and (220) lines are well resolved, but due to poor crystallization and/or strain the lines are rather broad; thus, in this instance, integrated intensities must be used. After the counts are collected, one of several different approaches can be employed for the estimation of the concentration of retained austenite. The simplest method is to establish a calibration curve of net peak count rate as a function of austenite concentration by using a number of standard samples. An alternative approach is to use the known structural parameters of martensite and austenite and calculate the relative concentrations [13,26] of the two phases from first principles.

13.8. QUANTITATIVE METHODS BASED ON USE OF THE TOTAL PATTERN

In most of the previous examples, quantitative measurements have been based on a single peak or, at best, a small subset of all of the peaks in the diffraction pattern. With the recent rapid growth in the use of versatile, inexpensive personal computers, it is much easier to solve the rather complicated mathematics involved in working with *all* of the lines in the pattern, or even the *total* pattern, including the background. The availability of powerful pattern decomposition techniques and the wide acceptance of the Rietveld method [27] have provided the analyst with most of the tools for working with *total* data. While at the time of completing this manuscript the application of such techniques is still in its infancy, there is little doubt that quantitative methods based on the use of the full pattern will become the basis of a new era in quantitative phase analysis.

Complete digital diffraction patterns provide the opportunity to perform quantitative phase analysis using all data in a given pattern rather than considering only a few of the strongest reflections. As the name implies, full-pattern methods involve fitting the entire diffraction pattern with a synthetic diffraction pattern. This synthetic diffraction pattern can either be calculated and fit dynamically from crystal structure data [28,29] or can be produced from a combination of observed or calculated standard diffraction patterns [30].

Accuracies in conventional X-ray diffraction quantitative analysis range from a few percent, in the best cases, to errors of several tens of percent, in the worst. The application of structural constraints to the problem and the additional statistical gain from using the whole pattern have produced quantitative analysis accurate to 0.1% in recent years [29]. This is one of the most promising applications of Rietveld analysis.

13.8.1. The Rietveld Method

Quantitative phase analysis using calculated full patterns is a natural outgrowth of the *Rietveld method* [27], originally conceived as a method of refining crystal structures using neutron powder diffraction data. Refinement is conducted by minimizing the sum of the weighted, squared differences between observed and calculated intensities at every step in a digital powder pattern. The Rietveld method requires a knowledge of the approximate crystal structure of all phases of interest (not necessarily all phases present) in a mixture [28, 29].

The input data to a refinement are similar to those required to calculate a diffraction pattern, i.e., space group symmetry, atomic positions, site occupancies, and lattice parameters, as dictated by Equation 13.1 for each phase in a mixture. In addition, in order to match the calculated pattern point by point with the observed digital pattern, parameters defining the profile shape, a background function, and a scale factor for each phase are also required. In a typical refinement, individual scale factors and profile, background, and lattice parameters are varied. In favorable cases, the atomic positions and site occupancies can also be successfully varied. The method consists of fitting the complete experimental diffraction pattern with calculated profiles and backgrounds, and obtaining quantitative phase information from the scale factors for each phase in a mixture. The quantity minimized in Rietveld refinements is the conventional least squares residual:

$$R = \sum_j w_j |I_{j(o)} - I_{j(c)}|^2 \tag{13.27}$$

where $I_{j(o)}$ and $I_{j(c)}$ are the intensity observed and calculated, respectively, at the jth step in the data, and w_j is the weight. Thus, it is more appropriate to consider the intensity at a given 2θ step rather than for a given reflection. In a single-phase powder pattern of pure α, the intensity at each step j is determined by summing the contributions from background and all neighboring Bragg reflections as

$$I_{j(c)} = S_\alpha \sum_{(hkl)} K_{(hkl)\alpha} G(\Delta\theta_{j,(hkl)\alpha}) P_{(hkl)} + I_{b(c)}, \tag{13.28}$$

where S_α is the conventional Rietveld scale factor that puts the computed intensities on the same scale as those observed; $P_{(hkl)}$ is a sometimes-used preferred orientation function for the ith Bragg reflection; $G(\Delta\theta_{j,(hkl)\alpha})$ is the profile shape function; and $I_{jb(c)}$ is the background [31]. For a simple single-line, integrated, background-corrected intensity we then have

$$S_\alpha = \frac{I_{(hkl)\alpha}}{K_{(hkl)\alpha}}. \tag{13.29}$$

Comparison of Equation 13.29 with Equation 13.2 for pure phase α shows that the Rietveld scale factor S_α includes the constant term K_e in Equation 13.2

along with the attenuation coefficient from Equation 13.1:

$$S_\alpha = \frac{K_e}{\mu_\alpha}. \qquad (13.30)$$

For a multiphase mixture, Equation 13.28 can be rewritten summing over the n phases in a mixture as

$$I_{ic} = I_{ib} + \sum_n S_n \sum_j K_{jn} G_{ijn}. \qquad (13.31)$$

Again, when one compares this equation with Equation 13.2, the scale factor for each phase may be seen to be

$$S_\alpha = K_e \frac{X_\alpha}{\rho_\alpha (\mu/\rho)_s}. \qquad (13.32)$$

In addition, a comparison of Equations 13.32 and 13.2 shows that Equation 13.29 is still valid, even in a polyphase mixture. Thus, the Rietveld scale factor is related to the actual observed intensity as it has been affected by the mass attenuation coefficient of the mixture. Because the scale factor acts as an intensity, it may be used in any of the techniques for quantitative analysis described in this chapter. In fact, the Rietveld scale factor acts in the role of a reference intensity ratio, permitting conventional RIR analysis. Since the sample mass attenuation coefficient is not known, one is forced to apply the usual internal standard analysis, calibrating the RIRs of the phases to be analyzed or, applying the normalization assumption, constraining the sum of the weight fractions of the phases considered to unity.

Internal Standard Analysis Using the Rietveld Method. When a known quantity of an internal standard X_β is added to a polyphase mixture and the entire pattern is fit using the Rietveld method of Equation 13.31, the ratio of the scale factors for any phase α with respect to the internal standard can be obtained from Equation 13.32:

$$\frac{S_\alpha}{S_\beta} = \frac{K_e \dfrac{X_\alpha}{\rho_\alpha (\mu/\rho)_s}}{K_e \dfrac{X_\beta}{\rho_\beta (\mu/\rho)_s}}. \qquad (13.33)$$

Therefore, just as in the case of the derivation of the internal standard equation (i.e., 13.18), the mass attenuation coefficient of the specimen cancels out, permitting the determination of X_α from

$$X_\alpha = \frac{\rho_\alpha S_\alpha}{\rho_\beta S_\beta} X_\beta. \tag{13.34}$$

The total weight fraction of any amorphous components can also be determined with this method by adjusting the Rietveld background polynomial to fit the broad amorphous scattering profile. The difference between the sum of the weight fractions of the crystalline components and 1.0 is the total weight fraction of the amorphous components. For example, O'Connor and Raven [32] used this method to conclude that a given quartz sample contained an 18% amorphous component.

Normalized Internal Standard Analysis Using the Rietveld Method. Just as in the conventional internal standard method, the normalization equation 13.24 may be applied to the refined Rietveld scale factors. The weight fraction relationship to the scale factor may be obtained from Equation 13.32:

$$X_\alpha = \frac{(\mu/\rho)_s}{K_e} S_\alpha \rho_\alpha. \tag{13.35}$$

In order to apply the normalized RIR approach in the Rietveld method, one can write the explicit equation for a weight fraction in the normalization equation 13.24 as

$$X_\alpha = \frac{X_\alpha}{X_\alpha + X_\beta + \cdots}. \tag{13.36}$$

On substituting Equation 13.35 into Equation 13.36, it is seen that the mass attenuation coefficient of the specimen cancels out just as it did previously, giving

$$X_\alpha = \frac{S_\alpha \rho_\alpha}{\sum_n S_n \rho_n}. \tag{13.37}$$

It will be seen that this procedure is exactly analogous to the normalization assumption in which RIRs are measured prior to analysis. Instead of measuring RIRs to put all intensities on an absolute scale, the Rietveld method calculates what may be called normalized RIR_N values in that they refer to the relative scale between the observed and calculated patterns. Comparison of

Equations 13.37 and 13.25 shows that the relationship between the Rietveld scale factor and the normalized RIR is

$$S_\alpha \rho_\alpha = \frac{1}{\text{RIR}_{N,\alpha}} \frac{I_{(hkl)\alpha}}{I^{\text{rel}}_{(hkl)\alpha}}. \tag{13.38}$$

The use of these normalized RIR Rietveld scale factors is as close as quantitative analysis can come to being "standardless" in that the use of Equation 13.37 permits the computation of weight fractions without the addition of an actual internal standard to the mixture. However, the crystal structure model used in computing normalized RIRs effectively acts as a standard. The disagreement between the crystal structure models and the observed pattern, as measured by the conventional least squares residual error (Equation 13.27), directly translates into error in the computed weight fractions; as such, the method should be treated as semiquantitative, except in cases when amorphous phases are known to be absent and the refinement converges to a low residual error.

In spite of the qualifications that must be made concerning the use and validity of the normalized Rietveld method, this technique offers an overwhelming number of advantages recommending its use. Among the most important of these are the following:

- Differences between the experimental standard and the phase in the unknown are minimized. This is particularly important when one is using mineral specimens with a high degree of variability.
- Phase-pure standards need not be available for the analysis. It is commonly the case that experimental standards are not obtainable.
- Overlapped lines and heavily overlapped patterns may be used with no difficulty.
- Lattice parameters for each phase are automatically produced, allowing the evaluation of solid solution effects in the phase.
- The use of the whole pattern rather than a few selected lines produces an accuracy and precision in the absence of amorphous phases that is significantly better than traditional methods.
- Preferred orientation effects are averaged over all of the crystallographic directions represented by the diffraction lines, and simple orientation may be modeled during the refinement.

As an example of the use of this method, D. L. Bish and J. E. Post (see Snyder and Bish [4]) analyzed a mixture of two volcanic feldspars with severely overlapped patterns. Figure 13.10 shows the observed and calculated

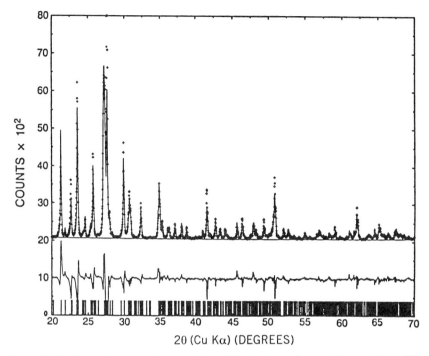

Figure 13.10. Observed (as points) and calculated (line) patterns for a sanidine and high albite mixture. The lower curve shows the difference between the observed and calculated patterns. Vertical marks at bottom indicate the positions of allowed lines.

patterns for the mixture of sanidine and high albite; the weighted pattern residual error (R_{wp}) is relatively good at 26.6%, considering the complexity of the pattern and the large number of reflections. Rietveld analysis yielded 93% sanidine and 7% high albite, agreeing with the results of an optical point count estimate of the true phase content that used 100 grains.

13.8.2. Full-Pattern Fitting with Experimental Patterns

The use of the whole pattern [30] involves collection of standard data on pure materials under fixed instrumental conditions. These standard patterns are processed to remove background and any artifacts, and the data may be smoothed. For standard materials that are unavailable in appropriate form, diffraction patterns can either be simulated from PDF data or calculated powder patterns. The next step in standardization is analogous to procedures used in conventional RIR quantitative analyses. Data are collected on each of the standard samples that have been mixed with a known amount of corundum, α-Al_2O_3, in order to determine the RIR. These patterns yield the basic calibration data used in the quantitative analysis. The final step involves collection of data for the unknown samples to be analyzed, under conditions identical to those used in obtaining the standard data. Background and artifacts are also removed from unknown-sample data.

RIRs are obtained from the whole patterns by initially assigning the standard material an RIR of 1.0. Then the values of "apparent" X_α and X_{std} are computed from Equation 13.22. The ratio of the "apparent" weight fractions of X_α to X_{std} gives the correct RIR value. For example [30], a standard mixture of 50% ZnO and 50% corundum yields results of 57.8% ZnO and 42.2% corundum. The RIR for ZnO is therefore $57.8/42.2 = 1.37$.

The digital patterns are fit to the standards with a least squares procedure, minimizing the expression

$$\Delta(2\theta) = I_{unk}(2\theta) - \sum_n X_n RIR_n I_n(2\theta), \qquad (13.39)$$

where $I_{unk}(2\theta)$ and $I_n(2\theta)$ are the diffraction intensities at each 2θ interval for the unknown and for each of the standard phases, respectively. Intensity ratios derived using this procedure are equivalent to peak height RIRs, rather than integrated-intensity RIRs, because the minimization is conducted on a step-by-step basis. This whole-pattern fitting procedure is simply a method for measuring RIRs from whole-pattern standards. Once this is accomplished, the method becomes a conventional RIR quantitative analysis procedure, which requires standards and calibration. The normalization assumption (Equation 13.24) may, of course, be used to eliminate the need for adding an internal standard.

13.9. DETECTION OF LOW CONCENTRATIONS

One question that may be asked in some quantitative analysis procedures is just what is the lower limit of detectability to which a given method is applicable. A commonly used criterion for the lower limit of detection is "that concentration equivalent to two standard deviations of the background" [33]. The ability to distinguish a peak standing above background depends on the number of counts in both the background and the peak. Whenever a peak is suspected in a low signal-to-background ratio situation, one should only rely on objective statistical analysis of the measurement and not on dangerous subjective evaluations. It may be seen in Equation 8.5 that as the divisor $N_p - N_b$ approaches zero, the counting error becomes infinite. It should also

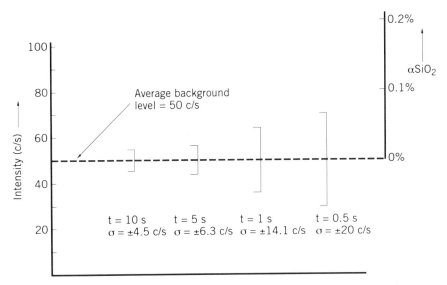

Figure 13.11. Statistical factors in the measurement of signal above background.

be apparent that as the value of N_b approaches the value of N_p, the net counting error increases. Since the number of counts accumulated, either in peak or background, is the product of the counting rate and the count time (see Equation 5.6), the lower limit of detection will clearly depend upon the square root of the total count time available.

As an example, the probability of detecting a signal above background is illustrated in Figure 13.11. In this example, the average value of the background is 50 c/s and the 2σ (95% probability) errors are shown for $t = 10, 5, 1,$ and 0.5 s, respectively. With an integration time of, for example, 5 s, any count datum greater than 55.3 c/s (i.e., 6.3 c/s above the background) would be statistically significant. To observe a count datum of only 4.5 c/s above background would require a count time of 10 s. In order to express these data in terms of phase concentration, one must convert the counts/second to percent by dividing by the sensitivity of the diffractometer for the appropriate phase in counts per second per unit concentration. For example, if one were determining the amount of α-SiO_2 in an airborne dust sample, and a standard containing 5% of α-SiO_2 gave 1550 c/s at the peak position of the (101) line, with background of 50 c/s, then for α-SiO_2 under these conditions one would have $(1500 - 50)/5 = 300$ c/s per unit concentration. In the example shown in Figure 13.11, this would correspond to a detection limit (2σ) of 0.015% for 10 s, 0.021% for 5 s, 0.047% for 1 s, and 0.067% for 0.5 s.

REFERENCES

1. Navias, A. L. Quantitative determination of the development of mullite in fired clays by and X-ray method. *J. Am. Ceram. Soc.* **8**, 296–302 (1925).
2. Clark, G. L., and Reynolds, D. H. Quantitative analysis of mine dusts. *Ind. Eng. Chem., Anal. Ed.* **8**, 36–42 (1936).
3. Alexander, L., and Klug, H.P. X-ray diffraction analysis of crystalline dusts. *Anal. Chem.* **20**, 886–894 (1948).
4. Snyder, R. L., and Bish, D. L. Quantitative analysis by X-ray powder diffraction. In *Modern Powder Diffraction* (D. L. Bish and J. E. Post, eds.), Rev. Mineral. Vol. 20, pp. 101–145. Mineral. Soc. Am., Washington, DC, 1989.
5. Lennox, D. H. Monochromatic diffraction absorption technique for direct quantitative X-ray analysis. *Anal. Chem.* **29**, 767–772 (1957).
6. Klug, H. P., and Alexander, L. E. *X-Ray Diffraction Procedures*, 2nd ed. Wiley, New York, 1974.
7. Copeland, L. E., and Bragg, R. H. Quantitative X-ray diffraction analysis. *Anal. Chem.* **30**, 196–206 (1958).
8. Snyder, R. L., and Hubbard, C. R. NBS*QUANT84: A System for Quantitative Analysis by Automated Powder Diffraction. National Bur. Stand. Spec. Publ., Washington, DC, 1981.
9. Snyder, R. L., Hubbard, C. R., and Panagiotopoulos, N. C. A second generation automated powder diffractometer control system. *Adv. X-Ray Anal.* **25**, 245–260 (1982).
10. Cline, J. P., and Snyder, R. L. The effects of extinction on X-ray powder diffraction intensities. *Adv. X-Ray Anal.* **30**, 447–456 (1987).
11. Gehringer, R. C., McCarthy, G. J., Garvey, R. G., and Smith D. K. X-ray diffraction intensity of oxide solid solutions: Application to qualitative and quantitative phase analysis. *Adv. X-Ray Anal.* **26**, 119–128 (1983).
12. Jenkins, R., and de Vies, J. L. *An Introduction to powder Diffractometry*. Philips, Eindhoven, The Netherlands, 1977 (No. 7000.02.3770.11).
13. Leroux, J., Lennox, D. H., and Kay, K. Applications of X-ray diffraction analysis in the environmental field. *Anal. Chem.* **25**, 740–748 (1953).
14. Sweatman, I. R., Norrish, K., and Durie, R. A. Misc. Rep. Div. Coal Res., CSIRO 177. Commonwealth Scientific and Industrial Research Organization, Adelaide, Australia, 1963.
15. Smith, S. T., Snyder, R. L., and Brownell, W. E. Quantitative phase analysis of Devonian shales by computer-controlled X-ray diffraction of spray dried samples. *Adv. X-Ray Anal.* **22**, 181–191 (1979).
16. Visser, J. W., and de Wolff, P. M. *Absolute Intensities*, Rep. 641.109. Technische Physische Dienst, Delft, The Netherlands, 1964.
17. Snyder, R. L. The use of reference intensity ratios in X-ray quantitative analysis. *Powder Diffr.* **7**, 186–193 (1992).

18. Hubbard, C. R., and Snyder, R. L. Reference intensity ratio—measurement and use in quantitative XRD. *Powder Diffr.* **3**, 74–78 (1988).
19. Chung, F. H. Quantitative interpretation of X-ray diffraction patterns. I. Matrix-Flushing method of quantitative multicomponent analysis. *J. Appl. Crystallogr.* **7**, 519–525 (1994).
20. Hubbard, C. R., Evans, E. H., and Smith, D. K. The Reference Intensity Ratio, I/I_c for computer simulated patterns. *J. Appl. Crystallogr.* **9**, 169–174 (1976).
21. Karlak, R. F., and Burnett, D. S. Quantitative phase analysis by X-ray diffraction. *Anal. Chem.* **38**, 1741–1745 (1966).
22. Chung, F. H. Quantitative interpretation of X-ray diffraction patterns. II. Adiabatic principle of X-ray diffraction analysis of mixtures. *J. Appl. Crystallogr.* **7**, 526–531 (1974).
23. Chung, F. H. Quantitative interpretation of X-ray diffraction patterns. III. Simultaneous determination of a set of reference intensities. *J. Appl. Crystallogr.* **8**, 17–19 (1975).
24. Davis, B. L., and Johnson, L. R. Sample preparation and methodology for X-ray quantitative analysis of thin aerosol layers deposited on glass fiber and membrane filters. *Adv. X-Ray Anal.* **25**, 295–300 (1982).
25. Hubbard, C. R., Robbins, C. R., and Snyder, R. L. (1983) XRD quantitative analysis using the NBS*QUANT82 system. *Adv. X-Ray Anal.* **26**, 149–157 (1983).
26. Ogilvie, R. E. Retained austenite by X-rays. *Norelco Rep.* **6**, 60–61 (1959).
27. Rietveld, H. M. A profile refinement method for nuclear and magnetic structures. *J. Appl. Crystallogr.* **2**, 65–71 (1969).
28. Hill, R. J., and Howard, C. J. Quantitative phase analysis from neutron powder diffraction data using the Rietveld method. *J. Appl. Crystallogr.* **20**, 467–474 (1987).
29. Bish, D. L., and Howard, S. A. (1988) Quantitative phase analysis using the Rietveld method. *J. Appl. Crystallogr.* **21**, 86–91 (1988).
30. Smith, D. K., Johnson, G. G., Jr., Scheible, A., Wims, A. N., Johnson, J. L., and Ullman, G. Quantitative X-ray powder diffraction using the full diffraction pattern. *Powder Diffr.* **2**, 73–77 (1987).
31. Wiles, D. B., and Young, R. A. A new computer program for Rietveld analysis of X-ray powder diffraction patterns. *J. Appl. Crystallogr.* **14**, 149–151 (1981).
32. O'Connor, B. H., and Raven, M. D. Application of the Rietveld refinement procedure in assaying powdered mixtures *Powder Diffr.* **3**, 2–6 (1988).
33. Kaiser, H., and Menzies, A. C. *The Limit of Detection of a Complete Analytical Procedure.* Adam Hilger, London, 1968.

APPENDICES

APPENDIX A: COMMON X-RAY WAVELENGTHS

The following values are taken from J. A. Bearden, *Rev. Mod. Phys.* **39**, 78, (1967), as reported in *International Tables for X-Ray Crystallography*, Vol. IV. Kynoch Press, Birmingham, England, 1974. The values are reported in Å* units, which are equal to 1 Å to 5 parts per million. As we explained in Chapter 1, current thinking suggests a value of 1.54060 for the Cu $K\alpha_1$ line.

Element	$K\alpha_1$	$K\alpha_1$	$K\beta$	K_{edge}
Cr	2.28970	2.293606	2.08487	2.07020
Fe	1.936042	1.939980	1.75661	1.74346
Co	1.788965	1.792850	1.62079	1.60815
Cu	1.540562	1.544390	1.392218	1.38059
Mo	0.709300	0.713590	0.632288	0.61978
Ag	0.5594075	0.563798	0.497069	0.48589

APPENDIX B: MASS ATTENUATION COEFFICIENTS

The table below lists values of the mass attenuation coefficients (μ/ρ) in square centimeters per gram for some of the common wavelengths. The data is abstracted from *International Tables for X-Ray Crystallography*, Vol. III. Kynoch Press, Birmingham, England, 1962.

Z		Cu Kα	Cr Kα	Fe Kα	Mo Kα	Z		Cu Kα	Cr Kα	Fe Kα	Mo Kα
1	H	0.435	0.545	0.483	0.380	51	Sb	270	288	472	33.1
2	He	0.383	0.813	0.569	0.207	52	Te	282	707	490	35.0
3	Li	0.716	1.96	1.25	0.217	53	I	294	722	506	37.1
4	Be	1.50	4.50	2.80	0.298	54	Xe	306	753	521	39.2
5	B	2.39	7.38	4.55	0.392	55	Cs	318	793	534	41.3
6	C	4.60	14.5	8.90	0.625	56	Ba	330	461	546	43.5
7	N	7.52	23.9	14.6	0.916	57	La	341	202	557	45.8
8	O	11.5	36.6	22.4	1.31	58	Ce	352	219	601	48.2
9	F	16.4	52.4	32.1	1.80	59	Pr	363	236	359	50.7
10	Ne	22.9	72.8	44.6	2.47	60	Nd	374	252	379	53.2
11	Na	30.1	95.3	58.6	3.21	61	Pm	386	268	172	55.9
12	Mg	38.6	121	74.8	4.11	62	Sm	397	284	182	58.6
13	Al	48.6	152	93.9	5.16	63	Eu	425	299	193	61.5
14	Si	60.6	189	117	6.44	64	Gd	439	314	203	64.4
15	P	74.1	229	142	7.89	65	Tb	273	329	214	67.5
16	S	89.1	272	170	9.55	66	Dy	286	344	224	70.6
17	Cl	106	318	200	11.4	67	Ho	128	359	234	73.9
18	Ar	123	366	232	13.5	68	Er	134	373	245	77.3
19	K	143	417	266	15.8	69	Tm	140	387	255	80.8
20	Ca	162	468	299	18.3	70	Ym	146	401	265	84.5
21	Sc	184	513	336	21.1	71	Lu	153	416	276	88.2
22	Ti	208	571	377	24.2	72	Hf	159	430	286	91.7
23	V	233	68.4	419	27.5	73	Ta	166	444	297	95.4
24	Cr	260	79.8	463	31.1	74	W	172	458	308	99.1
25	Mn	285	93.0	57.2	34.7	75	Re	179	473	319	103
26	Fe	308	108	66.4	38.5	76	Os	186	487	330	106
27	Co	313	125	76.8	42.5	77	Ir	193	502	341	110
28	Ni	45.7	144	88.6	46.6	78	Pt	200	517	353	113
29	Cu	53.0	166	103	50.9	79	Au	208	532	365	115
30	Zn	60.3	189	117	55.4	80	Hg	216	547	377	117
31	Ga	55.9	212	131	60.1	81	Tl	224	563	389	119
32	Ge	75.6	235	146	64.8	82	Pb	232	579	402	120
33	As	83.4	258	160	69.7	83	Bi	240	596	415	120
34	Se	91.4	281	175	74.7						
35	Br	99.6	305	190	79.8						
36	Kr	108	327	206	84.9						
37	Rb	117	351	221	90.0						
38	Sr	125	373	236	95.0						
39	Y	134	396	252	100.						
40	Zr	145	419	265	13.9						
41	Nb	153	441	284	17.1						
42	MO	162	163	300	18.4						
43	Tc	172	485	316	19.7						
44	Ru	183	509	334	21.1						
45	Rh	194	534	352	22.6						
46	Pd	206	559	371	24.1						
47	Ag	210	586	391	25.8						
48	Cd	281	613	412	37.5						
49	In	243	638	432	29.3						
50	Sn	256	662	451	31.1						

APPENDIX C: ATOMIC WEIGHTS AND DENSITIES

Z		Atomic Weight	Density (g/cm^3)	Z		Atomic Weight	Density (g/cm^3)
1	H	1.008	0.0000838	51	Sb	121.75	6.69
2	He	4.002	0.0001664	52	Te	127.60	6.25
3	Li	6.941	0.533	53	I	126.904	4.95
4	Be	9.012	1.85	54	Xe	131.29	0.005495
5	B	10.811	2.47	55	Cs	132.905	1.91
6	C	12.011	2.27	56	Ba	137.327	3.59
7	N	14.007	0.001165	57	La	138.906	6.17
8	O	15.999	0.001332	58	Ce	140.115	6.77
9	F	18.998	0.001696	59	Pr	140.908	6.78
10	Ne	20.180	0.0008387	60	Nd	144.24	7.00
11	Na	22.990	0.966	61	Pm	(147)	
12	Mg	24.305	1.74	62	Sm	150.36	7.54
13	Al	26.982	2.70	63	Eu	151.965	5.25
14	Si	28.086	2.33	64	Gd	157.25	7.87
15	P	30.974	1.82	65	Tb	158.925	8.27
16	S	32.066	2.09	66	Dy	162.50	8.53
17	Cl	35.452	0.003214	67	Ho	164.930	8.80
18	Ar	39.948	0.001663	68	Er	167.26	9.04
19	K	39.098	0.862	69	Tm	168.934	9.33
20	Ca	40.078	1.53	70	Yb	173.04	6.97
21	Sc	44.956	2.99	71	Lu	174.967	9.84
22	Ti	47.88	4.51	72	Hf	178.49	13.28
23	V	50.942	6.09	73	Ta	180.948	16.67
24	Cr	51.996	7.19	74	W	183.85	19.25
25	Mn	54.938	7.47	75	Re	186.207	21.02
26	Fe	55.847	7.87	76	Os	190.2	22.58
27	Co	58.933	8.8	77	Ir	192.22	22.55
28	Ni	58.69	8.91	78	Pt	195.08	21.44
29	Cu	63.546	8.93	79	Au	196.967	19.28
30	Zn	65.39	7.13	80	Hg	200.59	13.55
31	Ga	69.723	5.91	81	Tl	204.383	11.87
32	Ge	72.61	5.32	82	Pb	207.2	11.34
33	As	74.922	5.78	83	Bi	208.980	9.80
34	Se	78.96	4.81	84	Po	(210)	
35	Br	79.904	3.12	85	At	(210)	
36	Kr	83.80	0.003488	86	Rn	(222)	4.40
37	Rb	85.468	1.59	87	Fr	(223)	
38	Sr	87.62	2.68	88	Ra	226.026	5
39	Y	88.906	4.48	89	Ac	(227)	
40	Zr	91.224	6.51	90	Th	232.038	11.72
41	Nb	92.906	8.58	91	Pa	231.036	15.37
42	Mo	95.94	10.22	92	U	238.028	19.05
43	Tc	98.906	11.50	93	Np	237,048	20.45
44	Ru	101.07	12.36	94	Pu	(242)	19.81
45	Rh	102.906	12.42	95	Am	(245)	
46	Pd	106.42	12.00	96	Cm	(247)	
47	Ag	107.868	10.50	97	Bk	(247)	
48	Cd	112.411	8.68	98	Cr	(252)	
49	In	114.82	7.29	99	Es	(254)	
50	Sn	118.710	7.29	100	Fm	(259)	

APPENDIX D: CRYSTALLOGRAPHIC CLASSIFICATION

CRYSTAL SYSTEMS			CUBIC			
SPACE (BRAVAIS) LATTICES			PRIMITIVE		BODY	FACE
INTERNATIONAL ABBREVIATIONS			P		I	F
CRYSTAL CLASSES (ROGERS)			HEXOCTAHEDRAL	HEXTETRAHEDRAL	GYROIDAL	DIPLOIDAL	TETARTOIDAL
POINT GROUPS	INTERNATIONAL (HERMANN-MAUGUIN)		$\left(\frac{4}{m}\bar{3}\frac{2}{m}\right)$ m3m	$\bar{4}3m$	432	$\left(\frac{2}{m}\bar{3}\right)$ m3	23
	SCHOENFLIES		O_h	T_d	O	T_h	T
SPACE GROUPS	POINT SPACE GROUPS		•Pm3m ×Im3m √Fm3m	•P$\bar{4}$3m ×I$\bar{4}$3m √F$\bar{4}$3m	•P432 ×I432 √F432	•Pm3 ×Im3 √Fm3	•P23 ×I23 √F23
	GLIDE PLANE	P					
		C					
	SCREW AXIS	P			(432) •$4_1$32 $4_2$32 •$4_3$32	(m3)	(23) $2_1$3
		I			(432) $4_1$32		×(23) $2_1$3
		F			(432) $4_1$32		(23)
		C					
		R					
	GLIDE PLANE AND SCREW AXIS	P	(m3m) n3n n3m •m3n	($\bar{4}$3m) •$\bar{4}$3n		a3 n3	
		I	(m3m) a3d	($\bar{4}$3m) $\bar{4}$3d		(m3) a3	
		F	×(m3m) m3c d3m d3c	($\bar{4}$3m) ×$\bar{4}$3c		(m3) d3	
		C					
		A					
		R					
TOTAL			10	6	8	7	5

OF THE 230 SPACE GROUPS

	HEXAGONAL						TRIGONAL				
	BASE HEXAGONAL						PRIMITIVE RHOMBOHEDRAL			BASE HEXAGONAL	
	P						R			P	
	DIHEXAGONAL DIPYRAMIDAL	DITRIGONAL DIPYRAMIDAL	DIHEXAGONAL PYRAMIDAL	HEXAGONAL TRAPEZOHEDRAL	HEXAGONAL DIPYRAMIDAL	TRIGONAL DIPYRAMIDAL	HEXAGONAL PYRAMIDAL	HEXAGONAL SCALENOHEDRAL	DITRIGONAL PYRAMIDAL	TRIGONAL TRAPEZOHEDRAL	RHOMBOHEDRAL	TRIGONAL PYRAMIDAL
	$\left(\frac{6}{m}\frac{2}{m}\frac{2}{m}\right)$ $\frac{6}{m}mm$	$\bar{6}m2$	$6mm$	622	$\frac{6}{m}$	$\bar{6}$	6	$\left(\bar{3}\frac{2}{m}\right)$ $\bar{3}m$	$3m$	32	$\bar{3}$	3
	D_{6h}	D_{3h}	C_{6v}	D_6	C_{6h}	C_{3h}	C_6	D_{3d}	C_{3v}	D_3	(S_6) C_{3i}	C_3
	$P\frac{6}{m}mm$	$P\bar{6}m2$ $P\bar{6}2m$	$P6mm$	$P622$	$\bullet P\frac{6}{m}$	$\bullet P\bar{6}$	$\bullet P6$	$\bullet P\bar{3}m1$ $P\bar{3}1m$ × $R\bar{3}m$	$\bullet P3m1$ $P31m$ × $R3m$	$\bullet P312$ $\bullet P321$ × $R32$	$P\bar{3}$ $R\bar{3}$	$P3$ $R3$
		(6mm) 6cc						(3m1) (31m) √3c1 31c				
			(622) *6_122 *6_222 6_322 *6_422 *6_522	▼$6_3/m$		*6_1 *6_2 ▼6_3 *6_4 *6_5			(312) (321) *3_112 *3_212 *3_121 *3_221		*3_1 *3_2	
										(32)	(3)	(3)
(6/mmm) $\frac{6_3}{m}cm$ $\frac{6}{m}cc$ $\frac{6_3}{m}mc$	(6̄2m) (6̄m2) •6̄2c 6̄c2	6_3cm •6_3mc						(3̄1m) (3̄m1) √3̄c1 3̄1c				
								(3̄m) 3̄c	(3̄m) 3c			
4	4	4	6	2	1	6	6	6	7	2	4

(NOTE 1)

$(\frac{4}{m}mm)$ √ $\frac{4_2}{m}mc$
$\frac{4}{m}cc$ × $\frac{4_2}{m}cm$
$\frac{4}{n}bm$ $\frac{4_2}{n}bc$
$\frac{4}{n}nc$ $\frac{4_2}{n}nm$
$\frac{4}{n}bm$ $\frac{4_2}{n}bc$
○ $\frac{4}{n}nc$ ● $\frac{4_2}{n}nm$
$\frac{4}{n}mm$ $\frac{4_2}{n}mc$
$\frac{4}{n}cc$ $\frac{4_2}{n}cm$

(NOTE 2)

nma bam
bca ccn
mma bcm
nna bcn
mna mmn
cca bcn

BUERGER GROUPS

● ●
✱ ×
◆ √
▲ ○
▲ ▲
■ ■

ENANTIOMORPHIC PAIRS

✱ ✱

NOTE: SPACE GROUP SYMBOLS WHICH ARE CIRCLED ARE ALSO POINT SPACE GROUPS AND ARE THEREFORE LISTED TWICE.

(Cont'd.)

APPENDIX D (Cont'd.)

CRYSTAL SYSTEMS					TETRAGONAL					
SPACE (BRAVAIS) LATTICES					PRIMITIVE				BODY		
INTERNATIONAL ABBREVIATIONS					P				I		
CRYSTAL CLASSES (ROGERS)				DITETRAGONAL DIPYRAMIDAL	TETRAGONAL SCALENOHEDRAL	DITETRAGONAL PYRAMIDAL	TETRAGONAL TRAPEZOHEDRAL	TETRAGONAL DIPYRAMIDAL	TETRAGONAL DISPHENOIDAL	TETRAGONAL PYRAMIDAL
POINT GROUPS	INTERNATIONAL (HERMANN-MAUGUIN)			$\left(\frac{4}{m}\frac{2}{m}\frac{2}{m}\right)$ $\frac{4}{m}mm$	$\bar{4}2m$	$4mm$	422	$\frac{4}{m}$	$\bar{4}$	4
	SCHOENFLIES			D_{4h}	(V_d) D_{2d}	C_{4v}	D_4	C_{4h}	S_4	C_4
SPACE GROUPS	POINT SPACE GROUPS			$^*P\frac{4}{m}mm$ $I\frac{4}{m}mm$	$^*P\bar{4}2m$ $^*P\bar{4}m2$ $I\bar{4}2m$ $I\bar{4}m2$	$^*P4mm$ $I4mm$	$^*P422$ $I422$	$P\frac{4}{m}$ $I\frac{4}{m}$	$P\bar{4}$ $I\bar{4}$	$P4$ $I4$
	GLIDE PLANE	P			$(\bar{4}2m)$ $\bar{4}2c$	$(4mm)$ $4bm$ $4cc$ $\bigcirc 4nc$		$\frac{4}{n}$		
		C								
	SCREW AXIS	P			$(\bar{4}m2)$		4_12 4_22 $^*4_122$ $4_2 1_2$ 4_12_12 $^*4_32 2$ (422) $^*4_32_12$	$\frac{4_2}{m}$		$^*4_1$ 4_2 $^*4_3$
		I					$(\overline{422})$ 4_122		$(\bar{4})$	(4) 4_1
		F								
		C								
		R								
	GLIDE PLANE AND SCREW AXIS	P		(NOTE 1)	$\bar{4}2_1m$ $\bar{4}b2$ $\bar{4}2_1c$ $^\bullet\bar{4}n2$ $^X\bar{4}c2$	^X4_2cm $^\cdot4_2mc$ $^\bullet4_2nm$ 4_2bc		$\frac{4_2}{n}$		
		I		$(\frac{4_1}{a}mm)\frac{4}{m}cm$ $\frac{4_1}{a}md$ $\frac{4_1}{a}cd$	$(\overline{4m2})$ $\sqrt{4_2d}$ $(\overline{42m})$ $\bar{4}c2$	$(4mm)$ $^\checkmark 4_1md$ $4cm$ 4_1cd		$(\frac{4}{m})$ $\frac{4_1}{a}$		
		F								
		C								
		A								
		R								
TOTAL				20	12	12	10	6	2	6

ORTHORHOMBIC				MONOCLINIC		TRICLINIC	7
PRIMITIVE	ALL FACE	BODY	ONE FACE	PRIMITIVE	BASE	PRIMITIVE	14
P	F	I	C or A	P	C	P	
RHOMBIC DIPYRAMIDAL	RHOMBIC PYRAMIDAL	RHOMBIC DISPHENOIDAL	PRISMATIC	DOMATIC	SPHENOIDAL	PINACOIDAL	PEDIAL (ASYMMETRIC)	
$\left(\frac{222}{mmm}\right)$ mmm	(mm) $mm2$	222	$\frac{2}{m}$	m	2	$\bar{1}$	1	
(V_h) D_{2h}	C_{2v}	(V) D_2	C_{2h}	(C_{1h}) C_s	C_2	(S_2) C_i	C_1	
●Pmmm ×Immm √Fmmm ○Cmmm	●Pmm2 ×Imm2 √Fmm2 ○Amm2	●P222 ×I222 √F222 ○C222	●P$\frac{2}{m}$ C$\frac{2}{m}$	●Pm Cm	●P2 C2	P$\bar{1}$	P1	73
nnn ban ▲ccm	▲cc2 ma2 ba2 nc2 nn2		$\frac{2}{c}$	▼c				23
	$\overline{(mm2)}$ ■cc2			$\frac{\overline{m}}{×c}$				4
		222$_1$ 2,2,2$_1$ 2,2,2	√$\frac{2_1}{m}$		√2$_1$			45
		×$\overline{222}$ 2,2,2$_1$						11
		$\overline{222}$						4
		$\overline{222}$ 222$_1$			②			3
								3
(NOTE 2)	ca2$_1$ mc2$_1$ mn2$_1$ ■na2$_1$		$\frac{2_1}{c}$					65
$\overline{(mmm)}$ bam bca ▲mma	$\overline{(mm2)}$ ba2 ▲ma2							27
$\overline{(mmm)}$ ddd	$\overline{(mm2)}$ dd2							12
$\overline{(mmm)}$ ■ccm mma	cca ◀mcm ·mca	●mc2$_1$	$\overline{\left(\frac{2}{m}\right)}\overset{×}{\frac{2}{c}}$					9
	$\overline{(mm2)}$ bm2 ●ma2 , ba2							4
								4
28	22	9	6	4	3	1	1	230

(NOTE 1)

$\left(\frac{4}{m}mm\right)\overset{\downarrow}{\frac{4_2}{m}}mc$
$\frac{4}{m}cc \overset{×}{} \frac{4_2}{m}cm$
$\frac{4}{n}bm \quad \frac{4_2}{n}bc$
$\frac{4}{n}nc \quad \frac{4_2}{n}nm$
$\frac{4}{m}bm \quad \frac{4_2}{m}bc$
○$\frac{4}{n}nc$ ●$\frac{4_2}{n}nm$
$\frac{4}{m}mm \quad \frac{4_2}{m}mc$
$\frac{4}{n}cc \quad \frac{4_2}{n}cm$

(NOTE 2)

■nma bam
bca ccn
mma bcm
nna nnm
mna mmn
cca bcn

BUERGER GROUPS

● ●
  ×
▸ √
  ○
▲ ▲
■ ■

ENANTIOMORPHIC PAIRS

✸ ✸

NOTE: SPACE GROUP SYMBOLS WHICH ARE CIRCLED ARE ALSO POINT SPACE GROUPS AND ARE THEREFORE LISTED TWICE.

Source: From S. K. Dickinson, Jr., CRWPC, Air Force Cambridge Res. Labs., L. G. Hanscom Field, Bedford, MA.

INDEX

Absorption:
  by sample, 234
  of X-rays, 16, 79
Absorption–diffraction method, 362
Absorption edges, 18
Adiabatic principle, 374
Aerosols, 253
Air scatter, 157
Alignment checks, 216
Alignment quality, 219
Alignment tools, 205
α_2 stripping, 299
Amorphous silica, 25
Analysis:
  of air particulates, 365
  of thin samples, 374
Angular accuracy, 219
Angular dispersion, 153
Angular range, 287
Angular sensitivity, 219
Anisotropic temperature factor:
  β_{ij}, 69
  B_{ij}, 69
  U_{ij}, 69
Anomalous dispersion, 67
Anticoincidence unit, 136
APD (automated powder diffractometer) systems, 274
Aspect, 37
Atomic scattering factor, 15, 65
Auger effect, 6
Auger spectroscopy, 6
Axial divergence, 185
Axial divergence error, 187

Background, 157
  low-angle, 157
  measurement of, 141
  reduction, 157

  subtraction, 297
  threshold, 287
Backloading, 247
Band pass, 170
β-filter, 158
  calculation of thickness, 159
  placement, 162
β-radiation, 269
Binders, 240
Binding energy, 5
Body-centered unit cell, 31
Boolean search, 327, 345
Bragg angle, 48
Bragg–Brentano diffractometer, 180
Bragg–Pierce law, 17
Bragg's law, 48, 63
Bravais lattice, 30
Bremsstrahlung, 3
Brindley, G. W., 80
Broad-focus X-ray tube, 109
Broadening of diffraction lines, 141
Burger–Dorgelo rule, 10

Calibration standards, 281
Camera:
  Debye–Scherrer, 173
  Gandolfi, 173
  Guinier, 173, 177
  Seemann–Bohlin, 180
Camera methods, 173
Cavity mount, 247
CD–ROM, 330
Checking for contamination lines, 116
Chromium target tube, 108
Chung, F. H., 374
Cold finger, 132
Color centers, 234
Commercial Search/Match programs, 348
Computer-generated data, 265

Computer programs, 289
Contamination lines, 116
Continuous scanning, 278
Conversion errors, 305
Copper target tube, 108
Counting circuits, 138
Counting statistics, 142
Cristobalite, 24
Critical excitation potential, 100
Crystal data file, 326
Crystal system:
  anorthic, 29
  cubic, 29, 36
  hexagonal, 29, 38
  monoclinic, 29, 36
  orthorhombic, 29, 37
  rhombohedral, 29
  tetragonal, 29, 37
  triclinic, 29, 35, 53
  trigonal, 29, 38
Crystallite size analysis, 89
Crystallographic planes, 43

d-spacing:
  accuracy, 152
  large, 266
  range selection, 265
Darwin width, 89
Data acquisition, 261
Data display, 263
Data quality, 264, 310, 322
Data reduction, 287
  by computer, 288
  software, 289
Data smoothing, 292
Data treatment, 291
Database, crystal data, 33
Databases, 323
de Wolff figure of merit, 316
Dead time, 123
Dead-time correction, 124
Debye cone, 188
Debye rings, 60
Debye–Waller temperature factor, 68, 83
Demountable anode tube, 106
Detection limit, 384
Detection of X-rays, 121
Detector:
  cadmium telluride, 135

dead time, 123
gain, 126
gas counter configuration, 127
gas proportional, 127
Ge(Li), 132
ion pairs, 128
linearity, 123
mercuric iodide, 136
position sensitive, 130
properties, 121
proportional, 162
proportionality, 125
quantum counting efficiency, 122
resolution, 126, 128, 162
saturation, 124
scintillation, 131
Si(Li), 132
  advantages, 134
solid state, 163
xenon, 128
Determination:
  of anatase in rutile, 363
  of retained austenite, 376
Diffraction:
  peak intensity, 64
  peak location, 60
  powder pattern, 58
  single crystal, 57
Diffraction pattern calculation, 75
  for KCl, 82
Diffraction pattern distortions, 85
Diffractogram, contributions to, 154
Diffractometer, 178
  aberrations, 186
  alignment, 185, 205
  automated, 274
  automation, 275
  axial divergence, 188
  Bragg–Brentano, 180
  errors, 185
  horizontal, 180
  manual, 270
  optical arrangement, 184
  resolution, 13, 219
  resolving power, 167
  round-robin, 264
  Seemann–Bohlin, 180
Diffractometry, single-crystal, 81
Digital filter, 293
Discriminator, 136

Dispersion, 153
Divergence slit, 197, 208
  choice of, 157, 191, 199
Doublet, 8
Drift, 104
  characteristics, 104
Dynode, 132

EISI (Elemental and Interplanar Spacing Index), 327
Electromagnetic spectrum, 2
Electron diffraction, 57, 327
Encoder, 186
Energy-dispersive diffraction, 134
Equivalent positions, 39
Error window, 320
Euler coordinate system, 67, 72
Ewald, P. P., 49
Ewald sphere, 54
External calibration curve, 217
External standard, 308
Extinction:
  Bragg, 89
  primary, 78
  secondary, 79
  systematic, 74
Extra lines in a diffraction pattern, 155

Face-centered unit cell, 31
False peaks, 289
Fiber texture, 240
Fibrous specimens, 240
Field effect transistor, 132
Figures of merit, 310
  computer Search/Match, 350
  de Wolff, 316
  instrument, 224, 310
  pattern, 316
Film:
  blackening, 163
  as a detector, 121
Fine-focus X-ray tube, 109
Fink method, 339
Fixed-count method, 139
Fixed-time method, 139
Flat specimen error, 187, 191
Fluorescence yield, 6
Focal spot dimension, 109

Focal spot wander, 105
Focusing circle, 187
Forbidden transitions, 11
Formula units per unit cell, 43
Friedman, H., 2
Full-pattern fitting, 383

Gaussian distribution, 141
Gears:
  helical, 185
  herringbone, 185
  spiroid, 185
  spur, 185
  worm, 185
Ge(Li) detector, 135
Glide plane, 34
Goniometer, 97
  2:1, 206
  backlash, 280
  circle, 184
  focusing circle, 191
  gear systems, 186

Hanawalt method, 335
Hermann–Mauguin notation, 28
High-speed data acquisition, 131
High-voltage drift, 104
High-voltage generator, 99
High-voltage stability, 105
Huygens, 47

$I/I_{corundum}$, 372
ICDD (International Centre for Diffraction Data), 33, 319
Ideally imperfect crystals, 79
Improper rotation, 27
Indexing, 57, 62
Instrument checks, 117
Instrument performance, 311
Instrument reference standard, 226
Instrument-sensitive parameters, 357
Intensity:
  calculation of, 80
  integrated measurement with a timer-scaler, 360
  measurement, 357
  measurement of integrated, 141

Intensity (Cont'd.)
  of diffraction peaks, 64
  parameters in measurement, 356
  ratios of characteristic lines, 8
  relationship to concentration, 361
Interaxial angles, 30
Interference, 64
  constructive, 47
Internal standard, 309
  method, 370
  selection of, 371
Inversion, 27
Irradiation lengths, 240
Isowatt curve, 102
IUPAC (International Union of Pure and Applied Chemistry), 3
  notation, 10

Jenkins–Schreiner figure of merit, 224, 312
Johansson optics, 164
Johnson–Vand search, 347

$K\alpha_1 : K\alpha_2$ doublet dispersion, 153
Klug's equation, 365
Kramers' formula, 4
Kulbiki holder, 247

Lattice:
  body-centered, 31
  Bravais, 31
  Bravais types, 34
  face-centered, 31
  point, 30, 74
  primitive, 31
  reciprocal, 49, 50
  space, 30
  sub-, 33
  super-, 33
Laue camera, 142
Laue groups, 39
Laue, Max von, 1
Le Galley, D. P., 1
Lead mystearate, 214
Leroux, J., 367
Line asymmetry due to satellites, 13
Line overlaps, 357
Line voltage supply, 97

Line width, 114
Linear absorption coefficient, 17
Long-term drift, 105
Lorentz factor, 76
Lorentz polarization factor, 78
Low-absorbing materials, 194
Low concentrations, analysis of, 384

Macrostress, 91
Mass absorption law, 16
Mass attenuation coefficient, 16
  measurement, 367
Matrix flushing, 374
Max-d Index, 328
Maximum allowable count rates, 125
Mechanical zero, 206
Metals & Alloys Index, 328
Microabsorption, 80
Microfocus X-ray tube, 109
Microstress, 91
Miller indices, 43, 48
Molybdenum target tube, 109
Monochromatization, methods, 151
Monochromator, 164
  adjustment, 169
  antiparallel optics, 164
  comparison, 170
  crystal, 13
  diffracted beam, 80, 167, 180, 214
  parallel geometry, 164
  primary beam, 170
Moseley, H. G. J., 4
Multichannel analyzer, 136
Multiplicity, 40, 76

NBS*LATTICE, 33, 327
Net counting error, 141
Nichols search, 347
Niggli cell, 33
NIST (National Institute of Standards and Technology), 326
Nonparalyzable dead time, 124
Normal-focus X-ray tube, 109
Normalized RIR method, 373

Order, short-range, 26

INDEX

401

Paralyzable dead time, 124
Parrish, W., and Gordon, S. G., 1
Particle heterogeneity, 235
Particle size reduction, 244
Particle statistics, 240
Pattern indexing, 316
Pattern linearization, 298
Pattern reduction, 261
PDF (Powder Diffraction File), 322, 324
PDF-1, 330
PDF-2, 325
Peak location, 300
Peak overlap, 361
Peaks, maximum number of, 266
Pearson code, 329
Peltier cooling, 135
Phase identification, 319
   alphabetic method, 332
   Boolean search, 345
   computer methods, 345
   Fink method, 332
   Hanawalt method, 332
Phosphor, 132
Phosphor imaging plate, 145
Photoelectric absorption, 16
Photoelectric effect, 5
Photon, 2
Photopeak, 128
Pinhole pictures, 118
Planck's constant, 2
Point groups, 28
Poisson distribution, 140
Polarization, 80
Polarization factor, 64
Pole figure device, 89
Polychromatic wavelength problems, 299
Powder Diffraction File (PDF),
   See PDF
Power curve, 103
Power-rating curve, 113
Preferred orientation, 89, 236, 246
Primitive unit cell, 32
Probability, 141
Profile asymmetry, 188
Profile fitting, 12, 302, 361
Profile shape, 187
Pulse height selection, 136, 163
Pulse height selector window, 137
Pulse pileup, 124
Pulse shift, 124

Qualitative analysis, 319
   absorption–diffraction method, 363
   amorphous material in, 370
   of amorphous phase, 376
   with chemical constraints, 376
   Copeland–Bragg method, 375
   generalized internal standard method, 374
   internal standard method, 370
   normalized RIR method, 374, 375
   Norrish method, 367
   reference intensity ratio method, 372
   Rietveld method, 378
   RIR method, 372
   spiking method, 369
   standard addition method, 369
   "standardless," 374
   with structural constraints, 376
   use of computed mass attenuation
      coefficients, 367
   using the total pattern, 378
Quasicrystals, 28

Rachinger method, 300
Radiation safety, 19
   RAD, 21
   RBE, 21
   REM, 21
   röntgen, 19
Radiation:
   continuous, 3
   characteristic, 5
   continuum, 4
   white, 3
Radiography, 1
Rate-meter, 139, 270
Rate-meter setting, 270
Real-time image, 141
Receiving slit:
   effect of, 195
   selection, 197
Reciprocal lattice vector, 50
Reciprocal space, 49
Reduced cell, 33
Reduced pattern, 261, 287, 319
Reference file, 322
Reference intensity ratio:
   generalized, 372
   method, 372
Residual stress analysis, 91

Rietveld analysis, 380
Rietveld method, 305
Rietveld quantitative phase analysis, 380
Röntgen, W. C., 1

Safety, 19
Sample, definition, 231
Sample-sensitive parameters, 358
Satellites, 11
Scaler decades, 138
Scaler/timer, 138
Scan speed selection, 270
Scatter:
  anomalous, 67
  coherent, 14
  Compton, 15
  electron, 64
  low-angle, 200
Scatter σ, 16
Schönflies symbols, 37
Scherrer equation, 90
Schottky barrier, 133
Screw axis, 34
Search index, 323
Search/Match, 323
  computer methods, 344
  manual, 332
Seemann–Bohlin diffractometer, 180
Selection rules, 7
Separation methods, 240
Short-term drift, 105
Siegbahn notation, 10
Si(Li) detector, 163
SiO_2, 23
Site multiplicity, 40
Size and strain broadening, 91
Smith–Snyder figure of merit, 316
Solid solutions, 261
Solid substances, 23
Soller collimator, 116, 183, 188
  effect of, 188
Soller slits, 167
Source:
  constant potential, 100
  drift, 99
  duty cycle, 100
  flux, 102
  high-voltage generator, 100
    conditions, 102
  high-voltage stabilization, 100

line-voltage:
  supply, 98
  variations, 98
  spike, 99
  waveforms, 100
Space group notation, 35
Space group theory, 41
Space groups, 34
Special positions, 40
Specimen:
  changes on grinding, 232
  contamination, 240
  displacement, 254
  displacement error, 194
  fluorescence, 157
  holders, 246
  ideal, 231
  irradiation length, 200
  preparation, 231
  presentation, 254
  reactive, 257
  rotation, 255
  side loading, 247
  small, 257
  top loading, 249
  transparency, 192
  vs. sample, 233
Sphere of reflection, 55
Spiking method, 370
Spray drying, 250
Spreadsheet, 226
SRMs (Standard Reference Materials), 281
Standard addition method, 369
Standard deviation, 141
Standard, intensity, 309
Standards:
  alignment, 282
  external, 282
  internal, 283
  line profile, 285
  for quantitative analysis, 283
  selection of, 370
  sensitivity, 284
Statistically significant data, 298
Stepper motor, 275
Step scanning, 275, 277
Step size, 279
Step width selection, 280
Structure factor, 70
Structure of KCl, 41
Strukturbericht Symbol Index, 329

Symmetry, 26
  rotational, 26
Symmetry element, 26
Synchronous scanning, 270
Synchrotron, 106, 112
Systematic aberrations, 187
Systematic error, 205
Systematic extinction, 75

Takeoff angle, 103, 114, 185, 205
Temperature factor, 68
Texture, 240
Thermal motion, 68
Thompson equation, 66
Thompson, J. J., 64
Time constant, 140, 270
Transducer, 121, 125
Transients, 105
Translation vector, 30
Transparency error, 187
Triplet, 12
Troubleshooting, 226
True pattern, 261
Two-dimensional detectors, 142
2θ alignment, 219
2θ calibration methods, 308
  external standard, 308
  internal standard, 309
  zero background holder, 309
2θ range selection, 266
Types of lines in quantitative analysis, 356

Unique axes selection, 36
Unit cell, 30
Units:
  Ångstrom, 3
  kX, 3
  nanometer, 3
  RAD, 21
  RBE, 21
  REM, 21
  röntgen, 20

Vacancy, 6
Variable divergence slit, 201
Vegard's law, 42, 355
Vibration:
  anisotropic, 69
  mean square amplitude, 68
  thermal, 68
Visser and De Wolff, 372

Wavelength:
  minimum, 3
  nomenclature, 10
  selection, 266
Wehnelt cup, 106
Wet grinding, 240
Whole pattern fitting, 378

X-ray crystallography, 1
X-ray detectors, 121
  types of, 127
X-ray diffractometers, 178
X-ray generation, 97
X-ray penetration, 246
X-ray penetration depth, 234
X-ray region, 2
X-rays:
  nature of, 2
  scattering, 14
X-ray source intensity, 114
X-ray tube, 105
  anode pitting, 117
  care of, 113
  choice of, 126
  common targets, 152
  contamination, 116
  cooling system, 107, 114
  design, 106
  efficiency, 109
  failure, 114
  filament, 106
  foci, 108
  life, 103, 117
  loading, 103
  pulsed, 119
  rotating anode, 118
  running-in, 113
  sealed, 105
  shield, 97
  specific loading, 103, 109
  spectral purity, 157

Zero background holder, 194, 297
  calibration curve, 252
  calibration method, 249, 283, 309